T0281230

Metrics, Norms and Integrals

An Introduction to Contemporary Analysis

Metrics, Norms and Integrals

An Introduction to Contemporary Analysis

J J Koliha
University of Melbourne, Australia

World Scientific

NEW JERSEY · LONDON · SINGAPORE · BEIJING · SHANGHAI · HONG KONG · TAIPEI · CHENNAI

Published by

World Scientific Publishing Co. Pte. Ltd.

5 Toh Tuck Link, Singapore 596224

USA office: 27 Warren Street, Suite 401-402, Hackensack, NJ 07601

UK office: 57 Shelton Street, Covent Garden, London WC2H 9HE

British Library Cataloguing-in-Publication Data
A catalogue record for this book is available from the British Library.

METRICS, NORMS AND INTEGRALS
An Introduction to Contemporary Analysis

Copyright © 2008 by World Scientific Publishing Co. Pte. Ltd.

All rights reserved. This book, or parts thereof, may not be reproduced in any form or by any means, electronic or mechanical, including photocopying, recording or any information storage and retrieval system now known or to be invented, without written permission from the Publisher.

For photocopying of material in this volume, please pay a copying fee through the Copyright Clearance Center, Inc., 222 Rosewood Drive, Danvers, MA 01923, USA. In this case permission to photocopy is not required from the publisher.

ISBN-13 978-981-283-656-4
ISBN-10 981-283-656-X
ISBN-13 978-981-283-657-1 (pbk)
ISBN-10 981-283-657-8 (pbk)

Printed in Singapore.

To Milly,

my guiding star

Preface

The genesis of this book has its roots in the progressively developing set of lecture notes I have used for more than a decade in two semester long courses, one on metric and topological spaces, and one on introductory functional analysis and Lebesgue integration at the University of Melbourne. The book is intended as an introduction to contemporary analysis at the undergraduate level, but will be also useful to graduate students, and has been classroom tested over the years by many student populations.

Prerequisites A course in calculus and basic linear algebra, and some familiarity with proofs. The appendices serve to summarize background material, including sets and mappings, relevant results from calculus and real analysis, and a brief survey of transfinite cardinal numbers useful in various areas of mathematics.

Structure The book is mathematically rigorous, yet moderately paced, with ample motivation, and with proofs given in full detail. The book contains many worked examples, and every chapter concludes with a set of problems carefully designed to test and further develop the theory presented in the chapter. Most odd numbered problems are given detailed solutions in an appendix. This is a valuable resource for study and exam preparation. There is a comprehensive index of topics, which facilitates easy navigation in the book. The numbering of results is designed for easy reference. For instance, 1.14 refers to result 14 in Chapter 1, which happens to be Example; but it can be Definition, Theorem, Corollary, etc., as they are all subject to the same numbering system. Each chapter ends with a section containing problems numbered independently of Definitions, Theorems, Examples, etc. Thus Problem 3.12 refers to Problem 12 in Chapter 3.

The parts Each of the three parts of the book bears an imprint of one of the giants of modern analysis:

I. Felix Hausdorff – II. Stefan Banach – III. Henri Lebesgue.

These three mathematicians did not create their far reaching theories in isolation, but each of them became the focal figure, the leader of his field, after many before them had steadily contributed to the body of knowledge until a critical mass was attained for a breakthrough.

Part I—primarily concentrating on metric spaces—attempts to capture the beauty of a self-contained area of abstract mathematics which yields such rich and fruitful applications and pervades much of the contemporary mathematics to the point of being the preferred language of many branches of mathematics such as numerical mathematics, error analysis, differential equations, and much of modern physics including the theory of relativity and space-time. Metric and pseudometric spaces lead in a natural way to the introduction of topological spaces. In this part we encounter the five C's, which form the foundation of contemporary analysis: convergence, completeness, continuity, compactness and connectedness. Much of this material has been dealt with or foreshadowed in Hausdorff's 1914 seminal work *Grundzüge der Mengenlehre* on the foundations of 'point set theory'.

In Part II we encounter the concepts that shaped much of the mathematical development of the 20th century. The area of functional analysis crystalized in Banach's celebrated book *Théorie des Opérations Linéaires* on linear operators in 1932. While the classical analysis considered as its object a function of a real or complex variable and its behaviour, functional analysis views functions as anonymous points in a metric or normed space. Operators are the real kings of the domain, carrying points to points, functions to functions. This part culminates in the three basic principles of functional analysis, all authored or co-authored by Banach: the Hahn-Banach theorem, the uniform boundedness principle and the open mapping theorem.

The last part, Part III, is devoted to the theory of integral named after its founder, Henri Lebesgue. Lebesgue's 130 page doctoral dissertation *Intégrale, Longueur, Aire* of 1902 became arguably the most famous doctoral thesis of all times. It was quickly followed by *Leçons sur l'Integration et la Recherche des Fonctions Primitives* in 1904. Lebesgue's new theory of measure and integral underlined much of the subsequent development of modern analysis in the 20th century. In this book the order of exposition reverses the historical chronology, as we take advantage of the concept of the metric space completion to introduce the Lebesgue integral after we discussed metric space topology.

What is unique about 'Metrics, Norms and Integrals'? There are many textbooks covering a similar ground to this book. My goal was to write a user friendly book, one which would give enjoyment and afford glimpses of the beauty of mathematics at a higher level. Over the years I found presentations of certain topics that seemed clearer and more insightful than the standard approaches.

One such example is the completion of metric spaces. Rather than discard the original space and construct its isometric image as customary, we keep the space and add the so called ideal points to it. This makes the construction of the Lebesgue integral by the metric space completion into a logical, essentially one-step process, where the original simple functions are preserved, and the ideal points quickly linked to integrable functions. The completion based on norm-absolutely convergent series rather than Cauchy sequences is used to avoid the repeated extraction of subsequences.

The introduction of pseudometrics early on is intended to prepare students for the properties of non-Hausdorff topologies.

Integrable functions in this book are always complex valued. Rather than making the exposition more difficult, this choice often enables us to reach conclusions more easily, often in one step, where the real valued approach traditionally goes through several intermediate steps. A further innovation is that we introduce integrable functions without explicitly requiring measurability; this greatly simplifies the exposition.

The Lebesgue integrability of continuous functions on closed intervals and rectangles is established soon after the definition of the integral; this allows many interesting examples to be introduced early in the book. At this stage of the development many books rely on the Riemann integral for the existence and evaluation of integrals of elementary functions. While the Riemann integral is traditionally introduced in a first year calculus course, usually very little is proved in the class. Instead we use the Newton integral based on (generalized) primitives for functions on the real line, which makes evaluations of many integrals easy. An elementary proof of the fundamental theorem of calculus for the Lebesgue integral given in the book is based on the author's article in the *American Mathematical Monthly*.

Literature In writing this textbook I have drawn on many books, in particular in selection of problems. Inevitably, I have been influenced by the style of some of my former teachers at the Charles University in Prague, namely by the late Professor Vojtěch Jarník, whose meticulously written textbooks [23, 24] survived the test of time, and by the late Professor Jan Mařík for his ursurpassed clarity and precision.

My first encounter with metric spaces was through the excellent early monograph [13] of Eduard Čech originally written in Czech in 1926. Other texts which influenced my presentation of metric and topological spaces were the books by Kasriel [26], Gemignani [18], Cain [12], Pitts [37] and Sutherland [41].

The part on linear analysis owes a debt to the introductory texts by Brown and Page [11] and Kreyszig [29], and to the more advanced text [43] by Taylor and Lay. The section on self-adjoint compact operators is influenced by the classical text [40] of Riesz and Sz.-Nagy, and the treatment of the Sturm–Liouville system is fashioned after Pryce [38]. For further exploration of linear operators I strongly recommend the book [19] by Gohberg, Goldberg and Kaashoek.

The integration part of the book is probably the most original. The abstract Lebesgue integral is defined using the metric completion of the space of simple integrable functions. A similar approach to the integral is also adopted in Lang's monograph [31] and in the textbook [36] by the father and son team of Jan and Piotr Mikusiński. The treatment in the present book is substantially simplified into an essentially one-step process. Bartle's book [6] influenced the presentation of general measures. The preference for the Newton integral over the Riemann integral was reinforced by Dieudonné's book [16] and by the textbook [14] by Černý and Rokyta.

I would like to draw the reader's attention to an excellent internet resource, the history of mathematics MacTutor archive at the University of St. Andrews in Scotland at
`http://www-groups.dcs.st-and.ac.uk/~history`
for biographies of mathematicians and much more.

Notation In this book the symbols $\mathbb{N}$, $\mathbb{R}$, $\mathbb{Q}$ and $\mathbb{C}$ stand for the sets of natural, real, rational and complex numbers, respectively. Other symbols are listed in the general index.

J. J. Koliha   March 2008

Acknowledgments

I would like to thank all the people who contributed to the improvement of this book. To Associate Professor Craig Hodgson who read the first part and supplied several useful ideas, to Professors Vladimir Rakočević and Ivan Straškraba for their many helpful comments, to James Saunderson and Dr Raymond Lubansky who did a great job of proofreading, and to Jonathan Bowden for spotting mistakes in several arguments and helping to fix them.

My thanks are also due to all the students who attended my courses over the years and contributed corrections and improvements to the successive versions of the text. There are too many to mention by name.

And most of all I would like to thank my late wife Milly for her patience and sometimes a loss of it, without whose constant encouragement this book would not have been written. For a lifetime of joy, thank you. I can feel you smiling from where you are now.

.

Contents

PART 1
Metrics and Topologies

Chapter 1

Metric Spaces

When we study properties of functions of a real variable, we often refer to the distance $|x - y|$ between two real numbers. In calculus we say that a function $f \colon \mathbb{R} \to \mathbb{R}$ is continuous at a point a if the distance $|f(x) - f(a)|$ is getting small when the distance $|x - a|$ is getting small. The distance plays a crucial role in approximations: a rational number x is a good approximation to $\sqrt{2}$ when the distance $|x^2 - 2|$ is small. Another example is provided by a convergent sequence (x_n) of real numbers for which the distance $|x_n - x|$ is becoming small as n increases.

In the complex plane the distance between two complex numbers z_1, z_2 is again expressed as $|z_1 - z_2|$. We may proceed to more complicated sets, such as $\mathbb{R}^k$, and again we find that the notion of distance plays an important role, for instance in the approximate solution of systems of linear equations in k unknowns.

We extend the concept of distance to arbitrary sets and obtain a metric space. Elements of a metric space are called 'points'. They can be real numbers, vectors in $\mathbb{R}^k$, sequences of real or complex numbers, real valued functions defined on an interval $[a, b]$, certain sets in the plane, etc. Metric analysis is studied in abstract form, and then it can be applied to a wide range of concrete spaces important in mathematics, physics and engineering.

Metric spaces were first defined by Maurice Fréchet (1878–1973) in his doctoral thesis [17] in 1906 and further developed by Felix Hausdorff (1868–1942) in his 1914 classic [22]. Since then, hundreds of mathematicians have contributed to the theory, and metric space terminology has become the language of much of modern mathematics.

1.1   Metrics and pseudometrics

We start with a definition of metric which generalizes the concept of distance.

Definition 1.1. Let S be a nonempty set. A function $d\colon S \times S \to \mathbb{R}$ is said to be a *metric* on S if it satisfies the following axioms:

  M1. $d(x,y) \geq 0$ for all $x, y \in S$.

  M2. $d(x,y) = 0$ if and only if $x = y$.

  M3. $d(x,y) = d(y,x)$ for all $x, y \in S$.

  M4. $d(x,y) \leq d(x,z) + d(z,y)$ for all $x, y, z \in S$.

  A *pseudometric* $dS \times S \to \mathbb{R}$ satisfies the axioms **M1**, **M3** and **M4**; instead of **M2** we have merely

  M2′. $d(x,x) = 0$ for all $x \in S$.

  A *metric space* (S,d) consists of a nonempty set S and a metric d defined on S; a *pseudometric space* (S,d) consists of a nonempty set and a pseudometric.

  These axioms are often referred to as the *Hausdorff axioms*. Axiom **M4** is usually called the *triangle inequality*; in the Euclidean plane it says that the length of any side of a triangle does not exceed the sum of the lengths of the remaining two sides.

Example 1.2. A prototype of metric spaces is the set $\mathbb{R}$ with the Euclidean metric

$$d(x,y) = |x - y|.$$

The verification of **M1** – **M3** is left as an exercise. The triangle inequality **M4** follows from the basic inequality of calculus,

$$|a + b| \leq |a| + |b|, \tag{1.1}$$

known to be valid for any real numbers a, b.

Example 1.3. The set $\mathbb{C}$ of all complex numbers with the Euclidean metric

$$d(z,w) = |z - w|$$

is a metric space. Again, **M4** follows from the complex analogue of the inequality (1.1).

A comment about notation: Many spaces we consider consist of vectors with either finite or infinite number of coordinates. To avoid ambiguities, we use roman letters such as a, b, c, x, y, z for vectors, and Greek letters to denote their coordinates; for instance $a = (\alpha_1, \alpha_2, \alpha_3)$ for vectors in $\mathbb{R}^3$, $x = (\xi_1, \xi_2, \ldots, \xi_N)$ for vectors in $\mathbb{R}^N$.

Example 1.4. In the Euclidean space $\mathbb{R}^N$ the metric is defined by

$$d(x, y) = \left(\sum_{k=1}^{N} |\xi_k - \eta_k|^2 \right)^{1/2}, \tag{1.2}$$

where $x = (\xi_1, \ldots, \xi_N)$ and $y = (\eta_1, \ldots, \eta_N)$. It is not hard to verify that the function d has the properties **M1, M2, M3**. To prove the triangle inequality **M4**, we need the so called *Cauchy inequality* (which is a special case of the Schwarz inequality for inner products):

$$\sum_{k=1}^{N} |\xi_k \eta_k| \leq \left(\sum_{k=1}^{N} |\xi_k|^2 \right)^{1/2} \cdot \left(\sum_{k=1}^{N} |\eta_k|^2 \right)^{1/2}. \tag{1.3}$$

We use a method of proof which works equally well for $\mathbb{C}^N$. Set $A = (\sum_{k=1}^{N} |\xi_k|^2)^{1/2}$, $B = (\sum_{k=1}^{N} |\eta_k|^2)^{1/2}$. Without loss of generality, we may assume that $A > 0, B > 0$. Set $\alpha_k = \xi_k/A$, $\beta_k = \eta_k/B$. Note that $\sum_{k=1}^{N} |\alpha_k|^2 = 1 = \sum_{k=1}^{N} |\beta_k|^2$. Since $0 \leq (|\alpha_k| - |\beta_k|)^2 = |\alpha_k|^2 - 2|\alpha_k||\beta_k| + |\beta_k|^2$, we have $|\alpha_k||\beta_k| \leq \frac{1}{2}(|\alpha_k|^2 + |\beta_k|^2)$, and

$$\sum_{k=1}^{N} |\xi_k \eta_k| = AB \sum_{k=1}^{N} |\alpha_k||\beta_k| \leq AB \frac{1}{2} \left(\sum_{k=1}^{N} |\alpha_k|^2 + \sum_{k=1}^{N} |\beta_k|^2 \right) = AB.$$

This proves (1.3). Using the Cauchy inequality, we get

$$\sum_{k=1}^{N} |\xi_k + \eta_k|^2 \leq \sum_{k=1}^{N} (|\xi_k| + |\eta_k|)^2 = \sum_{k=1}^{N} |\xi_k|^2 + \sum_{k=1}^{N} |\eta_k|^2 + 2 \sum_{k=1}^{N} |\xi_k \eta_k|$$
$$\leq A^2 + B^2 + 2AB = (A + B)^2.$$

Taking square roots, we obtain

$$\left(\sum_{k=1}^{N} |\xi_k + \eta_k|^2 \right)^{1/2} \leq \left(\sum_{k=1}^{N} |\xi_k|^2 \right)^{1/2} + \left(\sum_{k=1}^{N} |\eta_k|^2 \right)^{1/2}. \tag{1.4}$$

When we replace ξ_k by $\xi_k - \zeta_k$ and η_k by $\zeta_k - \eta_k$, we finally get

$$d(x, y) \leq d(x, z) + d(z, y).$$

The space $\mathbb{C}^N$ can be also equipped with the Euclidean metric (1.2); the proof of the triangle inequality given above for real vectors has been set up so that it works without any change for complex vectors.

Example 1.5. If there is more than one metric defined on the same set, we may use different letters, such as d, ρ, σ, to distinguish between them, or we use subscripts, say d_1, d_2, d_3. If $S = \mathbb{R}^2$ and $x = (\xi_1, \xi_2)$, $y = (\eta_1, \eta_2)$, we may define metrics

$$d(x, y) = \left(|\xi_1 - \eta_1|^2 + |\xi_2 - \eta_2|^2\right)^{1/2},$$
$$\rho(x, y) = |\xi_1 - \eta_1| + |\xi_2 - \eta_2|,$$
$$\sigma(x, y) = \max\left(|\xi_1 - \eta_1|, |\xi_2 - \eta_2|\right).$$

Verification of the Hausdorff axioms for ρ and σ is left to the reader as an exercise. Here d is the Euclidean metric defined earlier; this is the familiar metric of elementary geometry. There are some sensible definitions of distances between plane points other than Euclidean. For example, ρ is known as the *taxi driver's metric*, because it gives the distance between two points travelled by a car in the grid of city streets. The metric σ can also be given a physical interpretation.

Example 1.6. On $S = \mathbb{R}^2$ define

$$p(x, y) = |\xi_1 - \eta_1|;$$

then p is a pseudometric on S (check the axioms). We note that p is not a metric: Set $x = (1, 4)$, $y = (1, -3)$, and observe that $p(x, y) = 0$ while $x \neq y$.

Example 1.7. On any nonempty set S we may define $d(x, y) = 0$ for all $x, y \in S$. This is the so called *trivial pseudometric*.

Example 1.8. The so called *discrete metric* is an important source of examples and counterexamples. Given a nonempty set S, we set $d(x, y) = 1$ if $x \neq y$, and $d(x, x) = 0$. It can be checked that d is a metric on S. The space (S, d) is called a *discrete metric space*.

Example 1.9. Let E be a nonempty set, and $S = B(E)$ the set of all bounded functions $f \colon E \to \mathbb{R}$. If $f, g \in B(E)$, then $f - g \in B(E)$, and we can define the *uniform metric*

$$d(f, g) = \sup_{t \in E} |f(t) - g(t)|.$$

Here $\sup M$ stands for the supremum, or the least upper bound, of a set M of real numbers; the set in question here is $M = \{|f(t) - g(t)| : t \in E\}$. (The definition and properties of the supremum are given in Appendix B; we return to them in due course after a discussion of further properties of metric spaces.)

We leave the proof of **M1** – **M3** as an exercise. To prove **M4**, assume $f, g, h \in B(E)$, and keep an element $s \in E$ fixed. Then

$$
\begin{aligned}
|f(s) - g(s)| &= |(f(s) - h(s)) + (h(s) - g(s))| \\
&\leq |f(s) - h(s)| + |h(s) - g(s)| \\
&\leq \sup_{t \in E} |f(t) - h(t)| + \sup_{t \in E} |h(t) - g(t)| \\
&= d(f, h) + d(h, g).
\end{aligned}
$$

Thus, the real number $d(f, h) + d(h, g)$ is an upper bound for $M = \{|f(s) - g(s)| : s \in E\}$. Since $\sup M = d(f, g)$ is the least upper bound, we have $d(f, g) \leq d(f, h) + d(h, g)$.

Definition 1.10. Given metric spaces (S_i, d_i), $i = 1, \ldots, k$, we define the so called *product metric* on the cartesian product $S = S_1 \times \cdots \times S_k$ by

$$
d(x, y) = \sum_{i=1}^{k} d_i(x_i, y_i),
$$

where $x = (x_1, \ldots, x_k)$, $y = (y_1, \ldots, y_k)$.

Then (S, d) is a metric space (Problem 1.10). An example of the product metric is the taxi driver's metric on $\mathbb{R} \times \mathbb{R}$.

Many of the metric spaces we consider are real or complex vector spaces, and metrics are often generated by norms. A *norm* $\| \cdot \|$ on a vector space X satisfies the following axioms:

N1. $\|x\| \geq 0$ for all $x \in X$.

N2. $\|x\| = 0$ implies $x = 0$.

N3. $\|\alpha x\| = |\alpha| \, \|x\|$ for all $x \in X$ and all scalars α.

N4. $\|x + y\| \leq \|x\| + \|y\|$ for all $x, y \in X$.

The axiom **N4** is the *triangle inequality*.

We observe that $\|0\| = 0$ and $\|-x\| = \|x\|$ in view of **N3.** A *normed space* $(X, \|\cdot\|)$ consists of a vector space X and a norm $\|\cdot\|$. The *metric induced by the norm* $\|\cdot\|$ is defined by

$$
d(x, y) = \|x - y\|.
$$

Verification of the Hausdorff axioms is left as an exercise. A *seminorm* satisfies the axioms for a norm except **N2**; a seminorm induces a pseudometric. A *seminormed space* consists of a vector space and a seminorm.

It is interesting to note that, with the exception of Example 1.8, all metric spaces discussed so far are in fact vector spaces, and the metrics are induced by norms. For instance, the Euclidean metric on $\mathbb{R}^N$ is induced by the so called Euclidean norm

$$\|x\| = \Big(\sum_{k=1}^{N} |\xi_k|^2\Big)^{1/2}.$$

The following example gives a natural generalization of the Euclidean space $\mathbb{R}^N$. In this example and elsewhere in this book we use the notation A^B, where A, B are nonempty sets, for the set of all functions $f\colon B \to A$. In particular, $A^{\mathbb{N}}$ describes all sequences $x = (\xi_1, \xi_2, \dots)$ with values in A, where $\xi_n = x(n)$ for each $x \in A^{\mathbb{N}}$ and all $n \in \mathbb{N}$.

Example 1.11. *Hilbert sequence space* ℓ^2. The space ℓ^2 consists of those vectors $x = (\xi_1, \xi_2, \dots) \in \mathbb{R}^{\mathbb{N}}$ for which the series $\|x\|^2 = \sum_{k=1}^{\infty} |\xi_k|^2$ converges. We show that ℓ^2 is a vector space by proving that it is closed under vector sums and under multiplication by scalars. Let $x = (\xi_k)$ and $y = (\eta_k)$ be elements of ℓ^2. Since $(a+b)^2 \le 2(a^2 + b^2)$ for any pair $a, b \ge 0$, we have

$$\sum_{k=1}^{N} |\xi_k + \eta_k|^2 \le 2\Big(\sum_{k=1}^{N} |\xi_k|^2 + \sum_{k=1}^{N} |\eta_k|^2\Big) \le 2(\|x\|^2 + \|y\|^2),$$

which ensures that $\sum_{k=1}^{N} |\xi_k + \eta_k|^2$ converges; so $x + y \in \ell^2$. We observe that also $\lambda x \in \ell^2$ whenever $x \in \ell^2$ and λ is a real number. From (1.4) we get

$$\Big(\sum_{k=1}^{\infty} |\xi_k + \eta_k|^2\Big)^{1/2} \le \Big(\sum_{k=1}^{\infty} |\xi_k|^2\Big)^{1/2} + \Big(\sum_{k=1}^{\infty} |\eta_k|^2\Big)^{1/2}$$

by taking the limit as $N \to \infty$. Hence

$$\|x\| = \Big(\sum_{k=1}^{\infty} |\xi_k|^2\Big)^{1/2}, \quad x \in \ell^2,$$

satisfies the triangle inequality; it is not difficult to se that it also satisfies **N1–N3**. Thus $(\ell^2, \|\cdot\|)$ is a normed space, and $d(x, y) = \|x - y\|$ induces a metric on ℓ^2. The space ℓ^2 also comes in the complex variety when the elements are $\mathbb{C}^{\mathbb{N}}$ vectors.

Example 1.12. *Sequence space* ℓ^1. In analogy with the preceding example we define ℓ^1 to be the set of those $\mathbb{R}^{\mathbb{N}}$ (or $\mathbb{C}^{\mathbb{N}}$) vectors $x = (\xi_1, \xi_2, \xi_3, \dots)$

for which the series $\sum_{k=1}^{\infty} |\xi_k|$ converges. Show as an exercise that ℓ^1 is a vector space and that

$$\|x\|_1 = \sum_{k=1}^{\infty} |\xi_k|$$

defines a norm on ℓ^1. We can then define the induced metric on ℓ^1.

Example 1.13. *The space ℓ^∞ consists of all $\mathbb{R}^{\mathbb{N}}$ (or $\mathbb{C}^{\mathbb{N}}$) vectors $x = (\xi_1, \xi_2, \xi_3, \dots)$ for which the sequence (ξ_k) of coordinates is bounded.* The set is a vector space, and

$$\|x\|_\infty = \sup_{k \in \mathbb{N}} |\xi_k|$$

defines a norm on ℓ^∞ (exercise).

Example 1.14. *Let $1 < p < \infty$. The space ℓ^p consists of the $\mathbb{R}^{\mathbb{N}}$ (or $\mathbb{C}^{\mathbb{N}}$) vectors $x = (\xi_1, \xi_2, \xi_3, \dots)$ for which the series $\sum_{k=1}^{\infty} |\xi_k|^p$ converges.* Again, ℓ^p is a vector space, and

$$\|x\|_p = \left(\sum_{k=1}^{\infty} |\xi_k|^p \right)^{1/p}$$

defines a norm on ℓ^p; a proof of these facts is a nontrivial exercise which requires Hölder's and Minkowski's inequalities (see Appendix C).

Example 1.15. The space $B(E)$ (Example 1.9) of all bounded functions $f \colon E \to \mathbb{R}$ with the uniform metric is a normed space when we define the *uniform norm*

$$\|f\| = \sup_{t \in E} |f(t)|;$$

the uniform metric is induced by the uniform norm. It is an interesting fact that every metric space (S, d) can be isometrically embedded into $B(S)$— see Problem 1.35. (An isometric embedding preserves the distance.)

1.2 Open and closed sets

Definition 1.16. Let (S, d) be a pseudometric space. Given a point $a \in S$ and a number $r \in \mathbb{R}, r > 0$, the *open ball* with centre a and radius r is the set

$$B(a; r) = \{x \in S : d(x, a) < r\}.$$

Example 1.17. In the Euclidean space $(\mathbb{R}, d)$, the open ball $B(a; r)$ is the set of all points $x \in \mathbb{R}$ with

$$d(x, a) = |x - a| < r, \quad \text{or} \quad -r < x - a < r, \quad \text{or} \quad a - r < x < a + r.$$

So $B(a; r)$ is the interval $(a - r, a + r)$.

Example 1.18. In the Euclidean space $(\mathbb{R}^2, d)$, an open ball takes the form of a *disc* without the circumference. See Figure 1.1 (a). In the Euclidean space $(\mathbb{R}^3, d)$, an open ball can be imagined as the space enclosed within a sphere, the sphere not included.

Example 1.19. Consider the metric space $(\mathbb{R}^2, \sigma)$ with the metric

$$\sigma(x, y) = \max\{|\xi_1 - \eta_1|, |\xi_2 - \eta_2|\}.$$

The ball $B((0, 0); 1)$ is the square with the vertices $(1, 1)$, $(-1, 1)$, $(-1, -1)$, $(1, -1)$, the sides excluded. See Figure 1.1 (b).

Example 1.20. Consider the metric space $(\mathbb{R}^2, \rho)$, with ρ the 'taxi driver metric'

$$\rho(x, y) = |\xi_1 - \eta_1| + |\xi_2 - \eta_2|.$$

You can check that the ball $B((0, 0); 1)$ is the square with the vertices $(1, 0)$, $(0, 1)$, $(-1, 0)$, $(0, -1)$, the sides excluded. See Figure 1.1 (c).

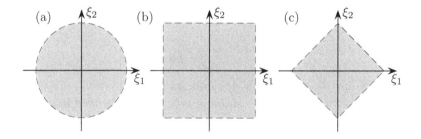

Figure 1.1

Example 1.21. Let (S, d) be the discrete space from Example 1.8. If $0 < r \leq 1$, then $B(a; r) = \{a\}$; if $r > 1$, then $B(a; r) = S$. Thus in a discrete metric space, an open ball is either a singleton or the whole space.

Example 1.22. Let p be the pseudometric of Example 1.6. The ball $B(a; r)$, where $a = (\alpha_1, \alpha_2)$, consists of all points $x = (\xi_1, \xi_2) \in \mathbb{R}^2$ satisfying

$$p(x, a) = |\xi_1 - \alpha_1| < r;$$

this is the infinite strip $(\alpha_1 - r, \alpha_1 + r) \times \mathbb{R}$ in the plane $\mathbb{R}^2$.

Definition 1.23. A set A in a pseudometric space (S, d) is said to be *bounded* if A is contained in some open ball $B(a; r)$. (See Problem 1.8 for an alternative definition.) In particular, every ball $B(a; r)$ is bounded.

From Example 1.22 we see that the strip $(\alpha_1 - r, \alpha_1 + r) \times \mathbb{R}$ is bounded in the pseudometric space $(\mathbb{R}^2, p)$. We note that boundedness is dependent on the pseudometric, and that it does not conform to our geometrical intuition.

Theorem 1.24. *Any subset of a bounded set is bounded, and the union of a finite number of bounded sets is bounded.*

Proof. Problem 1.9. Note that the theorem implies that every finite set in a pseudometric space is bounded. □

The concept of an interior point is one of the most fundamental in metric analysis. A failure to understand the following definition may render metric space topology quite incomprehensible. The concept will be illustrated by many examples and exercises; every effort should be made to understand it and be able to use it in logical arguments.

Definition 1.25. Let (S, d) be a pseudometric space and A a subset of S. A point $a \in A$ is called an *interior point of A* if there is an open ball $B(a; r)$ such that $B(a; r) \subset A$ (Figure 1.2).

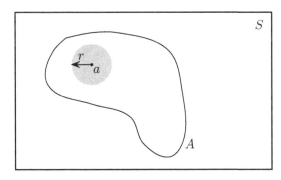

Figure 1.2

Example 1.26. Let A be the interval $A = (0, 1]$ in the Euclidean space $(\mathbb{R}, d)$. Then every point a with $0 < a < 1$ is an interior point of A: If $0 < r < \min\{a, 1 - a\}$, then

$$r < a \text{ and } r < 1 - a, \quad \text{which implies} \quad 0 < a - r \text{ and } a + r < 1;$$

so $B(a; r) = (a - r, a + r) \subset (0, 1) \subset A$. The point $a = 1$ belongs to A, but is not an interior point of A since *every* ball $B(1; r) = (1 - r, 1 + r)$ centred at 1 contains points $x > 1$ which are not in A.

Example 1.27. Let A be again the interval $A = (0, 1]$, this time in the metric space (T, d), where $T = (-\infty, 1]$ and $d(x, y) = |x - y|$. We show that every point of A is an interior point of A in (T, d). Indeed, the only dubious point is $a = 1$. From the definition of an open ball in T we have:

$$B(1; \tfrac{1}{2}) = \{x \in T : |x - 1| < \tfrac{1}{2}\} = (\tfrac{1}{2}, \tfrac{3}{2}) \cap (-\infty, 1] = (\tfrac{1}{2}, 1].$$

So $B(1; \tfrac{1}{2}) = (\tfrac{1}{2}, 1] \subset (0, 1] = A$, and $a = 1$ is also interior to A.

The preceding two examples show that the concept of interior point depends not only on the set A and the metric d, but also on the metric space in which the set A is embedded.

Definition 1.28. The set of all interior points of a set A in a pseudometric space S is called the *interior* of A, written

$$A^\circ \quad \text{or} \quad \text{int } A.$$

Example 1.29. In the Euclidean space $\mathbb{R}$, the interior of the interval $(0, 1]$ is the interval $(0, 1)$.

In general, the interior of a set does not coincide with the set itself. When it does, we say that the set is open.

Definition 1.30. A set $A \subset S$ is said to be *open* in (S, d) if every point of A is an interior point of A.

Example 1.31. Every open ball $B(a; r)$ in a pseudometric space (S, d) is an open set in S (Figure 1.3). If $b \in B(a; r)$, set $\alpha = r - d(b, a)$; since $b \in B(a; r)$, we have $\alpha > 0$. Now we prove that

$$B(b; \alpha) \subset B(a; r).$$

Figure 1.3

If $x \in B(b; \alpha)$, then

$$d(x, a) \le d(x, b) + d(b, a) < \alpha + d(b, a) = r,$$

and $x \in B(a; r)$ as required.

Theorem 1.32. *The interior $A°$ of a set A in a pseudometric space is an open set.*

Proof. Let a be a point of the interior $A°$ of A. Then there is a ball $B = B(a; r)$ contained in A. By the preceding example, every point of the ball B is interior to B, and hence also to A as $A \supset B$. $\square$

Example 1.33. In a discrete metric space (S, d) any subset A of S is open in S. Indeed, for any point $a \in A$, the open ball $B(a; 1) = \{a\}$ is a subset of A.

Example 1.34. The set

$$A = \{x = (\xi_1, \xi_2) : 2\xi_1 - \xi_2 + \xi_1 \xi_2 < 1\}$$

is open in the Euclidean space $(\mathbb{R}^2, d)$. We give a direct proof, but note that an easier proof can be obtained using continuity; this will be discussed in Chapter 3. Assume that $a = (\alpha_1, \alpha_2)$ is an element of A. Then

$$\beta = 2\alpha_1 - \alpha_2 + \alpha_1 \alpha_2 < 1.$$

Let x satisfy $d(x, a) < r$ for some $r > 0$ to be determined so that $x \in A$. Write

$$\xi_1 = \alpha_1 + \tau_1, \qquad \xi_2 = \alpha_2 + \tau_2.$$

We observe that $|\xi_i - \alpha_i| \le (|\xi_1 - \alpha_1|^2 + |\xi_2 - \alpha_2|^2)^{1/2} = d(x, a) < r$ for $i = 1, 2$. So $|\tau_i| \le d(x, a) < r$. Without loss of generality we may choose

$r \leq 1$; this will enable us to avoid quadratic inequalities in what follows. Then

$$
\begin{aligned}
2\xi_1 - \xi_2 + \xi_1\xi_2 &= 2(\alpha_1 + \tau_1) - (\alpha_2 + \tau_2) + (\alpha_1 + \tau_1)(\alpha_2 + \tau_2) \\
&= (2\alpha_1 - \alpha_2 + \alpha_1\alpha_2) + 2\tau_1 - \tau_2 + \alpha_1\tau_2 + \alpha_2\tau_1 + \tau_1\tau_2 \\
&\leq \beta + 2|\tau_1| + |\tau_2| + |\alpha_1|\,|\tau_2| + |\alpha_2|\,|\tau_1| + |\tau_1|\,|\tau_2| \\
&\leq \beta + (4 + |\alpha_1| + |\alpha_2|)r.
\end{aligned}
$$

The last inequality follows from $|\tau_1|\,|\tau_2| \leq r^2 \leq r$ (remember that $r \leq 1$). If we can choose r so that

$$
\beta + (4 + |\alpha_1| + |\alpha_2|)r < 1,
$$

then the ball $B(a;r)$ will be contained in A, and the proof will be finished. But such choice is possible: since $1 - \beta > 0$ $(a \in A)$,

$$
r' = \frac{1 - 2\alpha_1 + \alpha_2 - \alpha_1\alpha_2}{4 + |\alpha_1| + |\alpha_2|} = \frac{1 - \beta}{4 + |\alpha_1| + |\alpha_2|}
$$

is positive; then set $r = \min\left\{1, \frac{1}{2}r'\right\}$. Note that the radius r is a function of the point $a = (\alpha_1, \alpha_2)$.

The family of all open sets in a metric or pseudometric space plays an important role in the analysis of these spaces; we shall denote this family by $\mathcal{T}$, and call it the *topology* of S generated by the pseudometric d. The following theorem summarizes basic properties of open sets.

Theorem 1.35. *The following are true in a pseudometric space* (S, d).

(i) S *and* $\emptyset$ *are open sets.*

(ii)   *The union of any family of open sets is an open set.*

(iii)   *The intersection of a finite family of open sets is an open set.*

Proof. S is open because every ball in S lies in S. The empty set $\emptyset$ has no elements; so the implication $x \in \emptyset \implies x \in \emptyset^\circ$ is vacuously true, its assumption being false.

Let $\{A_\alpha : \alpha \in I\}$ be an arbitrary family of open sets with an index set I. Let $a \in A = \bigcup_{\alpha \in I} A_\alpha$. By the definition of union, there is an index β such that $a \in A_\beta$. Since A_β is open, there is a ball $B(a;r)$ contained in A_β. Then $B(a;r) \subset A$.

Finally, let the sets $A_1, A_2, \ldots, A_k$ be open in (S, d). Let $a \in A = \bigcap_{i=1}^k A_i$. Then a belongs to every A_i. Since A_i is open, there is $r_i > 0$ with $B(a;r_i) \subset A_i$. Define $r = \min\{r_i : i = 1, \ldots, k\}$. Then the ball $B(a;r)$ lies in A. $\square$

It is worth mentioning that the property (iii) breaks down if the family of open sets is infinite. This is illustrated by the next example.

Example 1.36. Each of the sets $A_n = (-1/n, 1 + 1/n)$ is open in the Euclidean space $(\mathbb{R}, d)$. However, the intersection of all the A_n is the interval $A = [0, 1]$, which is not open (the end points are not interior to A).

If a set is equipped with two different pseudometrics, we may ask when the two pseudometrics can be regarded as equivalent. A sensible definition requires that both pseudometrics lead to the same open sets.

Definition 1.37. Two pseudometrics d_1, d_2 on the same set S are called *equivalent*, or more precisely *topologically equivalent*, if they generate the same topology, that is, the same family of open sets.

Problem 1.24 gives a convenient criterion of equivalence of two pseudometrics: Two pseudometrics d and ρ on a set S are equivalent if and only if each ball $B_d(a; r)$ contains a ball $B_\rho(a; s)$, and each ball $B_\rho(a; u)$ contains a ball $B_d(a; v)$. (The subscripts specify which pseudometric is used in the definition of the ball.)

Example 1.38. Each of the metrics ρ and σ defined on $\mathbb{R}^2$ in Example 1.5 is equivalent to the Euclidean metric d. This follows from Problem 1.24 when we observe that, for every $r > 0$,

$$B_\rho(a; r) \subset B_d(a; r) \quad \text{and} \quad B_d(a; \tfrac{r}{\sqrt{2}}) \subset B_\rho(a; r),$$
$$B_\sigma(a; \tfrac{r}{\sqrt{2}}) \subset B_d(a; r) \quad \text{and} \quad B_d(a; r) \subset B_\sigma(a; r).$$

Example 1.39. The metric $\rho(x, y) = |f(x) - f(y)|$, where f is defined by $f(x) = x/(1 + |x|)$, is equivalent to the Euclidean metric $d(x, y) = |x - y|$ on $\mathbb{R}$. An easy proof of equivalence can be given with the help of Theorem 3.4 discussed later.

Definition 1.40. A set A in a pseudometric space (S, d) is said to be *closed* if its complement $A^c = S \setminus A$ is open.

Theorem 1.41. *The following are true in any pseudometric space* (S, d).

(i) *S and $\emptyset$ are closed sets.*

(ii) *The intersection of any family of closed sets is a closed set.*

(iii) *The union of a finite family of closed sets is a closed set.*

Proof. Follows from Theorem 1.35 and de Morgan's laws for sets:

$$\left(\bigcup_{\alpha\in I} A_\alpha\right)^{\mathrm{c}} = \bigcap_{\alpha\in I} A_\alpha^{\mathrm{c}}, \qquad \left(\bigcap_{\alpha\in I} A_\alpha\right)^{\mathrm{c}} = \bigcup_{\alpha\in I} A_\alpha^{\mathrm{c}}.$$

□

It is important to realize that sets do not behave like doors. We may have sets which are neither open nor closed, or sets which are simultaneously closed and open. In particular, proving that a set is not open does not establish that it is closed.

Example 1.42. The interval $(0, 1]$ in the Euclidean space $(\mathbb{R}, d)$ is neither open nor closed.

Example 1.43. Any set in a discrete space (S, d) is simultaneously open and closed.

Example 1.44. Any closed interval $A = [a, b]$ in the Euclidean space $(\mathbb{R}, d)$ is a closed set as the complement $A^{\mathrm{c}} = (-\infty, a)\cup(b, +\infty)$ is open (the union of two open sets). Similarly, the set $B = [a, +\infty)$ in the same metric space is closed.

The next example will demonstrate a difference in behaviour of closed sets in metric and pseudometric spaces.

Example 1.45. A singleton in a metric spaces is always closed. More generally, any finite set in a metric space is closed (Problem 1.19). However, in a pseudometric space, singletons need not be closed. For this we consider the pseudometric space $(\mathbb{R}^2, p)$, where $p(x, y) = |\xi_1 - \eta_1|$, and set $a = (1, 0)$. The set $\mathbb{R}^2 \setminus \{a\}$ is not open: the point $b = (1, 1)$ belongs to $\mathbb{R}^2 \setminus \{a\}$, yet any ball $B(b; r)$ (being the strip $|\xi_1 - 1| < r$) contains a. This shows that the singleton $\{a\}$ is not closed in $(\mathbb{R}^2, p)$.

1.3   Boundary and closure

Consider a set A in a pseudometric space (S, d). A point $u \in S$ which is neither in the interior of A nor in the interior of its complement A^{c} has a special position with respect to A. There is no ball centred at u which is contained either in A or A^{c}. We give a formal definition as follows.

Definition 1.46. A point $u \in S$ is a *boundary point of the set A* if every ball $B(u; r)$ contains at least one point $a \in A$ and at least one point $b \in A^{\mathrm{c}}$ (see Figure 1.4).

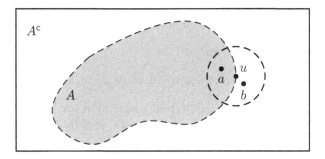

Figure 1.4

Example 1.47. Let A be the interval $(0, 1]$ in the Euclidean space $(\mathbb{R}, d)$. Each of the points 0, 1 is a boundary point of A. Indeed, each ball $B(0; r) = (-r, r)$ contains a point of A (e.g. $a = \min\{\frac{1}{2}r, 1\}$) and a point of A^c (e.g. $b = 0$). Similarly, each ball $B(1; r)$ contains a point of A (e.g. $a = 1$) and a point of A^c (e.g. $b = 1 + \frac{1}{2}r$). The set A has no other boundary points as each point of $(-\infty, 0) \cup (1, +\infty)$ is an interior point of A^c and each point of $(0, 1)$ an interior point of A.

Observe that, in the preceding example, one boundary point of A belongs to A while the other does not. So, boundary points of a set need not belong to that set.

The set of all boundary points of a set A is called the *boundary of A*. In symbols,

$$\partial A \quad \text{or} \quad \text{bd } A.$$

There is a symmetry in the definition of a boundary point which shows that

$$\partial A = \partial A^c.$$

Other important concepts in the theory of metric and pseudometric spaces are those of a limit point and an isolated point of a set.

Definition 1.48. Let A be a subset of a pseudometric space (S, d). A point $x \in S$ for which every ball $B(x; r)$ contains at least one point of A distinct from x is called a *limit point of A*. If $x \in A$ and there exists a ball $B(x; r)$ which contains only one point of A, namely x, we say that x is an *isolated point of A*. We write A' and iso A for the set of all limit and isolated points of A, respectively.

The next example shows that in pseudometric spaces our intuitive understanding of isolated and limit points can be wrong.

Example 1.49. Consider the set

$$A = \{(-1,0),(1,0),(0,0),(0,1),(0,2),(0,3)\}$$

in $\mathbb{R}^2$. In the Euclidean space $(\mathbb{R}^2, d_2)$ all points of A are isolated points of A. In the pseudometric space $(\mathbb{R}^2, p)$, where $p(x,y) = |\xi_1 - \eta_1|$, the points $(-1,0)$ and $(1,0)$ are isolated points of A, but $(0,0)$ is a limit point of A, as every ball $B_p((0,0);r)$ contains $(0,0)$ as well as $(0,1),(0,2),(0,3)$. (See also Problems 1.11, 1.12.)

Next we define the closure of a set; it is the set A together with all its boundary points.

Definition 1.50. Let A be a subset of a pseudometric space (S,d). The *closure* of A is defined as the set

$$\overline{A} = A \cup \partial A. \tag{1.5}$$

We often write cl A for the closure of A. Observe that $x \in \overline{A}$ if and only if every ball $B(x;r)$ contains at least one point of A. From the definition it follows that $\overline{A} = A \cup A'$. Further, the interior of A, interior of A^c and the boundary of A comprise all of S. So, $S = A^\circ \cup (A^c)^\circ \cup \partial A$, with the three sets mutually disjoint. Therefore $A \cup \partial A = A^\circ \cup \partial A$, which means that the closure of A is the disjoint union

$$\overline{A} = A^\circ \cup \partial A. \tag{1.6}$$

Theorem 1.51. *The closure $\overline{A}$ of a set A in a pseudometric space S is a closed set in S.*

Proof.   Follows from Theorem 1.32 and from $(\overline{A})^c = (A^c)^\circ$.   □

Example 1.52. If $A = (0,1]$ in the Euclidean space $(\mathbb{R}, d)$, then

$$A^\circ = (0,1), \quad \partial A = \{0\} \cup \{1\}, \quad \overline{A} = [0,1].$$

Example 1.53. Let $A = \{1/n : n \in \mathbb{N}\}$ in the Euclidean space $(\mathbb{R}, d)$. Then every point of A is a boundary point of A, and 0 is also a boundary point of A; any other point of $\mathbb{R}$ is an interior point of A^c. So $\partial A = A \cup \{0\}$, and $\overline{A} = A \cup \{0\}$. (Exercise.)

The following result is often useful in establishing that a set is closed without investigating its complement.

Theorem 1.54. *A set A is closed if and only if A contains all its boundary points, or, equivalently, $\overline{A} = A$.*

Proof. Assume that A contains all its boundary points. Then every point of A^c is interior to A^c since boundary points of A^c are also boundary points of A. So A^c is open, and A is closed by definition.

Assume that A is closed. Then A^c is open, and every boundary point of A^c lies in A. Since boundary points of A and A^c are the same, A contains all its boundary points. $\square$

Example 1.55. Let $A = \{1/n : n \in \mathbb{N}\}$ be a set in the Euclidean space $(\mathbb{R}, d)$. Then 0 is the only limit point of A as every ball $B(0; r) = (-r, r)$ contains all the points $1/n$ of A with $n > 1/r$. All other points of $\overline{A}$ are in A. Each point $1/n$ is isolated in A: If we choose $r = 1/(n(n+1))$, the ball $B(1/n; r) = (1/n - r, 1/n + r)$ contains only one point of A, namely, the center $1/n$ itself.

Definition 1.56. A subset A of a pseudometric space S is *dense* in S if $\overline{A} = S$.

For instance, the set $\mathbb{Q}$ of rational numbers in dense in the Euclidean space $\mathbb{R}$ (Theorem B.6).

1.4 Metric subspaces

If A is a nonempty subset of a metric space (S, d), we can consider the restriction d_A of the metric d to the set A. The pair (A, d_A) then becomes a metric space in its own right, with the metric *induced* by the metric d. (A, d_A) is called a *metric subspace* of (S, d). All these formulations are also applicable to pseudometric spaces.

For the sake of brevity, we often say that a set is *closed in A* when it is closed in the metric subspace (A, d_A); similarly we say *open in A*, or refer to the boundary with respect to A, etc. If U is a subset of A, it may be closed in A, but not in S; similarly, a set may be open in A, but not in S. We write $B_S(a; r)$ and $B_A(a; r)$ for balls in (S, d) and in (A, d_A), respectively.

Example 1.57. Let $S = \mathbb{R}$ and $A = [0, 1) \cup [2, 3]$. The set $[0, 1)$ is closed in A, but not in S; it is also open in A, but not in S. The set $[2,3]$ is closed in both A and S; it is open in A, but not in S.

Theorem 1.58. *A set U is open in a metric subspace A of S if and only if there is a set G open in S such that $U = G \cap A$.*

Proof. Suppose that U is open in A. Then for each $u \in U$ there is $r = r(u) > 0$ such that $B_A(u; r(u)) \subset U$. Define

$$G = \bigcup_{u \in U} B_S(u; r(u)).$$

Then G is open in S being the union of sets open is S. Then

$$A \cap G = \bigcup_{u \in U} \{A \cap B_S(u; r(u))\} = \bigcup_{u \in U} B_A(u; r(u)) = U.$$

Conversely, assume that $U = G \cap A$, where G is open in S. If $u \in U$, then also $u \in G$, and there is $r = r(u) > 0$ such that $B_S(u; r(u)) \subset G$. Therefore

$$A \cap B_S(u; r(u)) \subset A \cap G = U,$$

which means that $B_A(u; r(u)) \subset U$; so u is in the interior of U in A. □

Theorem 1.59. *A set U is closed in a metric subspace A of S if and only if there is a set F closed in S such that $U = F \cap A$.*

Proof. Follows from Theorem 1.58 using de Morgan's laws. □

It is often necessary to distinguish between the closure of a set in S and in a subspace A. We then write

$$\mathsf{cl}_S(U) \quad \text{or} \quad \mathsf{cl}_A(U).$$

Similarly we write $\mathsf{int}_S(U)$ and $\mathsf{int}_A(U)$; $\partial_S(U)$ and $\partial_A(U)$. We then have

$$\mathsf{cl}_A(U) = \mathsf{cl}_S(U) \cap A.$$

1.5   Problems for Chapter 1

1.   In a pseudometric space (S, d) prove the inequalities

M5. $|d(x, z) - d(y, z)| \leq d(x, y),$ **M6.** $|d(x, y) - d(a, b)| \leq d(x, a) + d(y, b).$

2.   Let $f : S \to \mathbb{R}$ be a function on S. If $d(x, y) = |f(x) - f(y)|$ for all $x, y \in S$, prove that d is a pseudometric on S. Show that d is a metric if and only if f is injective.

3. Which of the following are pseudometrics (metrics) on $\mathbb{R}$?

(i) $\rho(x,y) = |x/(1+|x|) - y/(1+|y|)|$ (ii) $\sigma(x,y) = |e^{-x} - e^{-y}|$

(iii) $\tau(x,y) = |\sin x - \sin y|$ (iv) $\varphi(x,y) = |x^2 - y^2|$

(v) $\psi(x,y) = |x-y|^2$ (vi) $\theta(x,y) = |x^3 - y^3|$

Repeat with $\mathbb{R}$ replaced by $\mathbb{C}$. (Use Problem 1.2 wherever possible.)

4. If d is a pseudometric on S, show that each of the following is also a pseudometric on S:

$$d_1 = k \cdot d \ (k > 0), \qquad d_2 = \frac{d}{1+d}, \qquad d_3 = \min(d, 1).$$

If d is a metric, show that each d_i is a metric, $i = 1, 2, 3$.

5. Let $S = C[a,b]$ be the set of all continuous real-valued functions on the interval $[a, b]$. Prove that

$$\rho(f, g) = \int_a^b |f(t) - g(t)|\, dt$$

defines a metric on S. (Continuity is needed for **M2**.)

6. In the space $S = \{(\xi_1, \xi_2) \in \mathbb{R}^2 : \max(|\xi_1|, |\xi_2|) \leq 1\}$ equipped with the Euclidean metric describe the following balls:

(i) $B((0,0); r)$ for varying values of $r > 0$.

(ii) $B((\xi_1, 0); \frac{1}{2})$ for varying values of ξ_1.

(iii) $B((1, 1); r)$ for varying values of $r > 0$.

Repeat with the pseudometric $\rho((\xi_1, \xi_2), (\eta_1, \eta_2)) = |\xi_1 - \eta_1|$.

7. Let $\rho(x, y) = |f(x) - f(y)|$ for all $x, y \in S = \mathbb{R}$, where $f(x) = x/(1 + |x|)$. In the metric space (S, ρ) find the balls:

(i) $B(0; \frac{1}{2})$, (ii) $B(0; 1)$, (iii) $B(1; \frac{1}{4})$, (iv) $B(1; \frac{1}{2})$, (v) $B(1; 1)$.

Repeat with the pseudometric $\sigma(x, y) = |x^2 - y^2|$.

8. Show that a set A in the pseudometric space (S, d) is bounded if and only if the set $M = \{d(x, y) : x, y \in A\}$ is a bounded set of real numbers. The number

$$\text{diam}\,(A) := \sup M = \sup_{x,y \in A} d(x, y)$$

is called the *diameter* of the set A.

9. Show that the union of a finite number of bounded sets is bounded.

10. Let $(S_1, d_1), \ldots, (S_k, d_k)$ be metric spaces and let $S = S_1 \times \cdots \times S_k$. Show that the function $d \colon S \to \mathbb{R}$ defined for $x = (x_1, \ldots, x_k)$, $y = (y_1, \ldots, y_k) \in S$ by

$$d(x, y) = d_1(x_1, y_1) + \cdots + d_k(x_k, y_k)$$

is a metric on S. Repeat for pseudometrics.

11.   Let A be a finite subset of a metric space (S, d). Prove that every point of A is isolated in A. Explain why this need not be true in a pseudometric space.

12.   Let a be a limit point of a set A contained in a metric space (S, d). Prove that each ball $B(a; r)$ contains infinitely many points of A. Show that this need not be true in a pseudometric space.

13.   Let A be a subset of a pseudometric space, and let $x \in A^c$ be a boundary point of A. Show that $x \in A'$.

14.   Let A be a subset of a pseudometric space, and let $x \in A^c$ be a limit point of A. Show that $x \in \partial A$.

15.   In each of the following cases decide whether the given set is open in the specified metric space, and find its boundary and closure:

(i) $\{(\xi_1, \xi_2) : \xi_1 + \xi_2 > 1\}$ in the Euclidean space $\mathbb{R}^2$

(ii) $\{(\xi_1, \xi_2) : \xi_1 \xi_2 < 1\}$ in the Euclidean space $\mathbb{R}^2$

(iii) $\{(\xi_1, \xi_2) : \xi_1 \xi_2 + \xi_1 > 1\}$ in $\mathbb{R}^2$ with the metric ρ given by

$$\rho((\xi_1, \xi_2), (\eta_1, \eta_2)) = \max_{i=1,2} |\xi_i - \eta_i|$$

(iv) $\{(\xi_1, \xi_2) : 2\xi_1 - 3\xi_2 < -1\}$ in $\mathbb{R}^2$ with the taxi driver's metric (Example 1.5)

(v) $(\frac{1}{2}, 1] \times (\frac{1}{2}, 1]$ in the space $S = (0, 1] \times (0, 1]$ with the Euclidean metric

16.   Sketch the set A and find iso A, A', $\overline{A}$, A°, ∂A in the Euclidean space $(\mathbb{R}^2, d)$:

(i) $A = \left\{ \left(\frac{1}{n}, \frac{n-1}{n} \right) : n \in \mathbb{N} \right\}$,   (ii) $A = \left\{ \left(1 - \frac{1}{m} \right) \left(\cos \frac{2\pi}{n}, \sin \frac{2\pi}{n} \right) : m, n \in \mathbb{N} \right\}$.

17.   Suppose S is the union of all intervals $(2k, 2k + 1]$, where k runs through all integers, and suppose d is the Euclidean metric $d(x, y) = |x - y|$. Show the following:

(i)   In the space (S, d), each of the sets $A_k = (2k, 2k + 1]$ is open and closed.

(ii)   In the space $(\mathbb{R}, d)$, the sets A_k are neither open nor closed.

18.   Let A be an open set in the metric space (S, d) and let K be a finite subset of A. Prove from first principles that the set difference $A \backslash K = A \cap K^c$ is also open in (S, d).

19.   Prove that any singleton in a metric space (S, d) is closed. Hence show that any finite set K is closed in (S, d). (This need not be the case in a pseudometric space: see Example 1.45.)

20.   *Alternative definition of infimum and supremum.* If A is a nonempty bounded subset of $\mathbb{R}$, show that $\alpha = \inf A$ is the least boundary point of A, and $\beta = \sup A$ is the greatest boundary point of A.

21. Prove the following properties of the interior in a pseudometric space:

(i) $S^\circ = S$ (ii) $A^\circ \subset A$ (iii) $A \subset B \implies A^\circ \subset B^\circ$

(iv) $(A \cap B)^\circ = A^\circ \cap B^\circ$ (v) $(A^\circ)^\circ = A^\circ$ (vi) $(A \cup B)^\circ \supset A^\circ \cup B^\circ$

22. Formulate and prove facts about closure analogous to the facts about interior from the preceding problem.

23. Prove that in a pseudometric space $\overline{A}\,^c = (A^c)^\circ \subset (A^\circ)^c = \overline{A^c}$.

24. Let d and ρ be two pseudometrics on a set S. Show that the two pseudometrics are equivalent if and only if each ball $B_d(a;r)$ contains a ball $B_\rho(a;s)$, and each ball $B_\rho(a;u)$ contains a ball $B_d(a;v)$. (The subscripts specify which pseudometric is used in the definition of the ball.)

25. Show that all the ℓ^p-metrics for $p \geq 1$ on $\mathbb{R}^N$ are equivalent, and that $d_\infty(x,y) = \lim_{p \to \infty} d_p(x,y)$ for all $x, y \in \mathbb{R}^N$.

26. Find the best possible $\alpha, \beta > 0$ such that $\alpha\|x\|_1 \leq \|x\|_2 \leq \beta\|x\|_1$ for all $x \in \mathbb{R}^N$. Here $\|\cdot\|_p$ is the ℓ^p norm on $\mathbb{R}^N$, $p = 1, 2$.

27. Let $A = \{1/n : n \in \mathbb{N}\}$ in the Euclidean space $(\mathbb{R}, d)$. Find $A^\circ, \partial A$ and $\overline{A}$. (See Example 1.55.)

28. In a pseudometric space (S, d) define a *closed ball* $K(a;r)$ with centre a and radius $r > 0$ as the set $\{x \in S : d(x,a) \leq r\}$. Prove that

(i) $K(a;r)$ is a closed set,

(ii) $\overline{B(a;r)} \subset K(a;r)$; give an example of a strict inclusion.

29. Let A be a set in a pseudometric space (S, d) and let $x \in S$. Show that

(i) $x \in \text{iso}\, A \iff B(x;r) \cap A = \{x\}$ for some $r > 0$.

(ii) $x \in \overline{A} \iff B(x;r) \cap A \neq \emptyset$ for each $r > 0$.

(iii) $x \in A' \iff (B(x;r) \setminus \{x\}) \cap A \neq \emptyset$ for each $r > 0$.

(iv) $x \in \partial A \iff B(x;r) \cap A \neq \emptyset$ and $B(x;r) \cap A^c \neq \emptyset$ for each $r > 0$.

30. For any point x and any set A in a pseudometric space (S, d) define $\text{dist}\,(x, A) = \inf\{d(x,z) : z \in A\}$. Prove the following:

(i) $x \in \text{iso}\, A \iff x \in A$ and $\text{dist}\,(x, A \setminus \{x\}) > 0$.

(ii) $x \in \overline{A} \iff \text{dist}\,(x, A) = 0$.

(iii) $x \in A' \iff \text{dist}\,(x, A \setminus \{x\}) = 0$.

(iv) $x \in \partial A \iff \text{dist}\,(x, A) = 0 = \text{dist}\,(x, A^c)$.

(v) $x \in A^\circ \iff \text{dist}\,(x, A^c) > 0$.

31. Let (S, d) and (T, ρ) be pseudometric spaces and let $S \times T$ be the product

space. If $A \subset S$ and $B \subset T$ are open sets, show that the set $A \times B$ is open in $S \times T$.

32.   Which of the following sets in the Hilbert sequence space ℓ^2 are open or closed?

(i) $\{x = (\xi_k) \in \ell^2 : \xi_k < 1 \text{ for all } k\}$

(ii) $\{x = (\xi_k) \in \ell^2 : \xi_k \geq -\frac{1}{2} \text{ for all } k\}$

(iii) $\{x = (\xi_k) \in \ell^2 : \xi_1 < 1 \text{ and } \xi_4 \geq 2\}$

(iv) $\{x = (\xi_k) \in \ell^2 : \xi_1 < 1\}$

(v) $\{x = (\xi_k) \in \ell^2 : \xi_3 \leq 1\}$

(vi) $\{x = (\xi_k) \in \ell^2 : \xi_1^2 - 2\xi_3 < 2\}$

(vii) $\{x = (\xi_k) \in \ell^2 : |\xi_k| \leq 1/k \text{ for all } k\}$ (the so called *Hilbert cube*)

33.   Consider ℓ^1 as a subspace of ℓ^∞ with the inherited norm $\|x\|_\infty = \sup_n |\xi_n|$. Prove that the closure of ℓ^1 in ℓ^∞ is the space $\mathbf{c}_0$ (defined in Appendix D).

34.   For $1 \leq p \leq \infty$ let X_p be the set of all $x \in \ell^p$ such that the series $\sum_{n=1}^\infty \xi_n$ converges and the sum is equal to 0. Show that X_p is a subspace of ℓ^p, and the following are true:

(i) X_1 is closed in ℓ^1.

(ii) X_2 is dense but not closed in ℓ^2, that is, $\overline{X}_2 = \ell^2 \neq X_2$.

(iii) X_∞ is not closed in ℓ^∞.

35.   *An isometric embedding of a metric space into a normed space.* Let (S, d) be a metric space and $B(S)$ the set of all bounded functions $f \colon S \to \mathbb{R}$ equipped with the equivalent norm. Select a point x_0 in S, and for each $a \in S$ define a function $f_a \colon S \to \mathbb{R}$ by $f_a(x) = d(x, a) - d(x, x_0)$, $x \in S$. Show that $f_a \in B(S)$, and that

$$\|f_a - f_b\| = d(a, b), \quad \text{for all } a, b \in S.$$

Chapter 2

Convergence and Completeness

In calculus we meet the concept of convergence, first for sequences of real numbers, later for sequences of functions. Intuitively, a sequence (x_n) of real numbers converges to a real number x if the distances $|x_n - x|$ are becoming small for large n. A rigorous definition requires a quantitative description of this process. Such definition was first introduced by French mathematician Augustin Cauchy and is discussed in Appendix B. Cauchy also gave a necessary and sufficient condition for convergence of real sequences that bears his name. A sequence (x_n) satisfies the Cauchy condition (or is a Cauchy sequence) if the distances $|x_m - x_n|$ are becoming small for large values of m and n. The result stating that a real sequence converges if and only if it satisfies the Cauchy condition is known as the Bolzano–Cauchy theorem (Appendix B). In many respects, the Bolzano–Cauchy theorem is one of the most important results about $\mathbb{R}$ on which much of real analysis is based. Analysis would be impossible in the set $\mathbb{Q}$ of rationals precisely because the Bolzano–Cauchy theorem fails in $\mathbb{Q}$.

We extend the concept of convergence to metric and pseudometric spaces. This will enable us to discuss in the same framework the convergence of numerical sequences, convergence of vectors in $\mathbb{R}^k$, pointwise and uniform convergence of functions and much more. The metric spaces which satisfy an analogue of the Bolzano–Cauchy theorem are called complete metric spaces, and have some pleasing mathematical properties.

Augustin Cauchy (1789–1857), an influential French mathematician
Bernhard Bolzano (1781–1848), a Jesuit priest living in Prague

2.1   Convergent and Cauchy sequences

A *sequence* in a pseudometric space is an element of $S^{\mathbb{N}}$, that is, a function $x \colon \mathbb{N} \to S$. We usually write $x(n) = x_n$ for each $n \in \mathbb{N}$, and denote the sequence by (x_n). As in the case of real or complex sequences, we can define convergence of a sequence in a metric or pseudometric space. We give the definition for pseudometric spaces as metrics are special cases of pseudometrics.

Definition 2.1. A sequence of points (x_n) in a pseudometric space (S, d) is said to *converge* to a point a of S if, for each $\varepsilon > 0$, there is an index $N = N(\varepsilon)$ such that

$$d(x_n, a) < \varepsilon \text{ for all } n > N.$$

Such a point a is called a *limit* of the sequence (x_n). We write

$$x_n \to a \qquad \text{or} \qquad x_n \xrightarrow{d} a \qquad \text{or} \qquad a = \lim_{n \to \infty} x_n.$$

Convergent sequences behave differently in metric and in pseudometric spaces.

Theorem 2.2. *A sequence in a metric space can have at most one limit.*

Proof.   Suppose that $x_n \to a$ and $x_n \to b$ in a metric space (S, d). Then

$$0 \le d(a, b) \le d(a, x_n) + d(x_n, b) \to 0.$$

Then $d(a, b) = 0$, and so $a = b$ by **M2**. □

In pseudometric spaces a sequence can have more than one limit.

Example 2.3. Let p be the pseudometric $p(x, y) = |\xi_1 - \eta_1|$ on $\mathbb{R}^2$ introduced in Example 1.6. The sequence $(n/(n+1), 2n)$ is convergent to every point of the form $(1, \alpha)$, $\alpha \in \mathbb{R}$.

Convergence of sequences is a topological property, that is, it can be rephrased solely in terms of open sets. An *open neighbourhood* of a is an open set U which contains a.

Theorem 2.4. *The sequence (x_n) in (S, d) converges to the point a if and only if for each open neighbourhood U of a there exists an index N such that*

$$x_n \in U \text{ for all } n > N.$$

Proof. This is left to the reader as an exercise. $\qquad\square$

Example 2.5. Let $x_n = (\xi_{n1}, \xi_{n2}, \ldots, \xi_{nN})$ be a sequence of points in the Euclidean space $(\mathbb{R}^N, d)$. The sequence converges to a point $a = (\alpha_1, \alpha_2, \ldots, \alpha_N)$ if

$$d(x_n, a) = \left(\sum_{k=1}^{N} |\xi_{nk} - \alpha_k|^2 \right)^{1/2} \to 0.$$

In fact, it can be shown (Problem 2.2), that $x_n \to a$ in $\mathbb{R}^N$ if and only if

$$\lim_{n \to \infty} |\xi_{nk} - \alpha_k| = 0 \quad \text{for every} \quad k = 1, \ldots, N,$$

in the sense of convergence in $\mathbb{R}$. This means that in the Euclidean space $\mathbb{R}^N$ the convergence of sequences is equivalent to coordinatewise convergence.

Example 2.6. Let (S, d) be a discrete space, and let (x_n) be a sequence in S. Then $x_n \to a$ if and only if there is an index N such that $x_n = a$ for all $n > N$ (exercise). Therefore a convergent sequence in a discrete space is one which assumes a constant value after a finite number of terms.

Definition 2.7. *Pointwise and uniform convergence of mappings.* Let E be a nonempty set, (T, ρ) a metric space, and f_n, $f : E \to T$, $n = 1, 2, \ldots$ The sequence (f_n) of mappings *converges pointwise on E* to the mapping f if for each $t \in E$, $\lim_{n \to \infty} \rho(f_n(t), f(t)) = 0$. The sequence (f_n) *converges uniformly on E* to f if

$$\lim_{n \to \infty} \sup_{t \in E} \rho(f_n(t), f(t)) = 0.$$

Clearly, the uniform convergence on E implies the pointwise convergence on E, but not vice versa. The next example deals with the case of scalar valued functions.

Example 2.8. Let E be a nonempty set and $B(E)$ the set of all bounded functions $f : E \to \mathbb{K}$, where $\mathbb{K}$ is the set of real or complex numbers. Then $B(E)$ is a vector space, and we can define the *uniform norm* on $B(E)$ by

$$\|f\|_\infty = \sup_{t \in E} |f(t)|.$$

A sequence (f_n) of functions $f_n : E \to \mathbb{K}$ converges uniformly on E to f if $\|f_n - f\|_\infty \to 0$. It is important to observe that the functions f_n and f need not be bounded; however, from a certain index on, $f_n - f$ must be bounded (that is, belong to $B(E)$) in order that $\|f_n - f\|_\infty$ be defined. The pointwise convergence means that $f_n(t) \to f(t)$ in $\mathbb{K}$ for each $t \in E$. The set $\mathbb{K}$ in this example can be replaced by a normed space $(X, \|\cdot\|)$.

Example 2.9. Consider the sequence $f_n(t) = t^n$ on the interval $[-\frac{1}{2}, \frac{1}{2}]$. The pointwise limit is found when we keep t fixed: $\lim_{n \to \infty} t^n = 0$ whenever $|t| < \frac{1}{2}$. So the pointwise limit is the function $f(t) = 0$. To determine whether the convergence is uniform:

$$d(f_n, f) = \sup_{t \in [-1/2, 1/2]} |t|^n = (\tfrac{1}{2})^n \to 0$$

as $n \to \infty$. So the convergence is uniform on $[-\frac{1}{2}, \frac{1}{2}]$.

If the same sequence is investigated on the interval $[0, 1)$, the conclusion is different: The pointwise limit is again a function $f(t) = 0$, but

$$d(f_n, f) = \sup_{t \in [0,1)} |t|^n = 1.$$

So $d(f_n, f) \not\to 0$, and convergence is merely pointwise but not uniform.

Example 2.10. Define $f_n(t) = 1/t + 1/n$ for $t \in (0, 1)$ and $n \in \mathbb{N}$. We note that the functions f_n and f are unbounded on $(0, 1)$, and that $f_n \to f$ pointwise on $(0, 1)$. Since $|f_n(t) - f(t)| = 1/n$ for all $t \in (0, 1)$, $\|f_n - f\|_\infty = 1/n \to 0$, and the sequence (f_n) converge uniformly to f on $(0, 1)$.

There is a useful characterization of the closure of a set in a pseudometric space in terms of convergence of sequences.

Theorem 2.11. *A point x in a pseudometric space (S, d) belongs to $\overline{A}$, where $A \subset S$, if and only if there is sequence (x_n) in A with $x_n \to x$.*

Proof.   See Problem 2.4. $\square$

Equivalence of pseudometrics can be conveniently characterized in terms of convergence of sequences:

Theorem 2.12. *Two pseudometrics d and ρ on the set S are equivalent if and only if for each sequence (x_n) in S*

$$x_n \xrightarrow{d} a \iff x_n \xrightarrow{\rho} a. \tag{2.1}$$

Proof.   Assume first that the pseudometrics d and ρ are equivalent and that $x_n \xrightarrow{d} a$. We want to prove that $x_n \xrightarrow{\rho} a$. If U is a ρ-open set, then U is also d-open. By Theorem 2.4 there is N such that $x_n \in U$ for all $n > N$. This means that $x_n \xrightarrow{\rho} a$. Interchanging the pseudometrics, we get $x_n \xrightarrow{\rho} a$ implies $x_n \xrightarrow{d} a$.

Conversely, suppose that (2.1) holds for every sequence in S. Using Theorems 2.11 and 1.54, we can show that a set $A \subset S$ is d-closed if and only if it is ρ-closed. The result follows on taking the complements. $\square$

The following definition is motivated by the Cauchy criterion for the convergence of real sequences.

Definition 2.13. A sequence (x_n) in a pseudometric space is said to be *Cauchy* if it satisfies the following condition: For each $\varepsilon > 0$ there is a positive integer N such that

$$m, n > N \implies d(x_m, x_n) < \varepsilon. \tag{2.2}$$

Theorem 2.14. *A Cauchy sequence in a pseudometric space is bounded.*

Proof. In the condition (2.2) choose $\varepsilon = 1$. Then there is an index N such that $d(x_m, x_n) < 1$ whenever $m, n > N$. This shows that the set $A = \{x_i : i > N\}$ is bounded. The set $\{x_n : n \in \mathbb{N}\}$ is bounded being the union of a finite set $B = \{x_1, \ldots, x_N\}$ and a bounded set A (see Theorem 1.24). $\square$

Theorem 2.15 (Bolzano–Cauchy theorem). *Any Cauchy sequence of real numbers is convergent.*

Proof. (Details given in Appendix B.) If (x_n) is Cauchy, it is bounded by the preceding theorem. We extract a monotonic subsequence (x_{k_n}) (Theorem B.7), which converges by the monotonic sequence theorem (Theorem B.8). Finally, we show that a Cauchy sequence with a convergent subsequence is itself convergent (Problem 2.8). $\square$

There is an equivalent way of expressing the Cauchy condition (2.2) in terms of a limit of a sequence as $m \wedge n := \min(m, n) \to \infty$:

$$\lim_{m \wedge n \to \infty} d(x_m, x_n) = 0.$$

The proof of this is left as an exercise.

A convergent sequence is always Cauchy (exercise). This means that every convergent sequence is bounded. However, in a general pseudometric or metric space, a Cauchy sequence is not always convergent.

Example 2.16. Consider the metric space (S, d) with $S = (0, 1]$ and $d(x, y) = |x - y|$. The sequence $x_n = 1/n$ is Cauchy: Given $\varepsilon > 0$, let N be the least positive integer such that $N > 2/\varepsilon$. Then, for any $m, n > N$,

$$d(x_m, x_n) = \left| \frac{1}{n} - \frac{1}{m} \right| \leq \frac{1}{m} + \frac{1}{n} \leq \frac{2}{N} < \varepsilon.$$

However, the sequence does not converge to any point of the given space S: If $a \in S$ and $n \geq 2/a$, then

$$d(x_n, a) = \left| \frac{1}{n} - a \right| = a - \frac{1}{n} \geq \frac{a}{2}.$$

This shows that (x_n) does not converge to a in (S, d).

Example 2.17. Let $S = \mathbb{R}$ and let $\rho(x, y) = |f(x) - f(y)|$, where $f(x) = x/(1 + |x|)$. Then the sequence $x_n = n$ is Cauchy, but does not converge to any point in the metric space $(\mathbb{R}, \rho)$. (Exercise.)

The Bolzano–Cauchy theorem motivates the following definition.

Definition 2.18. A pseudometric space (S, d) is said to be *complete* if every Cauchy sequence of points in S converges to a point in S. A space which is not complete is called *incomplete*.

Example 2.19. The Euclidean space $(\mathbb{R}, d)$ is complete in view of the Bolzano–Cauchy theorem. Completeness distinguishes the reals from the rationals.

Example 2.20. The Euclidean space $(\mathbb{R}^N, d)$ is complete (Problem 2.6).

Example 2.21. The pseudometric space $(\mathbb{R}^2, p)$ of Example 1.6 is complete.

Example 2.22. Let $S = B(E)$ be the set of all bounded real or complex valued functions on a nonempty set E equipped with the uniform metric d. Then the metric space (S, d) is complete (Problem 2.7).

Example 2.23. The metric space ℓ^∞ introduced in Section 1 of Chapter 1 is complete. This is a consequence of the fact that ℓ^∞ consists of bounded functions $x \colon \mathbb{N} \to \mathbb{R}$, that is, $\ell^\infty = B(\mathbb{N})$ with the uniform metric. This space is complete by the preceding example.

Examine the list of standard metric spaces given in Appendix D to see which are complete.

The following result is a useful criterion of completeness, provided the given space can be embedded into another pseudometric space known to be complete.

Theorem 2.24. *Let A be a closed set in a complete pseudometric space (S, d). Then the pseudometric space (A, d) is also complete.*

Proof. Let (x_n) be a Cauchy sequence in A. Then (x_n) is also Cauchy in (S, d), and so there is $a \in S$ such that $x_n \to a$. By the sequential characterization of closure (Theorem 2.11), the point a lies in the closure $\overline{A}$ of A. Since A is closed, $\overline{A} = A$, and so $a \in A$. □

Example 2.25. Let $\mathbf{c}$ be the set of all $\mathbb{R}^{\mathbb{N}}$ vectors $x = (\xi_1, \xi_2, \dots)$ for which the sequence (ξ_n) of coordinates converges in $\mathbb{R}$, equipped with the uniform metric

$$d(x, y) = \sup_{n \in \mathbb{N}} |\xi_n - \eta_n|.$$

In view of the preceding theorem it is enough to prove that $\mathbf{c}$ is closed in the metric space (ℓ^∞, d), which is known to be complete.

Let $x = (\xi_n)$ be a point of the closure of $\mathbf{c}$ in ℓ^∞. Given $\varepsilon > 0$, there exists at least one point $a = (\alpha_1, \alpha_2, \dots)$ in $\mathbf{c}$ with $d(x, a) < \frac{1}{3}\varepsilon$. By the definition of $\mathbf{c}$, the sequence (α_n) of real numbers converges to a real number; being convergent, it is also Cauchy, and there exists $N \in \mathbb{N}$ such that $|\alpha_m - \alpha_n| < \frac{1}{3}\varepsilon$ if $m, n \geq N$. Let $m, n \geq N$. Then

$$\begin{aligned}
|\xi_m - \xi_n| &= |(\xi_m - \alpha_m) + (\alpha_m - \alpha_n) + (\alpha_n - \xi_n)| \\
&\leq |\xi_m - \alpha_m| + |\alpha_m - \alpha_n| + |\alpha_n - \xi_n| \\
&\leq d(x, a) + |\alpha_m - \alpha_n| + d(x, a) \\
&= \tfrac{1}{3}\varepsilon + \tfrac{1}{3}\varepsilon + \tfrac{1}{3}\varepsilon = \varepsilon,
\end{aligned}$$

that is, (ξ_n) is Cauchy. In view of the Bolzano–Cauchy theorem the sequence (ξ_n) converges in $\mathbb{R}$, and so $x = (\xi_n) \in \mathbf{c}$. We proved that every point of the closure of $\mathbf{c}$ is in $\mathbf{c}$, which implies that $\mathbf{c}$ is closed in ℓ^∞.

2.2 Completions

There are many results in metric space theory that are true only when the metric space is complete, for example the contraction mapping theorem and Baire's theorem. It is therefore important to know that every incomplete metric space can be embedded into a complete metric space by adjoining new points to the original space.

The proof of the following theorem is quite straightforward, so the reader is asked to verify several technical details as their full discussion in print could obscure the basic simplicity of the idea. As most applications of completion are to metric spaces, we restrict the discussion to metric space, and leave the pseudometric case to the reader.

Theorem 2.26 (Completion theorem). *For every incomplete metric space (S, d) there exists a metric space $(S^\sharp, d^\sharp)$ satisfying the following conditions:*

(i) *$(S^\sharp, d^\sharp)$ is complete,*

(ii) *$S \subset S^\sharp$,*

(iii) *$d^\sharp$ is an extension of d,*

(iv) *S is dense in $(S^\sharp, d^\sharp)$, that is, $\overline{S} = S^\sharp$.*

Proof.   First we introduce some terminology. Two Cauchy sequences (x_n) and (y_n) in S are said to be *equivalent*, written $(x_n) \sim (y_n)$, if $d(x_n, y_n) \to 0$. It is not difficult to check that this is a true equivalence, that is, a reflexive, symmetric and transitive relation on Cauchy sequences in S. We observe that two equivalent Cauchy sequences in S either both converge in S (to the same point) or neither converges. This enables us to define *ideal points* of S as the equivalence classes $\alpha = [(x_n)]$ of Cauchy sequences on S where (x_n) has no limit in S. We denote the set of all ideal points of S by W, and will refer to the points of S as *ordinary points*.

Put $S^\sharp = S \cup W$. We define a metric $d^\sharp$ on $S^\sharp$ as follows:

(i) If $x, y \in S$, we set
$$d^\sharp(x, y) = d(x, y).$$

(ii) If $x \in S$ and α is an ideal point, we define
$$d^\sharp(\alpha, x) = \lim_{n \to \infty} d(x_n, x),$$
where (x_n) is a representative of α. We have to prove that the limit exists, and that it is independent of the choice of (x_n). This is left as an exercise.

(iii) Finally if α, β are two ideal points in S corresponding to Cauchy sequences (x_n), (y_n) in S, we define
$$d^\sharp(\alpha, \beta) = \lim_{n \to \infty} d(x_n, y_n).$$
Again we have to prove that the limit exists and is independent of the choice of the sequences (x_n) and (y_n).

We leave it to the reader as an exercise to prove that $d^\sharp$ satisfies the Hausdorff axioms.

Next we prove that S is dense in $S^\sharp$ and that $(S^\sharp, d^\sharp)$ is a complete space. The denseness of S in $S^\sharp$ follows from the fact that if α is an ideal point corresponding to a Cauchy sequence (x_n), then
$$d^\sharp(x_n, \alpha) \to 0.$$

Indeed, if $\varepsilon > 0$ and N is such that $d(x_n, x_k) \leq \varepsilon$ for $n, k \geq N$, then $d^\sharp(x_n, \alpha) = \lim_{k \to \infty} d(x_n, x_k) \leq \varepsilon$ whenever $n \geq N$.

To prove the completeness of $S^\sharp$, consider an arbitrary Cauchy sequence (x_n) in $S^\sharp$. The terms (x_n) can be ordinary or ideal points of S. We construct a sequence (y_n) of ordinary points of S as follows: If $x_n \in S$, put $y_n = x_n$. If $x_n \in W$, there is an ordinary point y_n of S such that $d^\sharp(x_n, y_n) < 1/n$ (the denseness of S in $S^\sharp$). The sequence (y_n) is Cauchy in (S, d) as

$$d(y_m, y_n) \leq d^\sharp(y_m, x_m) + d^\sharp(x_m, x_n) + d^\sharp(x_n, y_n) < \frac{1}{m} + d(x_m, x_n) + \frac{1}{n}.$$

So either the sequence (x_n) converges to a point x in S or there is an ideal point α of S corresponding to (y_n). In the latter case we show that $x_n \to \alpha$ in $(S^\sharp, d^\sharp)$:

$$d^\sharp(x_n, \alpha) \leq d^\sharp(x_n, y_n) + d^\sharp(y_n, \alpha) \leq \frac{1}{n} + d^\sharp(y_n, \alpha) \to 0.$$

This completes the proof of the theorem. □

Any metric space $S^\sharp$ satisfying the condition (i)–(iv) of Theorem 2.26 is called a *completion of S*. This definition does not specify the nature of the ideal points, that is, of the points of $S^\sharp \setminus S$. It is interesting to know that two different completions of the same metric space are metrically identical in some sense.

Definition 2.27. A mapping $f \colon (S_1, d_1) \to (S_2, d_2)$ between metric spaces is called an *isometry* if

$$d_2(f(x), f(y)) = d_1(x, y) \text{ for all } x, y \in S_1.$$

Two metric spaces S_1, S_2 are said to be *isometric* if there is a surjective isometry $f \colon S_1 \to S_2$. Any property which is preserved under an isometry is called a *metric property*.

Observe that any isometry between metric spaces is injective.

Theorem 2.28. *Any two completions of the same metric space (S, d) are isometric.*

Proof. Let (S_1, d_2) and (S_2, d_2) be completions of (S, d). Since S is dense in S_1, for every point $x \in S_1$, ordinary or ideal, there is a sequence (x_n) in S with $x_n \xrightarrow{d_1} x$. From the completeness of S_2 we can deduce that there is $y \in S_2$ with $x_n \xrightarrow{d_2} y$. Such point y is unique since both d_1 and

d_2 are extensions of d. The map $f \colon S_1 \to S_2$ which assigns to each point $x \in S_1$ the unique point $y \in S_2$ obtained in the foregoing construction is the desired isometry of S_1 onto S_2. □

From the definition of isometry it is clear that two isometric metric spaces are either both complete or both incomplete, and that the completions of two isometric spaces are isometric. The following theorem, whose proof is left to the reader as an exercise, provides a handy tool for constructing completions.

Theorem 2.29. *Let $f \colon (S, d) \to (T, \rho)$ be an isometry of S into T, that is, $\rho(f(x), f(y)) = d(x, y)$ for all $x, y \in S$. If S is incomplete and T complete, then the completion of (S, d) is isometric to $(\overline{f(S)}, \rho)$.*

Example 2.30. Complete the space $(\mathbb{R}, \rho)$, where $\rho(x_1, x_2) = |f(x_1) - f(x_2)|$ with f defined by $f(x) = x/(1 + |x|)$.

First we verify that $f \colon (\mathbb{R}, \rho) \to ((-1, 1), d)$ is an isometry, where d is the Euclidean metric on $\mathbb{R}$: The function $g \colon (-1, 1) \to \mathbb{R}$ defined by $g(y) = y/(1 - |y|)$ is the inverse of f, and $d(f(x_1), f(x_2)) = \rho(x_1, x_2)$ for all $x_1, x_2 \in \mathbb{R}$. The completion of $((-1, 1), d)$ is the closure of $(-1, 1)$ in $(\mathbb{R}, d)$, that is, $[-1, 1]$.

The completion of $(\mathbb{R}, \rho)$ will be isometric to $([-1, 1], d)$. Let α, β be two distinct elements not belonging to $\mathbb{R}$, and set $\mathbb{R}^\sharp = \mathbb{R} \cup \{\alpha, \beta\}$. The functions f and g can be extended to $f^\sharp \colon \mathbb{R}^\sharp \to [-1, 1]$ and $g^\sharp \colon [-1, 1] \to \mathbb{R}^\sharp$, respectively, by setting

$$f^\sharp(\alpha) = -1, \quad f^\sharp(\beta) = 1, \qquad g^\sharp(-1) = \alpha, \quad g^\sharp(1) = \beta;$$

the extended functions are isometries if we set $\rho^\sharp(x_1, x_2) = |f^\sharp(x_1) - f^\sharp(x_2)|$ for all $x_1, x_2 \in \mathbb{R}^\sharp$. The completion of $(\mathbb{R}, \rho)$ is the space $(\mathbb{R}^\sharp, \rho^\sharp)$.

Note. We may write $\alpha = -\infty$, $\beta = +\infty$; in this context, $-\infty$ and $+\infty$ cease to be mere symbols, and become actual elements of the completion $\mathbb{R}^\sharp$. In this notation we often write $\overline{\mathbb{R}} = \mathbb{R} \cup \{-\infty, +\infty\}$, and call $\overline{\mathbb{R}}$ the *extended real line*.

Example 2.31. The space $(\mathbb{Q}, d)$, where $\mathbb{Q}$ is the set of all rational numbers, and $d(x, y) = |x - y|$, has as its completion the Euclidean space $(\mathbb{R}, d)$. This is a consequence of the density of rationals in $\mathbb{R}$ (Appendix B.1.5). The ideal points of the completion are the irrational numbers, which can be always expressed as limits of Cauchy sequences of rationals.

Example 2.32. Another completion of the real line. First we embed the real line into the Euclidean plane $\mathbb{R}^2$ as the x axis. Then for each point x construct the line L_x joining the points $(0,1)$ and $(x,0)$. Define $f(x)$ as the point of intersection of L_x with the circle $C = \{(x,y) \in \mathbb{R}^2 : x^2 + y^2 = 1\}$. This construction is known as a stereographic projection of the x-axis from the point $(0,1)$. It can be checked that

$$f(x) = \left(\frac{2x}{1+x^2}, \frac{x^2-1}{1+x^2} \right).$$

We define the metric σ on $\mathbb{R}$ by

$$\sigma(x,x') = d(f(x),f(x')) = \frac{2|x-x'|}{\sqrt{1+x^2}\sqrt{1+x'^2}},$$

where d is the Euclidean metric in $\mathbb{R}^2$. Then f is an isometry of $\mathbb{R}$ onto the punctured circle $C \setminus \{(0,1)\}$. The completion of $(\mathbb{R}, \sigma)$ is isometric to the closure C of $C \setminus \{(0,1)\}$ in $\mathbb{R}^2$, equal to $(\mathbb{R} \cup \{\infty\}, \sigma_1)$, where σ_1 is an extension of σ satisfying $\sigma_1(x, \infty) = 2(1+x^2)^{-1/2}$. Note that the sequences (n), $(-n)$ and $((-1)^n n)$ are Cauchy in $(\mathbb{R}, \sigma)$, and determine the one and only ideal point ∞.

2.2.1 *Completion of a normed space*

Let $(X, \|\cdot\|)$ be a normed space with metric d in X induced by the norm, where $d(x,y) = \|x-y\|$ for all $x,y \in X$. The completeness of $(X, \|\cdot\|)$ means the completeness of (X,d). A series $\sum_{n=1}^{\infty} x_n$ whose terms x_n are elements of a normed space X *converges* to $x \in X$ if the sequence of partial sums converges to x, that is,

$$\lim_{n \to \infty} \left\| \sum_{k=1}^{n} x_k - x \right\| = 0.$$

A series $\sum_{n=1}^{\infty} x_n$ is *norm-absolutely convergent* if $\sum_{n=1}^{\infty} \|x_n\| < \infty$.

Theorem 2.33. *A normed space $(X, \|\cdot\|)$ is complete if and only if every norm-absolutely convergent series $\sum_{n=1}^{\infty} x_n$ converges in X.*

Proof. Suppose that the space is complete and $\sum_{n=1}^{\infty} \|x_n\| < \infty$. We set $s_n := \sum_{k=1}^{n} x_k$ and show that the sequence (s_n) is Cauchy. This follows from the estimate

$$\|s_n - s_m\| = \left\| \sum_{k=m+1}^{n} x_k \right\| \leq \sum_{k=m+1}^{n} \|x_k\| \leq \sum_{k=m+1}^{\infty} \|x_k\|$$

with $n > m$.

Conversely, assume that whenever $\sum_{n=1}^{\infty} \|x_n\| < \infty$, the series $\sum_{n=1}^{\infty} x_n$ converges in X. Let (s_n) be a Cauchy sequence. In view of Problem 2.8 it is enough to show that we can extract a convergent subsequence. We can select a strictly increasing sequence (k_n) of positive integers such that $\|s_{k_n} - s_{k_{n-1}}\| < (\frac{1}{2})^n$ for all $n \geq 2$. Set $x_1 = s_{k_1}$ and $x_n = s_{k_n} - s_{k_{n-1}}$ for $n \geq 2$. Then $\sum_{n=1}^{\infty} \|x_n\|$ converges, and consequently $s_{k_n} = \sum_{i=1}^{k_n} x_i$ converges in X.   □

The preceding theorem suggests an alternative method for the completion of an incomplete normed space using series instead of sequences. An ideal point α will be determined as the equivalence class of nonconvergent series $\sum_{n=1}^{\infty} x_n$ satisfying $\sum_{n=1}^{\infty} \|x_n\| < \infty$ under the equivalence

$$\sum_{n=1}^{\infty} x_n \sim \sum_{n=1}^{\infty} y_n \iff \lim_{n \to \infty} \left\| \sum_{k=1}^{n} (x_k - y_k) \right\| = 0.$$

The norm of α will be given by

$$\|\alpha\| = \lim_{n \to \infty} \left\| \sum_{k=1}^{n} x_k \right\|$$

for any series $\sum_{n=1}^{\infty} x_n$ determining the ideal point α.

2.2.2   *Lebesgue integrable functions in $\mathbb{R}^k$*

We devote the rest of the section to a construction by completion of the space L^1 of Lebesgue integrable functions on $\mathbb{R}^k$. This construction can be skipped on the first reading as the space L^1 described here is a special case of the space $L^1(S, \Sigma, \mu)$ discussed in Part III of this book. In order that L^1 can be used as an example of a metric space without reference to the general theory of integration, we sketch a simplified construction that takes advantage of special properties of $\mathbb{R}^k$. L^1 is obtained by the completion of the space of all step functions on $\mathbb{R}^k$ under an integral metric described below. A function $f \colon \mathbb{R}^k \to \mathbb{C}$ is called a *step function* if there is a closed k-cell K which can be expressed as the finite union of non-overlapping closed k-cells E_i $(i = 1, \ldots, n)$ such that f is constant on the interior of each cell E_i, and vanishes outside K. The integral of such a step function is defined by

$$\int f = \sum_{i=1}^{n} c_i \, \mathsf{vol} \, (E_i), \tag{2.3}$$

where $\mathsf{vol} \, (E_i)$ is the Euclidean k-volume of the cell E_i, and c_i is the constant value f takes on the interior of E_i. The set S of all step functions in $\mathbb{R}^k$ is a linear space with the property that $f \in S$ implies $|f| \in S$. A pseudometric d on S is generated by the so called L^1 seminorm,

$$d(f, g) = \|f - g\| = \int |f - g|. \tag{2.4}$$

We show that every ideal point of S can be identified with a certain function f on $\mathbb{R}^k$. For this we need the following concept. We say that a set $A \subset \mathbb{R}^k$ is a *null set* if for every $\varepsilon > 0$ there exists a sequence (E_n) of k-cells such that $A \subset \bigcup_{n=1}^{\infty} E_n$ and $\sum_{n=1}^{\infty} \text{vol}(E_n) < \varepsilon$. Examples of null sets in $\mathbb{R}^k$ are countable sets and faces of k-cells. In the context of integration, null sets are very small, and the complement A^c of a null set can be regarded as 'almost all' of $\mathbb{R}^k$. A *full set* is defined to be the complement of a null set in $\mathbb{R}^k$. We say that a statement holds *almost everywhere* if it holds on a full set in $\mathbb{R}^k$. By identifying any two step functions which are equal almost everywhere in $\mathbb{R}^k$, we convert the seminorm $\|f\| = \int |f|$ into a norm, called the L^1-norm. The following two results concerning step functions f_n are at the heart of our construction.

Theorem 2.34. *Let (f_n) be a sequence of step functions on $\mathbb{R}^k$ such that the series $\sum_{n=1}^{\infty} \|f_n\|$ converges. Then there exists a full set A such that the series $\sum_{n=1}^{\infty} |f_n(t)|$ converges for all $t \in A$.*

Theorem 2.35. *Let $\sum_n f_n$ be a norm-absolutely convergent series in the space (S, d) of step functions. Then*

$$\sum_{n=1}^{\infty} f_n = 0 \ \text{almost everywhere} \iff \lim_{n \to \infty} \left\| \sum_{k=1}^{n} f_n \right\| = 0.$$

We do not prove these results here as they will be proved later in Part III in a more general setting.

The construction of Lebesgue integrable functions is based on the completion of the normed space described in the preceding paragraph, and on Theorems 2.34 and 2.35. We say that a function f on $\mathbb{R}^k$ is *Lebesgue integrable* if there exists a series $\sum_n f_n$ of step functions such that $\sum_n \|f_n\| < \infty$ and $f(t) = \sum_n f_n(t)$ for all t in a full set. Such a series is called an *approximating series for f*. By $L^1 = L^1(\mathbb{R}^k)$ we denote the set of all Lebesgue integrable functions on $\mathbb{R}^k$. Note that every step function f is Lebesgue integrable with an approximating series $f + 0 + 0 + \cdots$.

By passing to a subsequence if necessary, we can convert every Cauchy sequence (f_n) of step functions to a norm-absolutely convergent series. An *approximating sequence* of a Lebesgue integrable function f is an L^1-norm Cauchy sequence of step functions (f_n) which converges almost everywhere to f. Theorems 2.34 and 2.35 then show that there is a one-to-one correspondence between points of the completion of (S, d) and Lebesgue integrable functions (up to equality almost everywhere).

If f is a Lebesgue integrable function on $\mathbb{R}^k$ with an approximating sequence (f_n), the integral of f is defined by

$$\int f = \lim_{n \to \infty} \int f_n.$$

The limit exists as the numerical sequence $\int f_n$ is Cauchy:

$$\left| \int f_m - \int f_n \right| = \left| \int (f_m - f_n) \right| \le \int |f_m - f_n| = \|f_m - f_n\|.$$

The definition of integral is independent of the choice of a Cauchy sequence (f_n) which converges pointwise to f on a full set A. This follows from Theorem 2.35 when we convert Cauchy sequences of step functions to norm-absolutely convergent series by passing to subsequences as in the preceding paragraph. The details are left to the reader.

According to the completion theorem, we can extend the integral metric $d(f,g) = \|f - g\|$ from (S, d) to L^1; clearly it is enough to extend the norm $\|\cdot\|$ by requiring $\|f\| = \lim_{n \to \infty} \|f_n\|$ for any approximating sequence (f_n) for f. Since $(|f_n|)$ is an approximating sequence for $|f|$, from the definition of the integral we get

$$\|f\| = \int |f| \text{ for all } f \in L^1.$$

The normed space $(L^1, \|\cdot\|)$ is complete as it is the completion of the space of step functions under the integral norm.

2.3   Contraction mapping theorem

The contraction mapping theorem is due to the celebrated Polish mathematician Stefan Banach (1892–1945). It is an existence theorem, which also gives a constructive procedure for finding a fixed point of a mapping. It has a wide range of applications to algebraic, integral and differential equations.

Definition 2.36. Let f be a *selfmap* of a nonempty set S, that is, let $f \colon S \to S$. A point u is a *fixed point* of f if $f(u) = u$. A selfmap f of a metric space (S, d) is called a *contraction* if there is constant α satisfying $0 \le \alpha < 1$ such that

$$d(f(x), f(y)) \le \alpha d(x, y) \text{ for all } x, y \in S.$$

The constant α is called a *coefficient of contraction* for f.

For a function $f \colon [a, b] \to \mathbb{R}$ to be a selfmap of $[a, b]$, the graph of f must be contained in the square $[a, b] \times [a, b]$ (Figure 2.1).

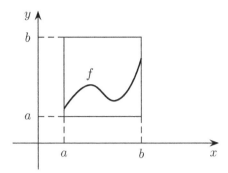

Figure 2.1

Theorem 2.37 (Contraction mapping theorem).

A contraction $f \colon S \to S$ on a complete metric space (S, d) has a unique fixed point u in S.

Proof. Let f be a contraction on S with a coefficient α. For any $x, y \in S$, $d(x, y) \le d(x, f(x)) + d(f(x), f(y)) + d(f(y), y) \le d(x, f(x)) + \alpha d(x, y) + d(f(y), y)$, and

$$d(x, y) \le \frac{1}{1 - \alpha}(d(x, f(x)) + d(f(y), y)). \tag{2.5}$$

This proves that there can be at most one fixed point for f.

Choose an arbitrary point x_0 in S, and define a sequence inductively by $x_{n+1} = f(x_n)$, $n = 0, 1, 2 \ldots$ First verify by induction that

$$d(x_k, x_{k+1}) \le \alpha^k d(x_0, x_1), \quad k = 0, 1, 2, \ldots \tag{2.6}$$

If $m, n \in \mathbb{N}$, then by (2.5),

$$d(x_n, x_m) \le \frac{1}{1 - \alpha}(d(x_n, x_{n+1}) + d(x_m, x_{m+1})) \le \frac{\alpha^n + \alpha^m}{1 - \alpha} d(x_0, x_1) \to 0$$

as $\min\{m, n\} \to \infty$, which means that (x_n) is a Cauchy sequence. From the completeness of S we deduce that there is a point $u \in S$ with $x_n \xrightarrow{d} u$. Since

$$d(u, f(u)) \le d(u, x_{n+1}) + d(f(x_n), f(u)) \le d(u, x_{n+1}) + \alpha \cdot d(x_n, u) \to 0$$

as $n \to \infty$, u is a fixed point of f in view of the Hausdorff axiom **M2**. $\square$

Note that the contraction mapping theorem has three hypotheses, each of which has to be verified before we can draw a conclusion:

(i) A mapping f is a selfmap of metric space (S, d), that is, $f \colon S \to S$.

(ii) f is a contraction on S.

(iii) The space (S, d) is complete.

From the proof of Theorem 2.37 we glean the following constructive procedure:

Theorem 2.38 (Picard iterations). *Under the hypotheses of Theorem 2.37 let x_0 be an arbitrary point in S, and let (x_n) be the sequence of the so called Picard iterations defined by*

$$x_{n+1} = f(x_n), \qquad n = 0, 1, 2, \ldots$$

Then (x_n) converges to the unique fixed point u of f.

Example 2.39. Show that the function $f(x) = 1/(1+x^2)$ satisfies the hypotheses of the contraction mapping theorem on the metric space $([0, 1], d)$, where d is the Euclidean metric. Relate this result to the solution of the equation

$$x^3 + x - 1 = 0$$

on the interval $[0, 1]$.

(i) To see that f is a selfmap of $[0, 1]$, we observe that f is decreasing on $[0, 1]$ since $g(x) = 1 + x^2$ is increasing and positive, and f is the reciprocal of g. So

$$f([0, 1]) = [f(1), f(0)] = [\tfrac{1}{2}, 1] \subset [0, 1],$$

and the graph of f is contained in the square $[0, 1] \times [0, 1]$.

(ii) To determine whether f is a contraction on $[0, 1]$, we utilize the result of Problem 2.18 applicable to differentiable selfmaps of $[0, 1]$: The function f is a contraction if and only if $\sup_{x \in [0,1]} |f'(x)| = \alpha$ is less than 1; the constant α is then the optimum coefficient of contraction. We calculate that

$$f'(x) = \frac{-2x}{(1+x^2)^2}, \qquad f''(x) = 2 \cdot \frac{3x^2 - 1}{(1+x^2)^3}.$$

From the sign of f'' we see that f' is decreasing on the interval $[0, 1/\sqrt{3}]$ and increasing on $[1/\sqrt{3}, 1]$. The absolute maximum of $|f'|$ on $[0, 1]$ is given by

$$\sup_{x \in [0,1]} |f'(x)| = \max\left(|f'(0)|, |f'(3^{-1/2})|, |f'(1)|\right) = \max(0, \tfrac{3\sqrt{3}}{8}, \tfrac{1}{4}) = \tfrac{3\sqrt{3}}{8},$$

where $(3\sqrt{3})/8 \approx 0.65$. So f is a contraction on $[0, 1]$ with the optimum coefficient of contraction $\alpha = (3\sqrt{3})/8$.

(iii) The space $([0, 1], d)$ is complete as $[0, 1]$ is a closed subset of the complete space $(\mathbb{R}, d)$.

Since the hypotheses of the contraction theorem are satisfied for f on $[0, 1]$, f has a unique fixed point $u \in [0, 1]$. As $u = f(u)$, we have

$$u = \frac{1}{1 + u^2} \qquad \Rightarrow \qquad u^3 + u - 1 = 0.$$

We conclude that the equation $x^3 + x - 1 = 0$ has a unique solution on the interval $[0, 1]$. We can approximate u by Picard iterations starting, say, with $x_0 = 0$: Define $x_{n+1} = 1/(1 + x_n^2)$ for $n = 0, 1, 2, \ldots$, to get

$$x_1 = 1, \qquad x_2 = \tfrac{1}{2}, \qquad x_3 = \tfrac{4}{5}, \qquad x_4 = \tfrac{25}{41}, \qquad x_5 = \tfrac{1681}{2306}.$$

Continuing with the iterations, we would get $x_9 \approx .69$, $x_{10} \approx .68$, etc.

The French mathematician Émile Picard used an iterative process to prove the existence of solutions to ordinary differential equations. We can recast the proof in terms of the contraction mapping theorem on a special complete metric space of continuous functions.

Theorem 2.40 (Picard's theorem). *Let f be a function of two variables x, y defined and continuous on a strip*

$$\Sigma = \{(x, y) \in \mathbb{R}^2 : |x - x_0| \leq a, \ y \in \mathbb{R}\},$$

satisfying the Lipschitz condition

$$|f(x, y_1) - f(x, y_2)| \leq K |y_1 - y_2|, \quad \text{for all } (x, y_1), \ (x, y_2) \in \Sigma,$$

for some $K > 0$. Then, for any given $y_0 \in \mathbb{R}$, the differential problem

$$\frac{dy}{dx} = f(x, y), \qquad y(x_0) = y_0, \tag{2.7}$$

has a unique solution $y = y(x)$ defined for $|x - x_0| \leq a$.

Proof.   The differential problem (2.7) is equivalent to the integral equation

$$y(x) = y_0 + \int_{x_0}^{x} f(t, y(t)) \, dt. \tag{2.8}$$

Consider first the interval $J = [x_0, x_0 + a]$. For a given constant $k > 0$ we define the metric ρ_k on the set $C(J)$ of all continuous functions on J by

$$\rho_k(y, z) = \sup_{x \in J} e^{-k(x - x_0)} |y(x) - z(x)|;$$

this metric is closely related to the uniform metric on $C(J)$, with the weight function $w(x) = e^{-k(x-x_0)}$ thrown in to obtain a contraction with a minimum of effort. It can be proved similarly as for the uniform metric that the space $(C(J), \rho_k)$ is complete (for any $k > 0$).

Define a mapping T on $C(J)$ by

$$(T(y))(x) = y_0 + \int_{x_0}^{x} f(t, y(t))\, dt, \quad y \in C(J).$$

Then $T(y)$ is a continuous function of the interval J, which means that T is a mapping of $C(J)$ into itself—a selfmap. Observe that a solution y to the fixed point equation $y = Ty$ is a solution to the original differential equation problem (2.7). We show that, with an appropriate choice of the constant k, the mapping T is a contraction on $(C(J), \rho_k)$. Let y_1, y_2 be two elements of $C(J)$. Then

$$|(T(y_1))(x) - (T(y_2))(x)| = \left| \int_{x_0}^{x} (f(t, y_1(t)) - f(t, y_2(t))\, dt \right|$$

$$\leq \int_{x_0}^{x} |f(t, y_1(t)) - f(t, y_2(t))|\, dt \leq K \int_{x_0}^{x} |y_1(t) - y_2(t)|\, dt$$

and

$$e^{-k(x-x_0)}|(T(y_1))(x) - (T(y_2))(x)|$$

$$\leq K \int_{x_0}^{x} e^{-k(x-t)} \{ e^{-k(t-x_0)} |y_1(t) - y_2(t)| \}\, dt$$

$$\leq K \int_{x_0}^{x} e^{-k(x-t)}\, dt \cdot \rho_k(y_1, y_2) \leq \frac{K}{k} \rho_k(y_1, y_2).$$

Taking the supremum over $x \in J$, we see that

$$\rho_k(T(y_1), T(y_2)) \leq \frac{K}{k} \rho_k(y_1, y_2).$$

So, if $\rho = \rho_{2K}$, we have

$$\rho(T(y_1), T(y_2)) \leq \tfrac{1}{2} \rho(y_1, y_2),$$

which shows that T is a contraction with a coefficient $\frac{1}{2}$. Then T has a unique fixed point $u \in C(J)$; that is, there is a unique solution u to the original differential problem (2.7).

Similarly we can show that there is a unique solution to the differential problem (2.7) on the interval $[x_0 - a, x_0]$, and then join the two solutions and show that the resulting function is a unique solution to the problem on the interval $[x_0 - a, x_0 + a]$. □

2.4   Problems for Chapter 2

1.   Describe convergent sequences in the metric space $(\mathbb{Z}, d)$, where d is defined by $d(a, b) = |a - b|$. Show that this space is complete.

2.   Show that in the Euclidean space $(\mathbb{R}^N, d)$ a sequence $x_n = (\xi_{n1}, \dots, \xi_{nN})$ converges to a point $y = (\eta_1, \dots, \eta_N)$ if and only if $\lim_{n \to \infty} |\xi_{nk} - \eta_k| = 0$ for each index $k \in \{1, \dots, N\}$.

3.   *Hilbert cube.* Let Q be the subset of ℓ^2 consisting of all $x = (\xi_1, \xi_2, \dots)$ with $|\xi_k| \leq 1/k$ for all $k \in \mathbb{N}$. Prove that $x_n \to y$ in Q if and only if $\xi_{nk} \to \eta_k$ for each k.

4.   *Limit characterization of closure.* Prove that a point a in a pseudometric space S is in the closure of $A \subset S$ if and only if there is a sequence (x_n) in A such that $x_n \to a$ in S.

5.   Find the limit function f of the sequence (f_n) and determine whether the convergence $f_n \to f$ is uniform on the interval indicated:

(i) $f_n(x) = -\dfrac{x^n}{n}$ on $[0, 1]$

(ii) $f_n(x) = \dfrac{nx}{1 + nx}$ on $[0, +\infty)$;   on $(a, +\infty)$, where $a > 0$

(iii) $f_n(x) = \dfrac{nx^2 + 1}{nx + 1}$ on $(0, 1)$;   on $(a, 1)$, where $0 < a < 1$

(iv) $f_n(x) = n^2 x (1 - x)^n$ on $[0, 1]$;   on $[a, 1]$, where $0 < a < 1$

6.   Using Problem 2.2 and the Bolzano–Cauchy theorem for real numbers, prove that the Euclidean space $(\mathbb{R}^N, d)$ is complete.

7.   Prove that the space $(B(E), d)$ of all bounded real (or complex) valued functions on a set E equipped with the uniform (supremum) metric d is complete.

8.   Prove that a Cauchy sequence in a pseudometric space which has a convergent subsequence is convergent.

9.   *Weierstrass M–test.* Let the functions $f_1, f_2, f_3, \dots$ on a set E be real or complex valued. Suppose that, for each n, there is a constant M_n such that $|f_n(x)| \leq M_n$ for all $x \in E$, where $\sum_{n=1}^{\infty} M_n$ is a convergent positive term series. Prove that the series $\sum_{n=1}^{\infty} f_n$ of functions converges uniformly on E.

Suggestion. Show that the sequence (S_N) of partial sums $S_N = f_1 + f_2 + \dots + f_N$ is Cauchy in the space $(B(E), d)$, where d is the uniform metric.

10.   *Interchange of limits.* Let (S, d) be a metric space, (T, ρ) a complete metric space, and (f_n) a sequence of functions $f_n \colon S \to T$ convergent uniformly to $f \colon S \to T$ on S. Suppose that for a given point $c \in S$, $\lim_{x \to c} f_n(x)$ exists in T for each n. Prove that

$$\lim_{x \to c} \lim_{n \to \infty} f_n(x) = \lim_{n \to \infty} \lim_{x \to c} f_n(x).$$

Suggestion. First show that the sequence (b_n), where $b_n = \lim_{x \to c} f_n(x)$, is Cauchy in T; write $b = \lim_{n \to \infty} b_n$ and consider the inequality

$$\rho(f(x), b) \leq \rho(f(x), f_n(x)) + \rho(f_n(x), b_n) + \rho(b_n, b).$$

11.   Prove that the metric space $(C[a,b], d)$ of all continuous real (or complex) valued functions on $[a,b]$ with the supremum metric is complete.

12.   Show that, for each real $p \geq 1$, the metric space (ℓ^p, d_p) is complete.

Suggestion. Let $x_n = (\xi_{n1}, \xi_{n2}, \xi_{n3}, \dots)$ be Cauchy in ℓ^p. Show first that $\xi_k = \lim_{n \to \infty} \xi_{nk}$ exists for each k. (For this note that $|\xi_{nk} - \xi_{mk}| \leq d_p(x_n, x_m)$.) Set $x = (\xi_1, \xi_2, \xi_3, \dots)$ and prove that $d_p(x_n, x) \to 0$.

13.   Prove that the Cartesian product of k metric spaces equipped with the product metric is complete if and only if each of the component spaces is complete.

14.   For any two vectors $x = (\xi_1, \xi_2, \xi_3, \dots)$, $y = (\eta_1, \eta_2, \eta_3, \dots)$ in $\mathbb{R}^\mathbb{N}$ define

$$d(x, y) = \sum_{k=1}^{\infty} (\tfrac{1}{2})^k \frac{|\xi_k - \eta_k|}{1 + |\xi_k - \eta_k|}.$$

Show that $x_n \to a$ in $(\mathbb{R}^\mathbb{N}, d)$ if and only if for each index k, $\lim_{n \to \infty} |\xi_{nk} - \alpha_k| = 0$. Conclude that this metric space is complete.

15.   Let E_0 be the space of all $\mathbb{R}^\mathbb{N}$ vectors $x = (\xi_1, \xi_2, \dots)$ with only a finite number of nonzero coordinates. Show that the metric space (E_0, d_∞), where d_∞ is the ℓ^∞ metric, is incomplete.

16.   In each of the following cases indicate why (S, ρ) is a metric space, show that it is incomplete, and find its completion (S_1, ρ_1):

(i) $S = (0, 1)$, $\rho(x, y) = |x^{-1} - y^{-1}|$

(ii) $S = (0, 1)$, $\rho(x, y) = |\tan \tfrac{1}{2}\pi x - \tan \tfrac{1}{2}\pi y|$

(iii) $S = (0, 1)$, $\rho(x, y) = |\sin x - \sin y|$

(iv) $S = (0, +\infty)$, $\rho(x, y) = |x^{-1} - y^{-1}|$

(v) $S = (0, +\infty)$, $\rho(x, y) = |e^{-x} - e^{-y}|$

(vi) $S = (0, +\infty)$, $\rho(x, y) = |x^2 - y^2|$

17.   Let $f : [0, 2] \to \mathbb{R}$ be defined by

$$f(x) = \begin{cases} x + 1 & \text{if } 0 < x < 1, \\ -x + 4 & \text{if } 1 < x < 2, \end{cases}$$

and $f(0) = 0$, $f(1) = 1$, $f(2) = 4$. Find the completion of the space $([0, 2], \rho)$, where $\rho(x, y) = |f(x) - f(y)|$.

18.   Let f be a differentiable function of a real variable which maps the interval J into itself. Show that f is a contraction on the metric space (J, d) with d the

Euclidean metric if and only if $\alpha = \sup_{x \in J} |f'(x)| < 1$. If this is the case, show that α is the coefficient of contraction for f.

19. Let $f: S \to S$ be a contraction in a complete metric space (S, d) with the coefficient of contraction α. For a given x_0 define a sequence (x_n) of Picard iterations by $x_{n+1} = f(x_n)$, $n = 0, 1, 2, \ldots$ If x^* is the fixed point of f, show that

$$d(x_m, x^*) \le \frac{d(x_0, x_1)}{1 - \alpha} \alpha^m, \qquad m = 0, 1, 2, \ldots$$

For any $\varepsilon > 0$, estimate the number of iterative steps needed to obtain an approximation x_m to x^* with $d(x_m, x^*) < \varepsilon$.

20. Suppose $S = [1, +\infty)$, $d(x, y) = |x - y|$ for $x, y \in S$, and $f(x) = \frac{1}{2}x + 1/x$ for $x \in S$. Show that the hypotheses of the contraction mapping theorem are satisfied, and identify the fixed point of f. Define a sequence (x_n) of Picard iterations by

$$x_0 = 1, \quad x_{n+1} = \tfrac{1}{2}x_n + \frac{1}{x_n}, \quad n = 0, 1, 2, \ldots$$

Explain why

$$|x_n - \sqrt{2}| \le (\tfrac{1}{2})^n.$$

21. Which hypotheses of the contraction mapping theorem are satisfied by the function $f(x) = x - \sqrt{x} + 2$ on the intervals $[0, 1]$; $[\frac{1}{4}, 5]$; $(\frac{1}{16}, 9]$; $[1, 3]$; $(1, 4]$ under the Euclidean metric? Can f have a fixed point on an interval on which some of the hypotheses are not satisfied?

Suggestion. Problem 2.18.

22. Let $S = [2, \infty)$ and let $d(x, y) = |x - y|$ for $x, y \in S$. For what real constants k does the function $f(x) = k + \log x$ become a selfmap of S? Verify that, for any such choice of k, f satisfies the hypotheses of the contraction mapping theorem on (S, d). Define inductively Picard iterations (x_n), starting with $x_0 = 4$, convergent to the solution $x^* \in [2, \infty)$ of the equation $x - \log x = 3$. Show that

$$|x_n - x^*| \le (\tfrac{1}{2})^n \quad \text{for } n = 0, 1, 2, \ldots$$

23. Let f be defined by $f(x) = x + 1/x$. Show that

(i) $f: [1, \infty) \to [1, \infty)$,

(ii) $|f(x) - f(y)| < |x - y|$ for all $x, y \in [1, \infty)$,

(iii) $[1, \infty)$ is complete under Euclidean metric,

but f has no fixed point on $[0, \infty)$. Is any hypothesis of the contraction mapping theorem violated?

24. *Solution to the equation $f(x) = 0$.* Let f be differentiable on $[a, b]$ satisfying $0 < k \le f'(x) \le K$ for all $x \in f[a, b]$ and let $f(a) < 0 < f(b)$. Define

$$g(x) = x - \lambda f(x), \quad x \in [a, b],$$

for some $\lambda > 0$. We observe that the fixed point of g is the solution to the equation $f(x) = 0$ (which exists in $[a, b]$ in view of our assumptions). Show the following.

(i) g is a self-map of $[a, b]$ if λ is chosen to satisfy

$$0 < \lambda \leq \min\{(b - a)/f(b), (b - a)/|f(a)|\}.$$

(ii) If, in addition, $\lambda < 2/K$, then g is a contraction on $[a, b]$.

Apply the preceding procedure to $f(x) = x^3 + x - 1$ on $[0, 1]$, and find a solution to $f(x) = 0$ correct to three decimal places.

25. Let $S = [\sqrt{3}, 2]$ and let $d(x, y) = |x - y|$ for $x, y \in S$. Show that the function $f(x) = (2 + x^{1/2})^{1/2}$ satisfies the hypotheses of the contraction mapping theorem on (S, d). Prove that the equation $x^4 - 4x^2 - x + 4 = 0$ has a unique root x^* in the interval $[\sqrt{3}, 2]$, and that the sequence (x_n) defined by

$$x_{n+1} = \sqrt{2 + \sqrt{x_n}}, \quad x_0 = \sqrt{3}$$

converges to x^*.

26. Let (S, d) be a complete metric space, and let $f: S \to S$ be such that f^p is a contraction for some $p \geq 1$. Show that f has a unique fixed point in S.

27. *Quasicontinuous functions.* Let X be a complete normed space. A function $f: [a, b] \to X$ is *quasicontinuous* if the one sided limits $f(x+)$ and $f(x-)$ exist in X at each $x \in [a, b]$ (postulating $f(a-) = f(a)$ and $f(b+) = f(b)$). Prove that:

(i) A function f on $[a, b]$ is quasicontinuous if and only if there exists a sequence (f_n) of step functions on $[a, b]$ such that $f_n \to f$ uniformly on $[a, b]$.

(ii) If f is quasicontinuous on $[a, b]$, then there exists a countable set $D \subset [a, b]$ such that f is continuous on $[a, b] \backslash D$. Give an example showing that this necessary condition is not sufficient for quasicontinuity.

(iii) Every monotonic real valued function on $[a, b]$ is quasicontinuous.

28. Let $QC([a, b], X)$ be the set of all quasicontinuous functions $f: [a, b] \to X$, where X is a complete normed space. Prove that $QC([a, b], X)$ is closed in the space $B([a, b], X)$ of all bounded functions $f: [a, b] \to X$ equipped with the uniform norm, and therefore $QC([a, b], X)$ is a complete normed space under this norm.

Chapter 3

Continuity in Metric Spaces

A function of a real variable is continuous if a small change of the variable produces a small change of the function. More precisely, if we want to approximate the value $f(a)$ by values $f(x)$ for x close to a, a continuous function allows us to achieve any degree of accuracy prescribed beforehand. A precise definition (due to Cauchy) is given in Appendix B. We can extend this concept to mappings between abstract metric spaces, so that we can cover in one definition functions from reals to reals, functions from complex numbers to complex numbers, functions from $\mathbb{R}^k$ to reals, mappings that carry functions (as points of metric spaces) onto functions, etc.

3.1 Continuous mappings

Unless specified otherwise, (S, d) and (T, ρ) will be two metric or pseudo-metric spaces with mappings usually going from S to T (source to target).

Definition 3.1. (See Figure 3.1.) The mapping $f \colon S \to T$ from (S, d) to (T, ρ) is said to be *continuous* at a point $a \in S$ if, for every $\varepsilon > 0$, there is $\delta > 0$ such that

$$d(x, a) < \delta \quad \Longrightarrow \quad \rho(f(x), f(a)) < \varepsilon. \tag{3.1}$$

As in the case of a function of real variable, this definition is often referred to as the *Cauchy definition of continuity*.

It is very important to rephrase this definition in terms of open sets. Recall that a set U is an *open neighbourhood* of a point x if U is an open set containing the point x.

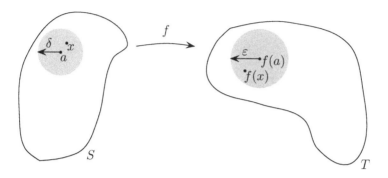

Figure 3.1

Theorem 3.2. *The mapping $f\colon S \to T$ is continuous at a point $a \in S$ if, for every open neighbourhood V of $f(a)$ in T, there is an open neighbourhood U of a in S such that*

$$f(U) \subset V.$$

The proof is left as an exercise. Observe that Definition 3.1 is subsumed in Theorem 3.2 if V is interpreted as an open ball centred at $f(a)$, and U as an open ball centred at a.

Definition 3.3. A property applicable to a pseudometric space is said to be *topological* if it can be expressed solely in terms of open sets without referring specifically to a pseudometric.

Theorem 3.2 shows that continuity is a topological property. This means that the pseudometrics d and ρ can be replaced by equivalent pseudometrics, and the mapping f will remain continuous.

Topological properties are exactly those that are preserved if we replace the pseudometric given in the space by a pseudometric equivalent to it. Other examples of topological properties: (i) a closed set; (ii) an open set; (iii) an isolated point (of a set); (iv) a limit point; (v) an interior point; (vi) a boundary point; (vii) a convergent sequence of points.

An example of a property which is not topological is completeness: Consider the Euclidean metric d on $\mathbb{R}$ and the metric $\rho(x,y) = |x-y|/(1+|x-y|)$ defined in Example 2.30. The metrics are equivalent, yet the space $(\mathbb{R}, d)$ is complete, while $(\mathbb{R}, \rho)$ is not. Another example of a property which is not topological is boundedness.

Theorem 3.4 (Heine's criterion of continuity). *A mapping* $f: S \to T$ *between pseudometric spaces is continuous at a point* $a \in S$ *if and only if for each sequence* (x_n) *in* S *we have*

$$x_n \xrightarrow{d} a \implies f(x_n) \xrightarrow{\rho} f(a).$$

Proof. Suppose first that f is continuous at a and that $x_n \xrightarrow{d} a$. Given $\varepsilon > 0$, there is $\delta > 0$ such that $\rho(f(x), f(a)) < \varepsilon$ whenever $d(x, a) < \delta$. By the definition of sequential convergence there is N such that $d(x_n, a) < \delta$ for all $n > N$. So,

$$n > N \implies d(x_n, a) < \delta \implies \rho(f(x_n), f(a)) < \varepsilon.$$

This shows that $f(x_n) \xrightarrow{\rho} f(a)$.

Conversely, assume that f is not continuous at a. To express this statement in a logically correct form, we write the definition of continuity in terms of logical quantifiers, and then negate it. The symbol $\forall$ means 'for all', and $\exists$ means 'there exists'. Continuity at a:

$$(\forall \varepsilon > 0)(\exists \delta > 0)(\forall x \in B(a; \delta)): \ \rho(f(x), f(a)) < \varepsilon.$$

Negated continuity at a:

$$(\exists \varepsilon > 0)(\forall \delta > 0)(\exists x \in B(a; \delta)): \ \rho(f(x), f(a)) \geq \varepsilon.$$

In words, there is $\varepsilon > 0$ such that for each $\delta > 0$ there is $x \in B(a; \delta)$ with $\rho(f(x), f(a)) \geq \varepsilon$. Choose δ in succession $1, \frac{1}{2}, \frac{1}{3}, \ldots$ Then for each $n \in \mathbb{N}$ there is $x_n \in B(a; \frac{1}{n})$ such that $\rho(f(x_n), f(a)) \geq \varepsilon$. So we have found a sequence (x_n) such that $x_n \to a$, but $f(x_n) \not\to f(a)$. $\quad\square$

Definition 3.5. A mapping $f: S \to T$ is *continuous on a set* A in S if it is continuous at every point of A. If $f: S \to T$ is continuous at each point of S, we say that it is *continuous* (without any qualification).

The following theorem gives a global characterization of continuity. The adjective 'topological' in its title refers to the fact that it is stated solely in terms of open sets.

Theorem 3.6 (Topological criterion of continuity). *A mapping* $f: S \to T$ *is continuous (on* S*) if and only if, for every open set* G *in* T*, the inverse image*

$$f^{-1}(G) = \{x \in S : f(x) \in G\}$$

is an open set in S.

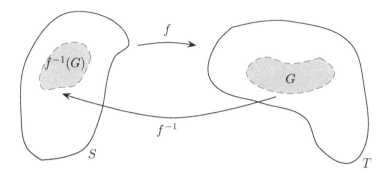

Figure 3.2

Proof. See Figure 3.2. Let f be continuous on S. If G is an open subset of T, set $A = f^{-1}(G)$. If $a \in A$, then $f(a) \in G$. By the definition of an open set, there is an open ball V centred at $f(a)$ contained in G. By the definition of continuity there is an open ball U centred at a such that $f(U) \subset V \subset G$. This means that every point of U lies in $f^{-1}(G) = A$, that is, $U \subset A$, and a is an interior point of A. Hence A is open.

Conversely, suppose that f has the property that, for every open set G in T, the inverse image $f^{-1}(G)$ is open in S. Let $a \in S$ and let V be an arbitrary open neighbourhood of $f(a)$ in T. By hypothesis, the set $U = f^{-1}(V)$ is open in S and it is an open neighbourhood of a. In addition, $f(U) = f(f^{-1}(V)) \subset V$. Hence, f is continuous at a. □

Note. In the preceding theorem, open sets can be replaced by closed sets. (Exercise.)

Theorem 3.7. *Let* $f : S \to T$ *and* $g : T \to U$ *be continuous mappings between pseudometric spaces. Then the composite mapping* $h = g \circ f : S \to U$ *is also continuous.*

Proof. Recall that continuous without any qualification means continuous on all of the space. The proof follows from the topological criterion of continuity since

$$(g \circ f)^{-1}(G) = f^{-1}(g^{-1}(G))$$

for every set $G \subset U$. □

With the concept of continuity there is the associated concept of the limit of a mapping. Let $f\colon S \to T$ be a mapping between pseudometric spaces. We say that $b \in T$ is a *limit of f at the point $a \in S$* if, for each $\varepsilon > 0$, there exists $\delta > 0$ such that

$$0 < d(x,a) < \delta \implies \rho(f(x), b) < \varepsilon.$$

We then write

$$\lim_{x \to a} f(x) = b.$$

It is important to note that the point a itself is excluded in this definition. The reason is that $f(a)$ need not be defined at all or may be different from b. We then have a handy criterion of continuity at a point which is obtained when we compare the definition of continuity and the definition of limit.

Theorem 3.8. *A mapping $f\colon S \to T$ between pseudometric spaces is continuous at a point $a \in S$ if and only if $\lim_{x \to a} f(x)$ exists and equals $f(a)$.*

3.2   Real and complex functions

Many functions encountered in analysis are real or complex valued. In this case the question of continuity can often be solved by using the algebra of continuous functions as expressed in the following theorem. Recall that $f + g$, $f \cdot g$, $|f|$, $\max\{f, g\}$, $\min\{f, g\}$, $1/f$ are the functions with functional values at x given by

$$f(x) + g(x), \;\; f(x) \cdot g(x), \;\; |f(x)|, \;\; \max(f(x), g(x)), \;\; \min(f(x), g(x)), \;\; \frac{1}{f(x)}.$$

We give formulations for metric spaces, and leave the extensions to pseudometric spaces to the reader.

Theorem 3.9 (Algebra of continuous real or complex functions).
Let f, g be mappings of an arbitrary metric space (S, d) into $\mathbb{R}$ or $\mathbb{C}$ both continuous at a point $a \in S$. Then the following are true:

(i)   *The functions $f + g$, $f \cdot g$ and $|f|$ are continuous at a.*

(ii)   *If $f, g \colon S \to \mathbb{R}$, the functions $\max\{f, g\}$, $\min\{f, g\}$ are continuous at a.*

(iii)   *If $f(a) \neq 0$, then f is bounded away from zero on some open neighbourhood U of a, and the reciprocal $1/f$, defined on U, is continuous at a.*

Proof.   Problem 3.1.   □

Example 3.10. Consider a function $f \colon \mathbb{R}^2 \to \mathbb{R}$ between Euclidean spaces defined below. We will write a variable point in $\mathbb{R}^2$ as $x = (\xi_1, \xi_2)$, and a point which is kept fixed as $a = (\alpha_1, \alpha_2)$. We want to find the points of continuity and discontinuity of f, where

$$f(\xi_1, \xi_2) = \begin{cases} \dfrac{\xi_1^2 + \xi_2^2}{|\xi_1| + |\xi_2|} & \text{if } (\xi_1, \xi_2) \neq (0, 0) \\ 0 & \text{if } (\xi_1, \xi_2) = (0, 0) \end{cases}$$

(a) The algebra of continuous functions can be used at any point $a = (\alpha_1, \alpha_2) \neq (0, 0)$. We define

$$f_1(x) = \xi_1, \quad f_2(x) = \xi_2, \quad f_3(x) = |\xi_1|, \quad f_4(x) = |\xi_2|.$$

The functions f_k $(k = 1, \ldots, 4)$ are continuous on $\mathbb{R}^2$:

$$|f_1(x) - f_1(a)| = |\xi_1 - \alpha_1| \leq \sqrt{|\xi_1 - \alpha_1|^2 + |\xi_2 - \alpha_2|^2} = d(x, a),$$

where d is the Euclidean metric on $\mathbb{R}^2$. Let $\varepsilon > 0$ be given. If we put $\delta = \varepsilon$, we have

$$d(x, a) < \delta \implies |f_1(x) - f_1(a)| \leq d(x, a) < \varepsilon.$$

A similar arguments will establish the continuity of f_2. To prove that f_3 is continuous, we need the inequality

$$|\,|\xi_1| - |\alpha_1|\,| \leq |\xi_1 - \alpha_1|.$$

A similar argument can be used for f_4.

If $a \neq (0, 0)$, then f is continuous at a by the algebra of continuous functions as

$$f = (f_1 \cdot f_1 + f_2 \cdot f_2) \cdot \frac{1}{f_3 + f_4},$$

where $(f_3 + f_4)(a) \neq 0$.

(b) Consider the point $0 = (0, 0)$. We cannot use the algebra of continuous functions as the denominator $f_3 + f_4$ takes value 0 at this point. We have to rely on Cauchy's definition of continuity (or other devices). We observe that $\sqrt{\xi_1^2 + \xi_2^2} \leq |\xi_1| + |\xi_2|$. Then, for any $x \neq 0$,

$$|f(x) - f(0)| = \frac{\xi_1^2 + \xi_2^2}{|\xi_1| + |\xi_2|} \leq \frac{\xi_1^2 + \xi_2^2}{\sqrt{\xi_1^2 + \xi_2^2}} = \sqrt{\xi_1^2 + \xi_2^2},$$

so that

$$|f(x) - f(0)| \leq d(x, 0).$$

Let $\varepsilon > 0$ be given. Choosing $\delta = \varepsilon$, we get

$$d(x, 0) < \delta \implies |f(x) - f(0)| \leq d(x, 0) < \varepsilon.$$

This proves that f is continuous at each point of $\mathbb{R}^2$.

Example 3.11. Repeat the previous example with

$$f(\xi_1, \xi_2) = \begin{cases} \dfrac{\xi_1 + \xi_2}{\xi_1} & \text{if } \xi_1 \neq 0 \\ 0 & \text{if } \xi_1 = 0 \end{cases}$$

(a) Similarly as in the preceding example we prove that f is continuous at each point $a = (\alpha_1, \alpha_2)$ with $\alpha_1 \neq 0$, using the algebra of continuous functions.

(b) Let $a = (0, \alpha_2)$. We use Heine's criterion (Theorem 3.4) to prove discontinuity of f at a: Choose a sequence of points on a horizontal line through a convergent to a, say $x_n = (1/n, \alpha_2)$. Then

$$f(x_n) = \frac{(1/n) + \alpha_2}{1/n} = 1 + n\alpha_2.$$

If $\alpha_2 \neq 0$, $(f(x_n))$ diverges to $+\infty$ or $-\infty$. If $\alpha_2 = 0$, then $f(x_n) = 1$ for all n. In either case $f(x_n) \nrightarrow f(a) = 0$. So we found a sequence (x_n) for which $x_n \to a$ and $f(x_n) \nrightarrow f(a)$. Thus f is discontinuous at a.

Example 3.12. When we consider functions $f \colon J \to \mathbb{R}$, where J is an interval, we can often use the fact known from Calculus, that a function f differentiable at a point a is continuous at a:

$$\lim_{x \to a} f(x) = \lim_{x \to a} \left(f(a) + \frac{f(x) - f(a)}{x - a} \cdot (x - a) \right) = f(a) + f'(a) \cdot 0 = f(a).$$

Then we can apply Theorem 3.8.

3.3   Homeomorphisms

Recall that two metric spaces (S, d) and (T, ρ) have the same metric properties if they are isometric. More generally, we can ask when they have the same topological properties. (Recall that a topological property is one which can be expressed solely in terms of open sets.) While every topological property is also a metric property, the converse is not true. For example, boundedness and completeness are metric, but not topological, properties. Just as isometries preserve metric properties, homeomorphisms, defined below, preserve topological properties.

Definition 3.13. A mapping $f \colon S \to T$ is called a *homeomorphism* if f is a bijection such that f is continuous on S and the inverse mapping $f^{-1} \colon T \to S$ is continuous on T. Metric spaces (S, d) and (T, ρ) are said to be *homeomorphic* if there is a homeomorphism $f \colon S \to T$.

It can be shown that every isometry is a homeomorphism, but not every homeomorphism is an isometry. Homeomorphisms are more general than isometries, and also more important.

Theorem 3.14. *Two homeomorphic metric spaces (S, d) and (T, ρ) have the same topological properties.*

Proof. Let $f\colon S \to T$ be a homeomorphism. Let P be a topological property of S, which is expressed in terms of open sets in S. If A is an open set in S, then the set $B = f(A)$ is open in T by the topological criterion of continuity, as $B = g^{-1}(A)$, where $g = f^{-1}$ is continuous. So the property P can be formulated for T by replacing A by $B = f(A)$. Similarly, every topological property of T can be formulated for S by replacing the open set U in T by the open set $f^{-1}(U)$ in S. □

The next theorem describes the relation between equivalent metrics and homeomorphisms.

Theorem 3.15. *Let d, ρ be two metrics on a nonempty set S. Then d and ρ are equivalent if and only if the identity mapping $\mathrm{id}\colon (S, d) \to (S, \rho)$ is a homeomorphism.*

Proof. The key is the topological criterion of continuity and the fact that the identity mapping id satisfies $\mathrm{id}^{-1} = \mathrm{id}$. Details of the proof are left as an exercise. □

3.4   Uniform continuity

Definition 3.16. A mapping $f\colon (S, d) \to (T, \rho)$ is said to be *uniformly continuous* on S if, for every $\varepsilon > 0$, there is $\delta > 0$ such that

$$d(x, y) < \delta \quad \Longrightarrow \quad \rho(f(x), f(y)) < \varepsilon.$$

Observe that in this definition δ depends merely on ε, not on the points x, y. Such δ is called a *uniform δ*. Loosely speaking, the points $f(x)$, $f(y)$ are close together whenever x, y are close together, regardless where in S are the points x, y located.

Clearly, every uniformly continuous function is continuous. The converse is not true, as we see from the following example.

Example 3.17. Let $f(x) = 1/x$ be the mapping from $S = (0, 1)$ to $\mathbb{R}$, where both sets are equipped with the Euclidean metric. Then f is continuous on S by the reciprocal rule of the algebra of continuous functions. We show that the continuity is not uniform; in fact we show that, for the same $\varepsilon > 0$, the corresponding $\delta > 0$ must be chosen smaller the closer we get to 0.

Choose $0 < \varepsilon < 1$ and keep it fixed; for any $a \in (0, 1)$ we determine the largest possible $\delta > 0$ to satisfy the Cauchy definition of continuity:

$$|x - a| < \delta \implies \left|\frac{1}{x} - \frac{1}{a}\right| < \varepsilon.$$

Suppose that $|x - a| < \delta$, where $0 < \delta < a$. From $x \geq a - |a - x| > a - \delta > 0$ we obtain

$$\left|\frac{1}{x} - \frac{1}{a}\right| = \frac{|x - a|}{ax} < \frac{\delta}{ax} \leq \frac{\delta}{a(a - \delta)}.$$

The function $\delta \mapsto \delta/[a(a - \delta)]$ is increasing, and so the largest δ to satisfy $\delta/[a(a - \delta)] \leq \varepsilon$ is given by

$$\delta = \frac{a^2 \varepsilon}{1 + a\varepsilon}.$$

This shows that, for ε kept fixed, this optimal δ converges to 0 as $a \to 0$. There is no uniform choice of δ that would work for all a in the interval $(0, 1)$. So f is not uniformly continuous on $(0, 1)$.

Example 3.18. In the previous example replace $(0, 1)$ by $M = (h, 1)$, where h is a constant $0 < h < 1$. Then the function $f(x) = 1/x$ is uniformly continuous on M: We apply the mean value theorem to f on $(h, 1)$. If $x, a \in (h, 1)$, there exists c between x and a such that

$$|f(x) - f(a)| = |f'(c)|\,|x - a| = \frac{1}{c^2}\,|x - a| \leq \frac{1}{h^2}\,|x - a|,$$

which shows that we can choose a uniform δ corresponding to ε with $0 < \varepsilon < 1$ to be

$$\delta = h^2 \varepsilon.$$

We give a sequential criterion for uniform continuity on the space:

Theorem 3.19. *A mapping* $f \colon (S, d) \to (T, \rho)$ *is uniformly continuous on* S *if and only if for each pair of sequences* (x_n), (y_n) *in* S *we have*

$$d(x_n, y_n) \to 0 \implies \rho(f(x_n), f(y_n)) \to 0.$$

Proof.   First we express uniform continuity in terms of logical quantifiers:

$$(\forall \varepsilon > 0)(\exists \delta > 0)(\forall x \in S)(\forall y \in S) : d(x,y) < \delta \implies \rho(f(x), f(y)) < \varepsilon.$$
(3.2)

The order of quantifiers is crucial here; if we interchange the second and third quantifier, we have the definition of the ordinary continuity at x.

Suppose that f is uniformly continuous on S, and that $d(x_n, y_n) \to 0$. Let $\varepsilon > 0$ be given. Then there exists a uniform $\delta > 0$ corresponding to this ε such that (3.2) is satisfied. There is N such that $d(x_n, y_n) < \delta$ for all $n > N$. So

$$n > N \implies d(x_n, y_n) < \delta \implies \rho(f(x_n), f(y_n)) < \varepsilon.$$

This shows that $\rho(f(x_n), f(y_n)) \to 0$ as $n \to \infty$.

Conversely, assume that the function f is not uniformly continuous on S. Negating (3.2), we get

$$(\exists \varepsilon > 0)(\forall \delta > 0)(\exists x \in S)(\exists y \in S) : d(x,y) < \delta \text{ and } \rho(f(x), f(y)) \geq \varepsilon.$$
(3.3)

Choose δ in succession to be $1, \frac{1}{2}, \frac{1}{3}, \dots$ Then for each $n \in \mathbb{N}$ there is a pair x_n, y_n of points of S with $d(x_n, y_n) < \frac{1}{n}$, and $\rho(f(x_n), f(y_n)) \geq \varepsilon$. Thus we have produced a pair of sequences (x_n), (y_n) such that $d(x_n, y_n) \to 0$ but $\rho(f(x_n), f(y_n)) \not\to 0$ as $n \to \infty$.   □

The preceding theorem is useful in demonstrating that a function is not uniformly continuous on the space. Returning to Example 3.17 we can choose sequences $x_n = 1/n$, $y_n = 1/(n+1)$. Then

$$|x_n - y_n| < \frac{1}{n}, \text{ and } |f(x_n) - f(y_n)| = |n - (n+1)| = 1$$

gives another—much easier—proof that f is not uniformly continuous on $(0,1)$.

Theorem 3.20. *Let a function $f : [a,b] \to \mathbb{R}$ be continuous on a closed bounded interval $[a,b]$. Then f is uniformly continuous on $[a,b]$.*

Proof.   If no metric is specified on subsets on $\mathbb{R}$, we understand it is equipped with the Euclidean metric. Let $\varepsilon > 0$ be given. Then for each point $x \in [a,b]$ there exists $\delta(x) > 0$ such that

$$\{z \in [a,b] \text{ and } |z - x| < \delta(x)\} \implies |f(z) - f(x)| < \tfrac{1}{2}\varepsilon.$$

We apply the Heine–Borel theorem (Theorem B.24) to the family $\{J_x : x \in [a,b]\}$ of the open intervals $J_x = (x - \frac{1}{2}\delta(x), x + \frac{1}{2}\delta(x))$. We note that the interval $[a, b]$ is contained in the union $\bigcup_{x \in [a,b]} J_x$. So, by the Heine–Borel theorem there is a finite subfamily

$$\{J_{x_1}, J_{x_2}, \ldots, J_{x_n}\} \tag{3.4}$$

whose union contains $[a, b]$. Let

$$\delta = \min\{\tfrac{1}{2}\delta(x_1), \tfrac{1}{2}\delta(x_2), \ldots, \tfrac{1}{2}\delta(x_n)\}.$$

We are going to show that this δ is a uniform δ for the given ε. Let u, v be two elements of $[a, b]$ with $|u - v| < \delta$. Then u lies in one of the intervals (3.4), say $u \in J_{x_k}$. Then

$$|v - x_k| \leq |v - u| + |u - x_k| < \delta + \tfrac{1}{2}\delta_k \leq \delta_k.$$

Therefore,

$$|f(u) - f(v)| \leq |f(u) - f(x_k)| + |f(x_k) - f(v)| < \tfrac{1}{2}\varepsilon + \tfrac{1}{2}\varepsilon = \varepsilon.$$

This completes the proof. $\qquad\qquad\square$

The preceding theorem can be probed using the sequential characterization of uniform continuity instead of the Heine–Borel theorem.

Uniform continuity is not a topological concept since it need not be preserved when we replace a metric in S by an equivalent metric. This is seen from the following example.

Example 3.21. The metrics $d(x, y) = |x - y|$ and $\rho(x, y) = |x^{-1} - y^{-1}|$ are equivalent on the space $S = (0, 1)$. The real valued function $f(x) = 1/x$ is uniformly continuous on (S, ρ) as

$$|f(x) - f(y)| = \left| \frac{1}{x} - \frac{1}{y} \right| = \rho(x, y).$$

However, we have seen that f is not uniformly continuous on (S, d).

3.5 Problems for Chapter 3

1. Let f, g be continuous real valued functions on a metric space (S, d). Prove:

(i) $f + g$ and and $f \cdot g$ are continuous on S.

(ii) $|f|$, $f \vee g = \max(f, g)$ and $f \wedge g = \min(f, g)$ are continuous on S.

(iii) If $f(a) \neq 0$, $1/f$ is defined in some neighbourhood of a, and is continuous at a.

Suggestion. (ii) $|\,|u| - |v|\,| \le |u - v|$; $f \vee g = \frac{1}{2}(f+g) + \frac{1}{2}|f-g|$, $f \wedge g = -(-f \vee -g)$
(iii) Show that there are $r, c > 0$ such that $|f(x)| \ge c$ for all $x \in B(a, r)$.

2. Prove that a function $f: S \to \mathbb{R}^n$ is continuous if and only if each component f_k of f is continuous. (The components f_k are defined by $f(x) = (f_1(x), \ldots, f_n(x))$.)

3. Show that the mapping $f: S \to T$ is continuous if and only if (i) for each set $A \subset S$ we have $f(\overline{A}) \subset \overline{f(A)}$, or (ii) for each set $B \subset T$ we have $f^{-1}(B^\circ) \subset (f^{-1}(B))^\circ$.

4. If $S = A \cup B$, and if $f: S \to T$ is such that the partial mappings $f|_A$, $f|_B$ are continuous at a point $a \in A \cap B$, show that f is continuous at a.

5. If a is an isolated point of a metric space S, show that any function $f: S \to T$ is continuous at a.

6. Prove that every contraction on a metric space S is uniformly continuous on S.

7. Let a be a point in a metric space (S, d). Prove that $f(x) = d(x, a)$ is uniformly continuous on S. Hence show that $f(x) = \|x\|$ is uniformly continuous on a normed space X.

8. Let (S, d) be a metric space. Is the function $g(x, y) := d(x, y)$ uniformly continuous on the product space $S \times S$? (See Problem 1.1).

9. Determine the points of continuity and discontinuity of the following real valued functions on Euclidean spaces:

(i) $f(\xi_1, \xi_2) = \dfrac{\xi_1}{\xi_2}$ if $\xi_2 \ne 0$, and $f(\xi_1, 0) = \xi_1$ otherwise

(ii) $f(\xi_1, \xi_2) = \dfrac{\xi_1 \xi_2}{1 - \xi_1}$ if $\xi_1 \ne 1$, $f(1, \xi_2) = 0$

(iii) $f(\xi_1, \xi_2, \xi_3) = \dfrac{\xi_1^2 + \xi_2^2 + \xi_3^2}{\log(1 + |\xi_1| + |\xi_2| + |\xi_3|)}$ if $(\xi_1, \xi_2, \xi_3) \ne (0, 0, 0)$, $f(0, 0, 0) = 0$

(iv) $f(\xi_1, \xi_2) = \dfrac{\xi_1^2 \xi_2}{2\xi_1^2 + \xi_2}$ if $2\xi_1^2 + \xi_2 \ne 0$, and $f(\xi_1, \xi_2) = 0$ otherwise

(v) $f(\xi_1, \xi_2) = \dfrac{|\xi_1| + \xi_2}{\xi_1^2 - \xi_2^2}$ if $|\xi_1| \ne |\xi_2|$, $f(\xi_1, \xi_2) = \dfrac{1}{2|\xi_1|}$ if $|\xi_1| = |\xi_2| \ne 0$, $f(0, 0) = 0$

10. Determine whether each given function is uniformly continuous on the given set.

(i) $f(x) = x^2$, $x \in [0, +\infty)$; $x \in [0, a]$ for some $a > 0$

(ii) $f(x) = \sqrt{x}$, $x \in [0, +\infty)$

(iii) $f(x) = \log x, \quad x \in [1, +\infty); \quad x \in (0, +\infty)$

(iv) $f(x) = \arcsin x, \quad x \in [-1, 1]$

(v) $f(\xi_1, \xi_2) = \xi_1^2 + 2\xi_2, \quad (\xi_1, \xi_2) \in [0, 1] \times [0, 1]$

(vi) $f(\xi_1, \xi_2) = \xi_2/\xi_1, \quad (\xi_1, \xi_2) \in (0, 1) \times (0, 1); \quad (\xi_1, \xi_2) \in [1, +\infty) \times [1, 2]$

11. Let $f : \mathbb{R}^2 \to \mathbb{R}$ be defined by $f(\xi_1, \xi_2) = \xi_1^3 \xi_2/(\xi_1^8 + \xi_2^2)$ for $(\xi_1, \xi_2) \neq (0, 0)$, and $f(0, 0) = 0$. Show that f is continuous along each straight line through the origin (in fact f has all directional derivatives at the origin equal to zero). However, the function is not continuous at the origin as a function of two variables, and is *unbounded* in every neighbourhood of $(0, 0)$.

12. Let $f, g : S \to T$ be continuous and let $A = \{x \in S : f(x) = g(x)\}$. Prove that the set A is closed in S.

Suggestion. Sequential criterion of continuity.

13. If $f, g : S \to \mathbb{R}$ are continuous on (S, d), show that $A = \{x \in S : f(x) < g(x)\}$ is open, and the set $B = \{x \in S : f(x) \leq g(x)\}$ is closed.

Suggestion. Topological criterion of continuity.

14. Use the topological criterion of continuity to show that the sets given in Problem 1.15 (i)–(iv) are open.

15. Suppose (S, d) is a discrete space and (T, ρ) an arbitrary metric space. Show that every function $f : S \to T$ is continuous.

16. Let $f : S \to \mathbb{R}$ be defined by $f(x) = x$ for $x < 0$ and $f(x) = x^2 + 1$ for $x \geq 0$. Decide whether f is continuous on the following metric spaces equipped with the Euclidean metric:

(i) $\mathbb{R}$, (ii) $(-1, 0) \cup (0, 1)$, (iii) $(-1, 0]$, (iv) $[0, 1)$.

17. In this problem we extend the concept of uniform convergence to mappings between metric spaces. We say that a sequence (f_n) of mappings $f_n : S \to T$ between metric spaces (S, d) and (T, ρ) *converges uniformly on S* to a mapping $f : S \to T$ if for every $\varepsilon > 0$ there exists $N = N(\varepsilon) \in \mathbb{N}$ such that $\rho(f_n(t), f(t)) < \varepsilon$ for every $t \in S$ whenever $n \geq N$. Here $N(\varepsilon)$ depends on ε but not on $t \in S$.

Let (S, d) and (T, ρ) be metric spaces, and (f_n) a sequence of mappings $f_n : S \to T$, each continuous on S, uniformly convergent on S to $f : S \to T$. Prove that f is continuous.

18. Let $f : (S, d) \to (T, \rho)$ be a homeomorphism between metric spaces. Define a function $\sigma : S \times S \to \mathbb{R}$ by $\sigma(x, y) = \rho(f(x), f(y))$ for all $x, y \in S$. Show that σ is a metric on S equivalent to d.

Chapter 4

Topological Spaces

The foundation of general topology was laid by Georg Cantor (1845–1918) in his work on set theory and properties of real numbers and by Felix Hausdorff (1868–1942) in 1914 in his classic book [22]. A topological space is a generalization of a pseudometric space in which only the open sets are specified. They have to obey the basic rules of topology: the empty set and the space itself are open, the union of any family of open sets is open, and the intersection of finitely many open sets is open.

In metric analysis it is of considerable interest whether a given property of a metric space is preserved under a homeomorphism. The properties that are preserved were named topological properties. Examples of topological properties are the closure, interior and boundary of a set, continuity of a function, compactness of a set, continuity of mappings, etc. All these concepts can be defined in topological spaces.

4.1 Topology

Topological properties of metric and pseudometric spaces are the ones that can be expressed solely in terms of open sets. In order to distill the essence of topological properties, we need a family of open sets without assuming that these sets were obtained from some metric or pseudometric, provided that they behave as dictated by Theorem 1.35. This brings us to the following definition.

Definition 4.1. Let X be a set and $\mathcal{T}$ a family of subsets of X. Then $\mathcal{T}$ is called a *topology on* X if it has the following properties:

   (i) X and $\emptyset$ belong to $\mathcal{T}$.

(ii)   The union of any subfamily of $\mathcal{T}$ belongs to $\mathcal{T}$.

(iii)   The intersection of any finite subfamily of $\mathcal{T}$ belongs to $\mathcal{T}$.

A *topological space* $(X, \mathcal{T})$ consists of a set X and a topology $\mathcal{T}$ on X. The sets which belong to $\mathcal{T}$ are called *open sets*.

Example 4.2. Any pseudometric or metric d on a set S induces a topology $\mathcal{T}$ on S consisting of all d-open sets in S.

Example 4.3. The *Euclidean topology* on $\mathbb{R}^k$ is the topology induced by the Euclidean metric on $\mathbb{R}^k$.

Example 4.4. Any nonempty set X can be equipped with two extreme topologies: The family $\mathcal{T} = \mathcal{P}(X)$, where $\mathcal{P}(X)$ is the power set of S (the family of all subsets of X), is the largest topology on X (often called the *discrete topology* on X). The family $\mathcal{T} = \{\emptyset, X\}$ is the smallest topology on X (often called the *indiscrete topology* on X). If X has at least two points, then the indiscrete topology cannot be generated by any metric (Problem 4.9).

We now try to retrieve topological concepts using only open sets. This is quite easy when we realize that the metric concept of an open ball with centre a has to be replaced by an open set containing a. It is worth introducing the following terminology.

Definition 4.5. An *open neighbourhood* of a point a is an open set that contains a. A *neighbourhood* of a is an arbitrary set which contains an open neighbourhood of a.

We may start with the concept of an interior point. Given a set A in X, a point $a \in A$ is an *interior point of A* if there is a neighbourhood of a contained in A. The set A° of all interior points of A is called the *interior of A*. A point $a \in X$ is a *boundary point of A* if every open neighbourhood of a contains at least one point of A and one point of A^c. The *boundary* ∂A of a set A is the set of all boundary points of A. In analogy with the pseudometric space situation we say that a set A is *closed* if its complement A^c in X is open. A point $x \in X$ lies in the *closure* $\overline{A}$ of a set A if each neighbourhood of x contains at least one point of A. Then $\overline{A} = A \cup \partial A$. We can prove the following result analogous to the situation in metric and pseudometric spaces.

Theorem 4.6. *The closure $\overline{A}$ of a set A in a topological space X is a closed set. A set A is closed if and only if A contains all its boundary points.*

Proof. Problem 4.8. □

Convergence of sequences in a topological space can be defined in terms of neighbourhoods. A sequence (x_n) *converges* to a point x if, for every neighbourhood U of x, there is an index k such that $x_n \in U$ for all $n \geq k$. Next, in analogy with pseudometric spaces, we hope to give a characterization of closures of sets in terms of sequences (Theorem 2.11). However, this dream is shattered by the following example.

Example 4.7. *Inadequacy of sequences in topological spaces.* We show that the closure of a set A in a topological space may be larger than the set of the limits of convergent sequences with terms in A.

Let $\mathcal{T}$ be the family consisting of $\emptyset$ and the complements of all countable sets in $\mathbb{R}$; $\mathcal{T}$ is a topology on $\mathbb{R}$ (Problem 4.3). Let $A = (a, b)$ be a finite interval in $\mathbb{R}$. First we observe that any point in A^c is a boundary point of A; so $\overline{A} = A \cup A^c = \mathbb{R}$. However, any convergent sequence (x_n) contained in A has its limit in A: The complement U of the set $\{x_n : n \in \mathbb{N}\}$ is open, contains A^c, but does not contain any of the x_n; the limit of (x_n) does not belong to U, and consequently to A^c. The points of A^c lie in $\overline{A}$, but are not the limits of sequences with terms in A.

The preceding example shows that sequences in topological spaces no longer have the privileged status they had in metric and pseudometric spaces. There is a remedy to this situation: Instead of sequences we can use the so called *generalized sequences* or *nets*. Whereas sequences have the natural numbers for the index set, nets are indexed by more general index sets with less restrictive ordering.

4.1.1   Nets and the Moore–Smith convergence

A *directed set* is a nonempty set D together with a relation $\geq$ which satisfies the following properties:

(a) $\gamma \geq \beta$ and $\beta \geq \alpha \implies \gamma \geq \alpha$ (transitive);

(b) $\alpha \geq \alpha$ (reflexive);

(c) for every pair α, β in D there exists $\gamma \in D$ such that $\gamma \geq \alpha$ and $\gamma \geq \beta$ (upper bound).

A *net* in a set X is a mapping $x \colon D \to X$ for some directed set D. We write x_α for $x(\alpha)$. If X is a topological space, we define convergence of nets as follows: A net $(x_\alpha : \alpha \in D)$ in X *converges to a point* $x \in X$ if for every neighbourhood U of x there exists $\alpha_0 \in D$ such that $x_\alpha \in U$ for all $\alpha \geq \alpha_0$. We write $x_\alpha \to a$. The convergence of nets is called the *Moore–Smith convergence*. We can now prove that a point a is in the closure of a set A if and only if there is a net (x_α) contained in A convergent to a in the sense of Moore–Smith (Problem 4.27 (i)).

For a given net (x_α), $c \in X$ is a *cluster point* for (x_α) if for each neighbourhood U of c and every $\alpha \in D$ there exists $\beta \in D$ such that $x_\beta \in U$. Subnets are notoriously difficult to define and work with, and we avoid them.

See Problem 4.27 for some results on nets.

The following definition is motivated by Theorem 3.6.

Definition 4.8. If $(X, \mathcal{T})$ and $(Y, \mathcal{U})$ are two topological spaces, we say that a mapping $f \colon X \to Y$ is *continuous* if

$$U \in \mathcal{U} \implies f^{-1}(U) \in \mathcal{T}.$$

The mapping f is a *homeomorphism* of X onto Y if f is bijective and both f and f^{-1} are continuous. Two topological spaces X and Y are said to be *homeomorphic* or *topologically equivalent* if there is a homeomorphism from X to Y. A *topological property* is one which is preserved under homeomorphism.

As in the case of pseudometric spaces, we can consider restricting the topology of $(X, \mathcal{T})$ to a subset A of X. Let $\mathcal{T}_A$ be the family of all subsets U of A of the form

$$U = A \cap G,$$

where G is open in X, that is, $G \in \mathcal{T}$. It is not difficult to verify that $(A, \mathcal{T}_A)$ is a topological space. This topology is known as the *subspace topology of A*. For metric spaces this coincides with the earlier defined concept of a (metric) subspace. We usually say that $B \subset A$ is *open in the subspace topology of A* if $B \in \mathcal{T}_A$, or more succinctly, that B *is open in A*. Similar terminology is used for other topological concepts in $(A, \mathcal{T}_A)$.

Definition 4.9. Let $\mathcal{T}_1$ and $\mathcal{T}_2$ be two topologies on X. If $\mathcal{T}_1 \subset \mathcal{T}_2$, we say that $\mathcal{T}_1$ is *weaker* (also *smaller* or *coarser*) than $\mathcal{T}_2$; then $\mathcal{T}_2$ is *stronger* (also *larger* or *finer*) than $\mathcal{T}_1$.

4.2   Bases for a topology

In metric and pseudometric spaces the open balls form a special system of open sets that generate all other open sets. In particular, a set A is open if, for each point $a \in A$ there exists a ball $B(a; \varepsilon)$ contained in A. This extends to topological spaces as follows.

Definition 4.10. Let $(X, \mathcal{T})$ be a topological space. A subfamily $\mathcal{B}$ of $\mathcal{T}$ is a *base for the topology* $\mathcal{T}$ if the following condition holds: for each $U \in \mathcal{T}$ and $a \in U$ there exists $W_a \in \mathcal{B}$ such that $a \in W_a \subset U$.

If $\mathcal{B}$ is a base for $\mathcal{T}$, then each set in $\mathcal{B}$ is open, and $\mathcal{B}$ is a cover of X. However, not every open cover of X is a base for the topology of X.

Example 4.11. Consider the set $X = \mathbb{R}^2$. Then the family $\mathcal{B}_1$ of all open discs is a base for the Euclidean topology in X. Similarly, the family $\mathcal{B}_2$ of all open squares with sides parallel to the axes is another base for the Euclidan topology of X.

A topology on a set X can be obtained by giving first a base.

Theorem 4.12. *Let X be a set and $\mathcal{B}$ a family of subsets of X which forms a cover for X. Then $\mathcal{B}$ is a base for a topology on X if and only if the following condition holds:*

If $U, V \in \mathcal{B}$ and $x \in U \cap V$, then there exists a set $W \in \mathcal{B}$ such that $x \in W \subset U \cap V$.

Proof. Suppose first that $\mathcal{B}$ is a base for a topology $\mathcal{T}$ on X. If $U, V \in \mathcal{B}$ and $x \in U \cap V$, then $U \cap V \in \mathcal{T}$. By the definition of base, there is $W \in \mathcal{B}$ such that $x \in W \subset U \cap V$.

Conversely, suppose that a family $\mathcal{B}$ satisfies the condition. Define the family $\mathcal{T}$ by including in it all unions of subfamilies of $\mathcal{B}$:

$$\mathcal{T} = \left\{ \bigcup_{A \in \mathcal{A}} A : \mathcal{A} \subset \mathcal{B} \right\}. \tag{4.1}$$

Then the family $\mathcal{T}$ satisfies the conditions of a topology. (The details are left as an exercise.) $\square$

It is often useful to consider more general families of sets than bases. A *subbase* for a topology $\mathcal{T}$ on X is a family $\mathcal{S} \subset \mathcal{T}$ such that the family of all finite intersections of elements of $\mathcal{S}$ forms a base for $\mathcal{T}$. The method of constructing the family $\mathcal{T}$ in the preceding proof is useful in the situation when we attempt to construct the smallest topology on X that contains a given cover $\mathcal{U}$. This topology $\mathcal{T}$ is called the *topology of X generated by $\mathcal{U}$*.

Theorem 4.13. *Let $\mathcal{U}$ be a nonempty family which forms a cover for X. Then there is a unique topology $\mathcal{T}$ on X such that $\mathcal{U}$ is a subbase for $\mathcal{T}$.*

Proof.   Define first the family $\mathcal{B}$ consisting of all finite intersections of elements of $\mathcal{U}$. It is enough to verify that $\mathcal{B}$ is a base and apply Theorem 4.12. To prove the uniqueness of $\mathcal{T}$, we show that any other topology containing $\mathcal{U}$ also contains the family $\mathcal{T}$ we have constructed. The details are left as an exercise. $\square$

4.3   Hausdorff spaces and separation properties

A convergent sequence in a metric space has a unique limit (Theorem 2.1.2), while it may have many limits in a pseudometric space (Example 2.3). The latter situation may persist in topological spaces.

Example 4.14. Let $\mathcal{T}$ be the family consisting of $\emptyset$ and of the complements of all finite sets in $\mathbb{R}$. Then $\mathcal{T}$ is a topology on $\mathbb{R}$. Define a sequence $x_n = n$ for all $n \in \mathbb{N}$. If $x \in \mathbb{R}$ and U is a neighbourhood of x, then $x_n \in U$ for all but a finite number of n because U^{c} is finite. This shows that (x_n) converges to x, and in fact to every point in the space.

Such a situation is often undesirable, and we would like to single out those topological spaces that behave in a more orderly manner.

Definition 4.15. A topological space X is called a *Hausdorff space* if for any pair of distinct points x, y in X there are disjoint open neighbourhoods U, V of x, y respectively.

Metric spaces are Hausdorff (Problem 4.13). Pseudometric spaces which are not metric are not Hausdorff; neither is the topological space defined in Example 4.7 as any two nonempty open sets have nonempty intersection being the complements of countable sets in $\mathbb{R}$. The next result indicates that in Hausdorff spaces the previously discussed 'pathology' of limits of sequences disappears.

Theorem 4.16. *In a Hausdorff space X any convergent sequence has a unique limit.*

Proof.   Problem 4.14. $\square$

In the preceding section we showed that sequences are not adequate to describe closures of sets in a topological space and gave an example in a non-Hausdorff space (Example 4.7). This inadequacy, however, persists also in Hausdorff spaces (see Problem 4.15).

Metric spaces are not only Hausdorff, but possess stronger separation properties. For instance, any two disjoint closed sets in a metric space can be separated by open sets (Problem 4.21); spaces with this property are called *normal* (see below). General topological spaces (including pseudometric spaces) need not satisfy even very weak separation properties. The topological space defined in Example 4.14 above is not Hausdorff. In the space constructed in Problem 4.23, there are points labelled 1 and 4 such that every open neighbourhood of 4 contains 1. It is therefore useful to classify topological spaces according to the separation properties they possess.

T_1-*space*: For any pair x, y of distinct points in X there is an open neighbourhood of x that does not contain y, and an open neighbourhood of y that does not contain x.

T_2-*space or Hausdorff space*: For any pair of disjoint points x, y in X there exists a pair of disjoint open neighbourhoods U_x, U_y of x and y.

Regular space: If A is a closed set and x a point not in A, then there exists a pair of disjoint open sets U_A and U_x containing A and x, respectively.

T_3-*space*: T_1 and regular.

Normal space: If A, B are disjoint closed sets, then there are disjoint open sets U_A, U_B containing A, B, repectively.

T_4-*space*: T_1 and normal.

Example 4.17. A pseudometric space need not be a T_1-space. Let p be the pseudometric on $\mathbb{R}^2$ defined by $p(x, y) = |\xi_1 - \eta_1|$. Then any open neighbourhood of the point $(2, 1)$ contains the point $(2, -3)$, and vice versa.

Metric spaces are standard examples of normal spaces (Problem 4.21). Normal topological spaces are of particular importance as they admit the following two theorems.

Theorem 4.18 (Urysohn's lemma). *Suppose that X is a normal topological space. For each pair of disjoint closed sets A, B in X there is a continuous function $f \colon X \to [0, 1]$ such that $f(A) = \{0\}$ and $f(B) = \{1\}$.*

Theorem 4.19 (Tietze's extension theorem). *Suppose that X is a normal topological space and A a nonempty closed subset of X. If a function $f \colon A \to [0, 1]$ is continuous, then there exist a continuous extension $F \colon X \to [0, 1]$ of f.*

As these two theorems are not going to be used in this book, their proofs are not given, and can be found, for instance, in [18] and [26].

4.4   Product spaces

In Section 1 of Chapter 1 we defined a product metric for the cartesian product $S = S_1 \times \cdots \times S_k$ of metric spaces (S_i, d_i), $i = 1, \ldots, k$, by

$$d(x, y) = \sum_{i=1}^{k} d_i(x_i, y_i), \tag{4.2}$$

where $x = (x_1, \ldots, x_k)$, $y = (y_1, \ldots, y_k)$. We may observe that other metrics may generate the same topology on S, such as

$$\rho(x, y) = \max_{i=1,\ldots,k} d_i(x_i, y_i),$$

$$\sigma(x, y) = \left(d_1(x_1, y_1)^2 + \cdots + d_k(x_k, y_k)^2 \right)^{1/2}.$$

The concept of product topology can be extended to topological spaces.

Definition 4.20. Let $(X_i, \mathcal{T}_i)$ $(i = 1, \ldots, k)$, be topological spaces, and let X be the cartesian product $X = X_1 \times \cdots \times X_k$. Let $\mathcal{B}$ be the family consisting of all sets $U = U_1 \times \cdots \times U_k$, where $U_i \in \mathcal{T}_i$ $(i = 1, \ldots, k)$. The topology $\mathcal{T}$ with $\mathcal{B}$ as a base is called the *product topology* on X generated by $\mathcal{T}_1, \ldots, \mathcal{T}_k$; $(X, \mathcal{T})$ is called the *product space* of $X_1, \ldots, X_k$.

It is important to understand that not every set U open in the product topology is of the form $U_1 \times \cdots \times U_k$ with $U_i \in \mathcal{T}_i$, $(i = 1, \ldots, k)$.

Theorem 4.21. *The product space $(X, \mathcal{T})$ of topological spaces $X_1, \ldots, X_k$ is a topological space in the sense of Definition 4.1.*

Proof.   Problem 4.16.   □

The topology defined by (4.2) on the cartesian product $S = S_1 \times \cdots \times S_k$ of metric spaces is exactly the product topology generated by $S_1, \ldots, S_k$ (Problem 4.17).

If $X = X_1 \times \cdots \times X_n$ is a product of topological spaces, we define the *projections* $\pi_i \colon X \to X_i$ $(i = 1, \ldots, n)$ by $\pi_i(x_1, \ldots, x_i, \ldots, x_n) = x_i$. The projections are continuous and open (Problem 4.20).

4.4.1 *Infinite product topologies*

Topological products can be defined for an infinite (countable or uncountable) number of spaces. A definition of these products and many results on product topologies were obtained by A. N. Tychonoff (1906–1993) in the 1930s. Given an infinite family of sets $\{X_\alpha : \alpha \in D\}$, we define the Cartesian product $X := \prod_\alpha X_\alpha$ as the set of all functions $x \colon D \to \bigcup_\alpha X_\alpha$ with the property that $x(\alpha) \in X_\alpha$ for all $\alpha \in D$. (See Section E.2 in Appendix E.) We usually write x_α instead of $x(\alpha)$. If $\{(X_\alpha, \tau_\alpha) : \alpha \in D\}$ is an indexed family of topological spaces and X is the Cartesian product of the X_α, we define a family $\mathcal{B}$ of subsets of X as follows:

$$U \in \mathcal{B} \iff U = \prod_{\alpha \in D} U_\alpha,$$

where $U_\alpha \in \tau_\alpha$ for all $\alpha \in D$, and $U_\alpha = X_\alpha$ for all but finitely many $\alpha \in D$. Then $\mathcal{B}$ is a base for a topology τ on X, called the *product topology* on X.

As for the finite case, we define the projections $\pi_\alpha \colon X \to X_\alpha$ by $\pi_\alpha(x) = x_\alpha$ for all $\alpha \in D$; these are again continuous and open in the product topology. We observe that the set $\prod_{\alpha \in D} U_\alpha$, where $U_\alpha = X_\alpha$ except for $\alpha = \alpha_1, \ldots, \alpha_n$, can be expressed as

$$\prod_{\alpha \in D} U_\alpha = \pi_{\alpha_1}^{-1}(U_{\alpha_1}) \cap \cdots \cap \pi_{\alpha_n}^{-1}(U_{\alpha_n}).$$

This shows that the family of all sets $\pi_\alpha^{-1}(U_\alpha)$, where U_α is open in X_α, is a subbase for the product topology. In fact, the product topology is the weakest topology for which each projection π_α is continuous.

4.5 Problems for Chapter 4

1. Let X be a nonempty set and $a \in X$. Decide whether the family $\mathcal{T} = \{\emptyset, \{a\}, X\}$ is a topology for X.

2. Let $X = \{a, b, c, d\}$ and let $\mathcal{T} = \{\emptyset, \{a\}, \{a, b\}, \{a, c\}, \{a, b, c\}, X\}$. Decide whether $\mathcal{T}$ is a topology for X.

3. Let X be an infinite set.

(i) *Finite complement topology.* If $\mathcal{T}$ is the family of subsets of X consisting of $\emptyset$ and the complements of all finite sets, show that $\mathcal{T}$ is a topology on X.

(ii) *Countable complement topology.* If $\mathcal{T}$ is the family of subsets of X consisting of $\emptyset$ and the complements of all countable sets, show that $\mathcal{T}$ is a topology on X.

4. Let X be a set and p a given point in X.

(i) *Particular point topology.* Let $\mathcal{T}$ be the family consisting of $\emptyset$ and all subsets of X containing p. Show that $\mathcal{T}$ is a topology.

(ii) *Excluded point topology.* Let $\mathcal{T}$ be the family consisting of X and all subsets of X which do not contain p. Show that $\mathcal{T}$ is a topology.

5. *Compact complement topology.* Let $\mathcal{T}$ be the family consisting of $\emptyset$ and all subsets of $\mathbb{R}$ whose complement is bounded and closed in the Euclidean topology. Prove that $\mathcal{T}$ is a topology on $\mathbb{R}$. Prove further that $\mathcal{T}$ is stronger than the finite complement topology on $\mathbb{R}$, and weaker than the Euclidean topology.

6. Let $(X, \mathcal{T})$ be a topological space. Prove that $\mathcal{T}$ is the discrete topology on X if and only if all singletons belong to $\mathcal{T}$.

7. Let (X, d) be a metric space with $\mathcal{T}$ the topology induced by d, and let $\mathcal{U}$ be the finite complement topology on X. Prove that $\mathcal{T}$ is stronger than $\mathcal{U}$.

8. Prove that the following facts are valid in a topological space.

(i) The interior of a set is an open set.

(ii) The closure of a set is a closed set.

(iii) A set is open if and only if every point of the set is an interior point of the set.

(iv) A set is closed if and only if it contains all its boundary points.

(v) The boundary of a set is a closed set.

9. Let X be a set with at least two points. Show that the indiscrete topology of X cannot be generated by any metric.

10. Show that a mapping $f \colon X \to Y$ between topological spaces is continuous if and only if one of the following conditions is satisfied:

(i) $f^{-1}(F)$ is closed in X for every F closed in Y.

(ii) $f(\overline{A}) \subset \overline{f(A)}$ for any $A \subset X$.

(iii) $f^{-1}(B^\circ) \subset (f^{-1}(B))^\circ$ for any $B \subset Y$.

11. Determine which of the topologies defined in Problems 4.1–4.3 are Hausdorff.

12. Show that two composable continuous mappings between topological spaces have a continuous composition.

13. Prove that the topology generated by a metric is Hausdorff.

14. Show that any sequence in a Hausdorff space can have at most one limit.

15. Let $X = \mathbb{R}^{\mathbb{R}}$ be the set of all functions $f \colon \mathbb{R} \to \mathbb{R}$, and let $\mathcal{T}$ consist of $\emptyset$ and all sets $A \subset X$ with the property that, for each $f \in A$, A contains a set of the form

$$U(f; x_1, \ldots, x_n; \varepsilon) = \{g \in X : |g(x_i) - f(x_i)| < \varepsilon \text{ for } i = 1, \ldots, n\}.$$

(i) Show that $\mathcal{T}$ is a Hausdorff topology on X.

(ii) Show that sequences are not adequate to describe the closures of sets in $(X, \mathcal{T})$.

16. Prove that the product topology on the product $X = X_1 \times \cdots \times X_k$ of topological spaces X_i is a topology in the sense of Definition 4.1.

17. Prove that the topology induced by the product metric on the cartesian product $S = S_1 \times \cdots \times S_k$ of metric spaces coincides with the product topology on S.

18. Let $X \times Y$ be the product space of topological spaces X, Y and let $A \subset X$, $B \subset Y$. Prove the following:

(i) $\overline{A \times B} = \overline{A} \times \overline{B}$.

(ii) $(A \times B)^\circ = A^\circ \times B^\circ$.

(iii) $\partial(A \times B) = (\partial A \times \overline{B}) \cup (\overline{A} \times \partial B)$.

(iv) $A \times B$ is open if and only if both A and B are open.

(v) $A \times B$ is closed if and only if both A and B are closed.

(vi) A point (a, b) is isolated in $A \times B$ if and only if a is isolated in A and b is isolated in B.

19. Let $X = X_1 \times \cdots \times X_k$ be the product space of topological spaces X_i. Prove the following statements.

(i) X is Hausdorff if and only if each X_i is Hausdorff. (True also for infinite products.)

(ii) X is separable if and only if each X_i is separable. (True also for countably infinite products. X is *separable* if it contains a countable subset dense in X.)

20. Let $X = X_1 \times \cdots \times X_k$ be the product space of topological spaces X_i and let $\pi_i \colon X \to X_i$ be the projection map $\pi_i(x) = x_i$ for all $x = (x_i, \ldots, x_k)$.

(i) Show that each projection π_i is continuous.

(ii) Show that each projection π_i is open. (True also for infinite products. A mapping $f \colon X \to Y$ between topological spaces is *open* if the direct image $f(U)$ of every open set U in X is open in the subspace $f(X)$.)

21. Let A, B be subsets of a metric space (S, d) such that $\overline{A} \cap B = \emptyset$ and $A \cap \overline{B} = \emptyset$. Prove that A, B are separated by disjoint open sets. (Any T_1 topological space with this property is called *completely normal.*) Conclude that any metric space is completely normal.

22. Let $\{(S_n, d_n) : n \in \mathbb{N}\}$ be a sequence of pseudometric spaces, $S := \prod_n S_n$,

and let

$$d(x,y) = \sum_{n=1}^{\infty} (\tfrac{1}{2})^n d_n(x_n, y_n), \quad x, y \in S.$$

Show that d is a pseudometric on S which induces the product topology on S.

23. Let $X = \{1, 2, 3, 4\}$ and let $\mathcal{T} = \{\{1\}, \{1, 2\}, \{1, 3\}, \{1, 2, 3\}, \emptyset, X\}$. Show that $\mathcal{T}$ is a topology on X which does not satisfy the T_1 separation axiom.

24. Show that the excluded point topology on a set X with at least two elements is a non-Hausdorff topology on X.

25. Show that a topological space X is regular if and only if for each point $x \in X$ and each open set U containing x there exists an open set V such that $x \in V \subset \overline{V} \subset U$.

26. Show that a topological space X is normal if and only if for each closed set A and each open set U containing A there exists an open set V such that $A \subset V \subset \overline{V} \subset U$.

27. *Moore–Smith convergence.* Let X, Y be topological spaces. Prove the following:

(i) Let $A \subset X$. Then $a \in \overline{A}$ if and only if there exists a net (x_α) in A such that $x_\alpha \to a$.

(ii) X is Hausdorff if and only if each net in X converges to at most one point.

(iii) A mapping $f \colon X \to Y$ is continuous on X if and only if $x_\alpha \to x$ implies $f(x_\alpha) \to f(x)$ for every net (x_α) in X.

(iv) A metric space X is net-complete if and only if it is (sequentially) complete.

Chapter 5

Compactness

The properties of closed bounded intervals discussed in Appendix B may serve as the motivation for this chapter. It may be helpful to read Section B.4 of Appendix B before proceeding.

The key fact is that closed bounded intervals satisfy the Heine–Borel theorem (Theorem B.24):

Every family $\{J_\alpha\}$ of open intervals whose union contains a closed bounded interval $[a, b]$ has a finite subfamily whose union contains $[a, b]$.

The property expressed in the Heine–Borel theorem is usually referred to as *topological compactness*. It was used to prove that a continuous real valued function on a closed bounded interval is uniformly continuous. We observe that many theorems concerning the behaviour of continuous functions on closed bounded intervals can be proved using Theorem B.16:

Every sequence contained in $[a, b]$ has a subsequence convergent to a point in $[a, b]$.

The property expressed in this theorem is usually called *sequential compactness*. The result enables us to prove that a continuous real valued function on a closed bounded interval is bounded, and that it attains its maximum and minimum on that interval. This need not be true for functions defined on intervals which are not closed or not bounded. Both these properties have their generalizations to metric and topological spaces, and as it turns out, they are equivalent in metric spaces. They are not in general equivalent in topological spaces. Since the Heine–Borel characterization applies in both topological and metric spaces, we use it for our definition of compactness.

Some key players in the development of the concept of compactness:

73

Karl Weierstrass (1815–1897)
Eduard Heine (1821–1881)
Felix Hausdorff (1868–1942)
Émile Borel (1871–1956)
Maurice Fréchet (1878–1973)
Marshall Stone (1903–1989)
Andrey Tychonoff (1906–1993)

5.1   Compact sets

First we introduce some terminology to help us with formulation of the definition of compactness. For greater generality, all sets considered are subsets of a topological space $(X, \mathcal{T})$.

Definition 5.1. A *cover* for a set A is a family $\{U_\alpha : \alpha \in D\}$ of subsets of X such that

$$A \subset \bigcup_{\alpha \in D} U_\alpha.$$

(The index set D may be finite or infinite; in the latter case it can be either countable or uncountable.)   A *subcover* $\{U_\beta : \beta \in E\}$ of a given cover $\{U_\alpha : \alpha \in D\}$ is a subfamily of this cover which is still a cover for A; here E is a subset of the index set D. A cover is *finite* if there are only finitely many sets in the family (that is, if the index set is finite).

We can now give our definition of (topological) compactness.

Definition 5.2. A set A in a topological space X is *compact* if every open cover of A has a finite subcover.

Example 5.3. In view of the Heine–Borel theorem of real analysis (Theorem B.24), every closed bounded interval $[a, b]$ is compact in the Euclidean space $(\mathbb{R}, d)$.

Example 5.4. The interval $(0, 1)$ in the Euclidean line is not compact. Consider the open intervals $U_n = (1/n, 1)$, $n = 1, 2, \ldots$ The family $\{U_n : n \in \mathbb{N}\}$ is an open cover for $(0, 1)$ since for every $a \in (0, 1)$ there exists a positive integer k such that $k > 1/a$ (Archimedean property of integers), and therefore $a \in U_k$. However, no finite subfamily of this family can cover $(0, 1)$ since the union of any such subfamily $U_{n_1}, \ldots, U_{n_s}$ is equal to $(1/m, 1)$, where $m = \max\{n_1, \ldots, n_s\}$.

Example 5.5. Any topological space X that has only finitely many points is compact. (Exercise.)

We derive some general properties of compact spaces.

Theorem 5.6. *A closed subset of a compact topological space is compact.*

Proof. Let A be a closed subset of a compact topological space $(X, \mathcal{T})$. Suppose that $\mathcal{U}$ is an open cover for A. Add the complement A^c of A to that cover. The resulting family consists of open sets in X, and is in fact a cover for X ($\mathcal{U}$ covers A, A^c covers the rest). By the compactness of X, there exists a finite subcover of X consisting of $U_1, \ldots, U_s$. This is a finite cover for A. If one of the U_i is equal to A^c, we omit it, and the remaining $s - 1$ sets still cover A. This gives a finite subcover of $\mathcal{U}$ for A. □

A question arises whether a compact subset of a topological space must be closed. Interestingly enough, the answer depends on whether the space is Hausdorff or not. Problem 5.13 shows that (in a non-Hausdorff space) a set can be compact but not closed.

Theorem 5.7. *Let A be a compact subset of a Hausdorff space X. Then A is closed.*

Proof. We show that the complement A^c of A is open. Suppose that A^c is nonempty and pick $a \in A^c$. By the definition of a Hausdorff space, for each $x \in A$ there are disjoint open neighbourhoods $U(x)$ and $V(x)$ of x and a, repectively. The family $\{U(x) : x \in A\}$ is an open cover of A. Since A is compact, there is a finite subcover $U(x_1), \ldots, U(x_n)$. The set $V = \bigcap_{i=1}^n V(x_i)$ is an open neighbourhood of a. Since V is disjoint with each set $U(x_i)$ and since $A \subset \bigcup_{i=1}^n U(x_i)$, V is disjoint with A, so that $V \subset A^c$. □

Theorem 5.8. *The product space $X = X_1 \times \cdots \times X_k$ of topological spaces X_i is compact if and only if each X_i is compact.*

Proof. We first consider two compact spaces X_1, X_2 and prove that the product space $X_1 \times X_2$ is compact. Let $\mathcal{U}$ be an open cover for $X_1 \times X_2$, and let $\mathcal{A}$ be the family consisting of those open sets $A \subset X_1$ such that $A \times X_2$ can be covered by a finite subfamily of $\mathcal{U}$.

We show that $\mathcal{A}$ is an open cover for X_1. Let $x \in X_1$. For each $y \in X_2$ there is $U(y) \in \mathcal{U}$ such that $(x, y) \in U(y)$. Since $U(y)$ is open, by the definition of the product topology there are open sets $P(y)$, $Q(y)$ in

X_1, X_2, respectively, such that $(x, y) \in P(y) \times Q(y) \subset U(y)$. The family $\{Q(y) : y \in X_2\}$ is an open cover for X_2, and so it has a finite subcover $Q(y_1), \ldots, Q(y_s)$. The set $A(x) = P(y_1) \cap \cdots \cap P(y_s)$ is open being the intersection of a finite number of open sets. Moreover,

$$A(x) \times X_2 = A(x) \times \bigcup_{i=1}^{s} Q(y_i) \subset \bigcup_{i=1}^{s} U(y_i),$$

so that $A(x) \times X_2$ can be covered by a finite subfamily of $\mathcal{U}$. Thus we proved that every point $x \in X_1$ has a neighbourhood $A(x) \in \mathcal{A}$.

Since X_1 is compact and $\mathcal{A}$ is an open cover for X_1, $\mathcal{A}$ has a finite subfamily $\{A(x_1), \ldots, A(x_s)\}$ that covers X_1. By the definition of $\mathcal{A}$, each set $A(x_i) \times X_2$ can be covered by a finite subfamily $\mathcal{U}_i$ of $\mathcal{U}$. Then $\mathcal{U}_1 \cup \cdots \cup \mathcal{U}_s$ is a finite subfamily of $\mathcal{U}$ covering $X_1 \times X_2$. This proves that the product space $X_1 \times X_2$ is compact.

An extension of the previous result to the product of finitely many spaces by induction is left as an exercise.

Conversely, suppose that the product space $X_1 \times X_2$ is compact. Each projection map $\pi_i(x_1, \ldots, x_k) = x_i$ associated with the cartesian product is continuous (Problem 4.20), and $X_i = \pi_i(X)$. To conclude our proof, we need to anticipate a result that will be proved in Section 5.3: the direct image of a compact set under a continuous mapping is compact. (This is Theorem 5.21).   □

Example 5.9. From the preceding theorem and Example 5.3 it follows immediately that a closed cell $A = [a_1, b_1] \times \cdots \times [a_N, b_N]$ in the Euclidean space $\mathbb{R}^N$ is compact.

Note. The infinite product of compact spaces is compact; this celebrated result is called *Tychonoff's theorem*.

5.2 Compactness in metric spaces

The property of closed bounded intervals we referred to in the introduction to the chapter motivates the following definition.

Definition 5.10. A set A in a metric space S is *sequentially compact* if every sequence (x_n) in A has a subsequence (x_{k_n}) which is convergent to some point of A.

Example 5.11. Any closed bounded interval in the Euclidean space $(\mathbb{R}, d)$ is sequentially compact (Theorem B.16).

Example 5.12. The extended real line $\overline{\mathbb{R}}$ with the metric $\rho(x, y) = |f(x) - f(y)|$, where $f(x) = x/(1 + |x|)$ for real x and $f(-\infty) = -1$, $f(+\infty) = +1$, is a sequentially compact metric space. This is true since $(\overline{\mathbb{R}}, \rho)$ is isometric to the interval $[-1, +1]$ equipped with the Euclidean metric (see Example 2.30 and Note 2.2).

Before we prove the equivalence of the two concepts of compactness in metric spaces, we need to discuss a strong version of boundedness.

Definition 5.13. A set A in a metric space (S, d) is *totally bounded* if for each $\varepsilon > 0$ there exists a finite family of balls $B(a_1; \varepsilon), B(a_2; \varepsilon), \ldots, B(a_k; \varepsilon)$ which is a cover for A, that is,

$$A \subset \bigcup_{i=1}^{k} B(a_i; \varepsilon). \tag{5.1}$$

The set $\{a_1, a_2, \ldots a_k\}$ of the centres is called an *ε-net for A*. So, a set is totally bounded if for every $\varepsilon > 0$ there exists a finite ε-net for A.

Every totally bounded set is also bounded, but in general, total boundedness is a much stronger property than boundedness. However, in certain metric spaces the two concepts coincide.

Example 5.14. In a Euclidean space $(\mathbb{R}^N, d)$, every bounded set is totally bounded.

Suppose that a set A is bounded. Then there is $r > 0$ such that the set A is contained in the cube $K = [-r, r] \times \cdots \times [-r, r]$ in $\mathbb{R}^N$. Let $\varepsilon > 0$ be given. Pick $k \in \mathbb{N}$ such that $\sqrt{N}/k < \varepsilon$. Consider the set $E(\varepsilon)$ of all points $(c_1/k, \ldots, c_N/k)$, where c_i are integers with $|c_i| \leq kr$. Then $E(\varepsilon)$ is a finite ε-net for A. (Sketch an appropriate diagram.)

The next theorem gives a useful property of sequential compactness.

Theorem 5.15. *Every sequentially compact set A in a metric space (S, d) is totally bounded (therefore bounded).*

Proof. We may assume that A is a nonempty sequentially compact set. For a proof by contradiciton, supppose that A is not totally bounded. Then there is $\varepsilon > 0$ such that no finite family of balls of radius ε covers A. Choose $x_1 \in A$ arbitrarily. This is possible as $A \neq \emptyset$. By our hypothesis, A is not a subset of $B(x_1; \varepsilon)$, so we can choose $x_2 \in A \backslash B(x_1; \varepsilon)$. Similarly, there is $x_3 \in A$ not contained in the union $B(x_1; \varepsilon) \cup B(x_2; \varepsilon)$. Continuing the construction this way, we obtain a sequence (x_n) in A with $d(x_m, x_n) \geq \varepsilon$ for all $m \neq n$. This sequence has no Cauchy subsequence, and therefore no convergent subsequence. □

Now we can present the main theorem.

Theorem 5.16. *A set A in a metric space (S, d) is compact if and only if it is sequentially compact.*

Proof.   First assume that A is sequentially compact. Let $\{U_\alpha : \alpha \in D\}$ be an open cover of A. We show that there is a number $\varepsilon > 0$ such that every ball with centre in A and radius ε is contained in at least one of the open sets U_α. (This number is often referred to as a *Lebesgue number* of the open cover $\{U_\alpha\}$.) Suppose such a number does not exist. Then for any $\varepsilon > 0$ there is a ball centred in A of radius ε which is not contained in any of the open sets U_α. So, for any positive integer n, there is a ball $B(a_n; 1/n)$ with $a_n \in A$, which is not contained in any of the U_α. Since A is sequentially compact, we can select a subsequence (a_{k_n}) which converges to a point $a \in A$. One of the sets of the open cover contains a, say U_β. Since U_β is open, there is a ball $B(a; r) \subset U_\beta$. Among the positive integers k_n there is one, say $m = k_s$, such that $m > 2/r$ and $a_m \in B(a; r)$. Then $1/m < \frac{1}{2}r$, and

$$B(a_m; 1/m) \subset B(a; r) \subset U_\beta.$$

This contradicts the definition of a_m.

Let $\varepsilon > 0$ be a Lebesgue number for the open cover $\{U_\alpha\}$ whose existence we have just established. By Theorem 5.15, the set A is totally bounded. So there is a finite cover for A by balls of radius ε. Each of these balls is contained in some U_α, which shows that $\{U_\alpha\}$ has a finite subcover for A.

Conversely assume that A is not sequentially compact; then A contains a sequence (x_n) which has no subsequence convergent to a point in A. Then every point of A is a centre of a ball which contains x_n for only finitely many n. These balls form an open cover of A. We cannot select a finite subcover, since one of the balls would have to contain x_n for infinitely many n. □

From this point on, we can use sequential compactness interchangeably with compactness for metric spaces; this often leads to simplification of proofs. From Theorem 5.16 and properties of compact sets derived in the preceding section we obtain the following useful result.

Corollary 5.17. *A compact subset A of a metric space (S, d) is closed and bounded.*

However, the following example shows that the converse of the preceding corollary is false.

Example 5.18. Consider the set

$$A = \{x = (\xi_k) \in \ell^\infty : |\xi_k| \leq 1\}$$

in the space ℓ^∞. The set A is closed and bounded (exercise). The sequence

$$e_1 = (1, 0, 0, 0, \ldots), \quad e_2 = (0, 1, 0, 0, \ldots), \quad e_3 = (0, 0, 1, 0, \ldots), \ldots$$

has the property that $d(e_n, e_m) = 1$ for any $m \neq n$, and so it cannot contain a convergent subsequence. So A is not sequentially compact.

Theorem 5.19 (Borel–Lebesgue theorem). *A subset A of a Euclidean space $(\mathbb{R}^N, d)$ is compact if and only if it is closed and bounded.*

Proof. Compact sets in metric spaces are closed and bounded by the preceding corollary. Suppose A is a bounded subset of the Euclidean space $(\mathbb{R}^N, d)$. Then A lies in some ball $B(0; r)$ centred at the origin 0. For each point $x = (\xi_1, \xi_2, \ldots, \xi_N) \in A$ we have

$$|\xi_k| \leq \sqrt{|\xi_1|^2 + \cdots + |\xi_N|^2} < r, \ k = 1, \ldots N.$$

This means that $x \in [-r, r] \times \cdots \times [-r, r] = K$. Therefore A is a closed subset of a compact metric space (K, d) (Theorem 5.8), and therefore compact (metric spaces are Hausdorff). □

The following criterion of compactness gives the third equivalent formulation of compactness in metric spaces.

Theorem 5.20 (Fréchet's criterion of compactness). *A metric space (S, d) is compact if and only if it is totally bounded and complete.*

Proof. Suppose that the space S is compact. Then S is sequentially compact, and therefore totally bounded by Theorem 5.15. From any Cauchy sequence in S we can extract a convergent subsequence. Since a Cauchy sequence with a convergent subsequence is itself convergent (exercise), S is complete.

Conversely, suppose that S is complete and totally bounded. According to Problem 5.32, every sequence (x_n) in a totally bounded space S has a Cauchy subsequence; since S is complete, this subsequence is in fact convergent. So S is sequentially compact, and therefore compact. □

5.3   Continuity and compactness

One of the reasons for our interest in compactness is the behaviour of continuous functions on compact sets. In Section 3 of Appendix B it is shown that real valued functions on closed bounded intervals, which are the prototypes of compact sets, are bounded and attain their maxima and minima. There is also Theorem 3.20, which states that a function continuous on a closed bounded interval is uniformly continuous. These and other properties generalize to continuous functions on compact sets in topological spaces.

Theorem 5.21. *Let $f\colon X \to Y$ be a continuous mapping between topological spaces, and let A be a compact subset of X. Then the direct image $f(A) = \{f(x) : x \in A\}$ of A is a compact subset of Y.*

Proof.   Let $\mathcal{V} = \{V_\alpha : \alpha \in D\}$ be an open cover for the set $f(A)$ in Y. Each set $U_\alpha = f^{-1}(V_\alpha)$ is open being the inverse image of an open set under a continuous mapping, and the family $\mathcal{U} = \{U_\alpha : \alpha \in D\}$ is an open cover for A. By the compactness of A, there is a finite subcover $U_{\alpha_1}, \ldots, U_{\alpha_s}$ of A. It can then be verified that $V_{\alpha_1}, \ldots, V_{\alpha_s}$ is an open cover for $f(A)$, which is a finite subcover of $\mathcal{V}$.   □

Theorem 5.22 (Inverse mapping theorem). *Let X be a compact space, Y a Hausdorff space, and $f\colon X \to Y$ a continuous bijection. Then f is a homeomorphism.*

Proof.   Let $g\colon Y \to X$ be the inverse mapping of f. We show that g is continuous by proving that each closed set V in X has a closed inverse image $g^{-1}(V)$ under g. If V is a closed subset of X, it is compact by Theorem 5.6. The direct image $f(V)$ of V under f is compact by the preceding theorem, and therefore closed in the Hausdorff space Y by Theorem 5.7. The conclusion follows as $f(V) = g^{-1}(V)$.   □

The inverse mapping theorem can be applied to prove the continuity of inverses of continuous functions in real analysis. Recall that a continuous function maps intervals onto intervals (Theorem B.21).

Example 5.23. Let $f\colon I \to \mathbb{R}$ be a continuous injective function on an arbitrary interval $I \subset \mathbb{R}$. Then the inverse f^{-1} defined on the interval $J = f(I)$ is also continuous. (Problem 5.17.)

We say that a mapping $f\colon X \to Y$ from a topological space X into a metric space Y is *bounded* if the range $f(X)$ is a bounded set in Y. (We need Y to be a metric space, as bounded sets are not defined in general topological spaces.) From Theorem 5.21 we see that any continuous mapping on a compact topological space into a metric space is bounded.

Theorem 5.24. *If a mapping $f\colon X \to \mathbb{R}$ is continuous and X is a compact topological space, then f is bounded and attains its maximum and minimum on X.*

Proof. By Theorem 5.21 the set $f(X)$ is compact in $\mathbb{R}$. This implies that $f(X)$ is closed and bounded in $\mathbb{R}$. So the real numbers $u = \inf f(X)$ and $v = \sup f(X)$ exist and belong to $f(X)$ as this set is closed. Therefore $u = f(a)$ for some $a \in X$ and $v = f(b)$ for some $b \in X$. ☐

In Chapter 3 we showed that a function continuous on a closed bounded interval in uniformly continuous. The proof used the Heine–Borel theorem. Virtually the same proof works for a continuous mapping on a compact metric space. Here we give an alternative proof based on the sequential criterion of uniform continuity (Theorem 3.19). (Uniform continuity is not a topological concept, and therefore it cannot be defined on arbitrary topological spaces.)

Theorem 5.25. *If $f\colon (S, d) \to (T, \rho)$ is a continuous mapping between metric spaces and S is compact, then f is uniformly continuous on S.*

Proof. Suppose f is a continuous mapping which is not uniformly continuous on S. Then there is $\varepsilon > 0$ and a pair of sequences (x_n), (y_n) in S with $d(x_n, y_n) \to 0$ and $\rho(f(x_n), f(y_n)) \geq \varepsilon$ for all $n \in \mathbb{N}$. Since S is compact, there is a subsequence (x_{k_n}) of (x_n) such that $x_{k_n} \xrightarrow{d} a$ for some point $a \in S$. Then $d(y_{k_n}, a) \leq d(y_{k_n}, x_{k_n}) + d(x_{k_n}, a)$, which shows that $y_{k_n} \xrightarrow{d} a$. However, $\rho(f(x_{k_n}), f(y_{k_n})) \geq \varepsilon$, which means that at least one of the sequences $(f(x_{k_n}))$, $(f(y_{k_n}))$ cannot converge to $f(a)$. So f cannot be continuous. ☐

5.4 Separable sets

Definition 5.26. A set A in a topological space $(X, \mathcal{T})$ is said to be *separable* if there is a countable set B which is dense in A, that is,

$$B \subset A \subset \overline{B}.$$

Example 5.27. The Euclidean space $(\mathbb{R}, d)$ is separable as the countable set $\mathbb{Q}$ is dense in $\mathbb{R}$. Similarly, the Euclidean space $(\mathbb{R}^N, d)$ is separable as $\mathbb{Q}^N$ is a countable dense subset.

Example 5.28. Let $p \geq 1$. The space ℓ^p is separable: Let A be the set of all points $a = (\alpha_1, \alpha_2, \ldots) \in \ell^p$ such that (i) each α_n is rational, and (ii) there is an index N such that $\alpha_n = 0$ for all $n > N$. Then the set A is countable (verify). Given any point $x = (\xi_n) \in \ell^p$ and any $\varepsilon > 0$, there is $N \in \mathbb{N}$ such that $\sum_{n=N+1}^{\infty} |\xi_n|^p < \frac{1}{2}\varepsilon^p$. For $1 \leq n \leq N$ there are rational numbers α_n such that $\sum_{n=1}^{N} |\xi_n - \alpha_n|^p < \frac{1}{2}\varepsilon^p$. Set $a = (\alpha_1, \alpha_2, \ldots, \alpha_N, 0, 0, \ldots)$. Then $a \in \ell^p$, and

$$d(x, a) = \left(\sum_{n=1}^{\infty} |\xi_n - \alpha_n|^p \right)^{1/p} < \varepsilon.$$

This means that the set A is dense in ℓ^p.

Example 5.29. The metric space ℓ^∞ is not separable.

Let $A = \{x_n : n \in \mathbb{N}\}$ be a countable subset of ℓ^∞, with $x_n = (\xi_{n1}, \xi_{n2}, \ldots)$. We show that ℓ^∞ contains a point x which is not in the closure of A. Define x as follows: If $|\xi_{nn}| \leq 1$, set $\xi_n = \xi_{nn} + 1$, otherwise set $\xi_n = 0$. Then

$$d(x_k, x) = \sup_{n \in \mathbb{N}} |\xi_{kn} - \xi_n| \geq |\xi_{kk} - \xi_k| \geq 1.$$

So A cannot be dense in ℓ^∞. (This is an example of the so called diagonal argument pioneered by Georg Cantor.)

Theorem 5.30. *Every totally bounded set A in a metric space is separable. Every compact set is separable.*

Proof. Let A be totally bounded. For each $n \in \mathbb{N}$ there is a finite $(1/n)$-net E_n for A contained in A. The union $E = \bigcup_{n \in \mathbb{N}} E_n$ is a countable subset of A. It can be shown that $\overline{E} \supset A$. The second part of the theorem follows from the fact that compact sets are totally bounded. $\qquad\square$

5.5 The Ascoli–Arzelà theorem

If (S, d) is a compact metric space, then the set $C(S)$ of all real valued continuous functions on S is a real vector space. Since continuous functions on compact spaces are bounded, $C(S)$ can be equipped with the norm

$$\|f\| = \sup_{t \in S} |f(t)| \quad (= \max_{t \in S} |f(t)|).$$

The metric $d(f,g) = \|f - g\|$ induced by this norm is the familiar uniform metric on $C(S)$. The space $C(S)$ is of great importance to analysis. It is a closed subset of the space $B(S)$ of all bounded functions on S. Since $B(S)$ is complete, so is $C(S)$. In this section we investigate when a subset A of the space $C(S)$ is compact.

Definition 5.31. Let A be a nonempty subset of $C(S)$. The set A is said to be *equicontinuous* at a point $a \in S$ if, for each $\varepsilon > 0$, there is $\delta = \delta(a; \varepsilon) > 0$ such that

$$\{f \in A \text{ and } x \in B(a; \delta)\} \implies |f(x) - f(a)| < \varepsilon.$$

Observe that δ depends on a and ε, but not on the function $f \in A$. We say that a set $A \subset C(S)$ is equicontinuous if it is equicontinuous at each point $a \in S$.

Theorem 5.32 (Ascoli–Arzelà theorem). *Let S be a compact metric space. A nonempty set $A \subset C(S)$ is totally bounded if and only if A is bounded relative to the norm in $C(S)$ and equicontinuous.*

Proof. Suppose that A is bounded and equicontinuous. Let $\varepsilon > 0$ be given. For each $a \in S$ there is a ball $B(a; \delta(a))$ such that

$$\{f \in A \text{ and } x \in B(a; \delta(a))\} \implies |f(x) - f(a)| < \tfrac{1}{3}\varepsilon.$$

Applying the Heine–Borel property to the space S, we conclude that there are balls $B(x_k; \delta_k)$ with $\delta_k = \delta(x_k)$, $k = 1, \ldots, N$, that cover S. Then, for every k,

$$\{f \in A \text{ and } x \in B(x_k; \delta_k)\} \implies |f(x) - f(x_k)| < \tfrac{1}{3}\varepsilon.$$

Let us define a mapping $P \colon A \to \mathbb{R}^N$ by setting

$$P(f) = (f(x_1), \ldots, f(x_N)), \quad f \in A.$$

The set $P(A)$ is bounded in $\mathbb{R}^N$: Since A is bounded, there is a positive constant c such that

$$\|f\| \leq c \text{ for all } f \in A.$$

Then, for each $f \in A$,

$$\|P(f)\|_\infty = \max_k |f(x_k)| \leq \|f\| \leq c.$$

Bounded sets in $\mathbb{R}^N$ are totally bounded. Let $P(f_1), \ldots, P(f_p)$ be a $\tfrac{1}{3}\varepsilon$-net for $P(A)$. We show that $f_1, \ldots, f_p$ is an ε-net for A.

Let $f \in A$. Then $P(f) \in P(A)$, and there is f_j such that $\|P(f) - P(f_j)\|_\infty < \frac{1}{3}\varepsilon$. For any $x \in S$ there is x_k with $x \in B(x_k; \delta_k)$. Then

$$|f(x) - f_j(x)| \leq |f(x) - f(x_k)| + |f(x_k) - f_j(x_k)| + |f_j(x_k) - f_j(x)|$$
$$\leq \tfrac{1}{3}\varepsilon + \|P(f) - P(f_j)\|_\infty + \tfrac{1}{3}\varepsilon$$
$$< \tfrac{1}{3}\varepsilon + \tfrac{1}{3}\varepsilon + \tfrac{1}{3}\varepsilon = \varepsilon.$$

Since the inequality is true for any $x \in S$, we have $\|f - f_j\| < \varepsilon$.

The converse is left as an exercise. $\qquad\square$

If A is a bounded equicontinuous subset of $C(S)$, then the preceding theorem implies that *every* sequence in A has a convergent subsequence, but the limit need not lie in A.

Corollary 5.33. *Let S be a compact metric space. A nonempty set $A \subset C(S)$ is compact if and only if A is bounded, closed and equicontinuous.*

Proof. Follows from the preceding theorem and Fréchet's criterion of compactness (Theorem 5.20). $\qquad\square$

5.6   The Stone–Weierstrass theorem

The Weierstrass approximation theorem states that every continuous real valued function on a compact interval $[a, b]$ can be uniformly approximated by real polynomials. Stone's far reaching generalization of this result is one of the most important theorems in approximation theory. Let S be a compact Hausdorff space, and $C(S, \mathbb{R})$ the space of all continuous real valued functions on S with the uniform norm $\|f\| = \sup_{t \in S} |f(t)|$. We say that a subset $\mathcal{F}$ of $C(S, \mathbb{R})$ *separates points of S* if, for any pair of distinct points x, y in S, there is $f \in \mathcal{F}$ such that $f(x) \neq f(y)$. A subset $\mathcal{A}$ of $C(S, \mathbb{R})$ is an *algebra* if it is a real vector space with operations of pointwise addition and scalar multiplication, and if $f, g \in \mathcal{A}$ implies that $fg \in \mathcal{A}$. In particular, $C(S, \mathbb{R})$ is an algebra. In the real algebra $C(S, \mathbb{R})$ we often use the notation

$$f \vee g = \max\{f, g\}, \quad f \wedge g = \min\{f, g\},$$

which refers to the partial order structure in the algebra.

Theorem 5.34 (Stone–Weierstrass theorem). *Let S be a compact Hausdorff space. If $\mathcal{A}$ is an algebra in $C(S, \mathbb{R})$ which separates points of S and contains the constant function 1, then $\mathcal{A}$ is dense in $(C(S, \mathbb{R}), \|\cdot\|)$.*

Proof. There exists a sequence of real polynomials (p_n) such that $p_n(t) \to \sqrt{t}$ uniformly on the interval $[0,1]$ (Problem 5.25). We deduce that if $f \in \mathcal{A}$, then $|f| \in \overline{\mathcal{A}}$: Let $a = \|f\| \neq 0$. Then the sequence $\big(ap_n(f^2/a^2)\big)$ belongs to the algebra $\mathcal{A}$ and converges uniformly to $|f|$ on S. Next we observe that if $f, g \in \overline{\mathcal{A}}$, then $f \vee g$ and $f \wedge g$ belong to $\overline{\mathcal{A}}$:

$$f \vee g = \tfrac{1}{2}(f + g + |f - g|), \qquad f \wedge g = \tfrac{1}{2}(f + g - |f - g|).$$

We make the following observation: If x, y are distinct points in S and α, β are distinct real numbers, then there exists a function $g \in \overline{\mathcal{A}}$ such that

$$g(x) = \alpha, \qquad g(y) = \beta \tag{5.2}$$

(Problem 5.26).

Let $f \in C(S)$, $x \in S$ and let $\varepsilon > 0$. Then there exists a function $g \in \overline{\mathcal{A}}$ such that $g(x) = f(x)$ and $g(t) > f(t) - \varepsilon$ for all $t \in S$. Indeed, by (5.2), for each $s \in S$ there exists $g_s \in \overline{\mathcal{A}}$ such that $g_s(x) = f(x)$ and $g_s(s) > f(s) - \frac{1}{2}\varepsilon$. Since g_s is continuous, there exists an open neighbourhood U_s of s such that $g_s(t) > f(t) - \varepsilon$ for all $t \in U_s$. The sets U_s form an open cover for S. Since S is compact, there exists a finite subcover $U_{s_1}, \ldots, U_{s_n}$. Define $g = g_{s_1} \vee \cdots \vee g_{s_n}$. Then $g \in \overline{\mathcal{A}}$ by the preceding argument, $g(x) = f(x)$ and $g(t) > f(t) - \varepsilon$ for all $t \in S$.

Finally, let $f \in C(S)$ and let $\varepsilon > 0$. By the preceding argument, for each $s \in S$ there exists $h_s \in \overline{\mathcal{A}}$ such that $h_s(s) = f(s)$ and $h_s(t) > f(t) - \varepsilon$ for all $t \in S$. In view of the continuity of f and h_s, there exists an open neighbourhood V_s of s such that $h_s(t) < f(t) + \varepsilon$ for all $t \in V_s$. We can cover the compact space S with a finite number of the neigbourhoods $V_{s_1}, \ldots, V_{s_k}$. Set $g = g_{s_1} \wedge \cdots \wedge g_{s_k}$. Then $g \in \overline{\mathcal{A}}$ and $|f(t) - g(t)| < \varepsilon$ for all $t \in S$. Hence $\|f - g\| < \varepsilon$, and f belongs to $\overline{\overline{\mathcal{A}}} = \overline{\mathcal{A}}$. $\qquad\square$

We have a complex analogue of the preceding theorem. A subset $\mathcal{A}$ of $C(S, \mathbb{C})$ is called a *self-adjoint algebra* if it is a complex algebra and if $f \in \mathcal{A}$ implies $\overline{f} \in \mathcal{A}$.

Theorem 5.35 (Complex form of Stone–Weierstrass theorem).
Let S be a compact Hausdorff space. If $\mathcal{A}$ is a self-adjoint algebra in $C(S, \mathbb{C})$ which separates points of S and contains the constant function 1, then $\mathcal{A}$ is dense in $(C(S, \mathbb{C}), \|\cdot\|)$.

Proof. Problem 5.28. $\qquad\square$

If K is a compact subset of $\mathbb{R}^k$, then the set of all real valued polynomials in k real variables is an algebra in $C(K, \mathbb{R})$; it contains the constant function

1 and separates points of K (Problem 5.27). A similar argument applies to the set of all complex valued polynomials in k real variables; in addition, the complex valued polynomials form a self-joint algebra. Applying the Stone–Weierstrass theorems to this setting we obtain the following classical result.

Theorem 5.36 (Weierstrass approximation theorem I). *Let $K \subset \mathbb{R}^k$ be compact. Then the real valued polynomials in k real variables are uniformly dense in $C(K, \mathbb{R})$, and the complex valued polynomials in k real variables are uniformly dense in $C(K, \mathbb{C})$.*

From the Weierstrass approximation theorem we can deduce that the spaces $C(K, \mathbb{R})$ and $C(K, \mathbb{C})$ are separable.

Let us consider the set $C([0, 2\pi], \mathbb{C})$ of all continuous complex valued functions defined on the interval $[0, 2\pi]$ and the subset $\mathcal{T}$ of $C([0, 2\pi], \mathbb{C})$ consisting of the so called *trigonometric polynomials*, that is, functions of the form

$$\sigma(t) = \sum_{k=-n}^{n} a_k (e^{it})^k = \sum_{k=-n}^{n} a_k e^{ikt}, \quad a_k \in \mathbb{C}.$$

Then $\mathcal{T}$ is a self-adjoint algebra which contains the constant function 1, however, $\mathcal{T}$ does not separate points in $[0, 2\pi]$ (Problem 5.31). We note that $\sigma(0) = \sigma(2\pi)$ for any $\sigma \in \mathcal{T}$. This suggests that we form a topological space S by identifying the points 0 and 2π in $[0, 2\pi]$; then S is a compact Hausdorff space and $\mathcal{T}$ separates points of S. Functions continuous on S can be then regarded as functions continuous on $[0, 2\pi]$ satisfying $f(0) = f(2\pi)$. These considerations lead to the following result.

Theorem 5.37 (Weierstrass approximation theorem II).
Let $C_{2\pi} = C_0([0, 2\pi], \mathbb{C})$ be the set of all continuous complex valued functions f on $[0, 2\pi]$ such that $f(0) = f(2\pi)$. Then the set of all trigonometric polynomials $\sigma(t) = \sum_{k=-n}^{n} \alpha_k e^{ikt}$ is uniformly dense in $C_0([0, 2\pi], \mathbb{C})$.

5.7   Hausdorff hyperspace and fractals

Let (S, d) be a complete metric space. We recall that the (infimum) distance between a point $x \in S$ and a set $A \subset S$ is defined by

$$\text{dist}(x, A) = \inf_{z \in A} d(x, z).$$

(See Problem 1.30.) If the set A is compact, then the infimum in this definition can be replaced by the minimum. (Exercise.)

By $\mathcal{H}(S)$ we denote the set whose points are compact subsets of S other than $\emptyset$. For any two points A, B of $\mathcal{H}(S)$ let

$$\rho(A, B) = \max_{x \in A} \text{dist}\,(x, B).$$

It is left as an exercise to check that the maximum exists and that there are points $x^* \in A$ and $y^* \in B$ such that

$$\rho(A, B) = d(x^*, y^*).$$

In general, $\rho(A, B) \neq \rho(B, A)$. To get symmetry in A and B we set

$$h(A, B) = \max\,(\rho(A, B), \rho(B, A)).$$

The function h then becomes a metric on $\mathcal{H}(S)$ (Problem 5.35). The metric h is known as the *Hausdorff metric* and the space $(\mathcal{H}(S), h)$ is often called the *Hausdorff hyperspace of S*. In applications it is important to know that the Hausdorff hyperspace is complete (Problem 5.37).

The Hausdorff hyperspace $(\mathcal{H}(S), h)$ is a space in which we can study the so called fractal geometry. A French mathematician Benoit Mandelbrot introduced *fractals* in 1975 to study self-similarity of natural structures. It is a widespread phenomenon in nature that certain shapes repeat themselves within the same object. A typical example is a fern leaf; other examples include clouds, forests, flowers, mountains, coastlines, etc.

Theorem 5.38. *Let (S, d) be a complete metric space and $f_1, \ldots, f_s$ contraction mappings on S with contraction coefficients $\alpha_1, \ldots, \alpha_s$. Define F to act on subsets of S by*

$$F(E) = \bigcup_{i=1}^{s} f_i(E).$$

Then F is a contraction mapping on the Hausdorff hyperspace $(\mathcal{H}(S), h)$ with the contraction coefficient $\alpha = \max\,(\alpha_1, \ldots, \alpha_s)$.

Proof. Problem 5.39. □

Fixed points of contraction mappings F in the Hausdorff hyperspace $\mathcal{H}(S)$ of the kind described in the preceding theorem are examples of *deterministic fractals*. Using the contraction mapping theorem, we can produce deterministic fractals by starting with an arbitrary point E_0 in $\mathcal{H}(S)$, that is a compact set in the space (S, d), and iterate:

$$E_{n+1} = F(E_n), \quad n = 0, 1, \ldots$$

The sequence (E_n) converges to a limit A in $\mathcal{H}(S)$, which is the fixed point of F, a deterministic fractal. For more details on this topic, which leads to fractal image compression, see Barnsley's illuminating books [4] and [5].

5.8   Problems for Chapter 5

(Problems marked with an asterisk are optional.)

1. Show that every finite set in a topological space is compact.

2. Show that every compact set in a discrete topological space is finite.

3. Prove that a finite union of compact sets in a topological space is compact.

4. Show that an arbitrary intersection of compact sets in a Hausdorff space is compact.

5. Prove that an infinite set X equipped with the finite complement topology ($\emptyset$ and the complements of all finite sets in X) is compact.

6. Prove that an infinite set X equipped with the discrete topology is not compact.

7. Prove that a topological space X is compact if and only if every family $\mathcal{F}$ of closed subsets of X with the finite intersection property has nonempty intersection. (The *finite intersection property* means that any finite subfamily of $\mathcal{F}$ has nonempty intersection.)

8. Which of the following sets is compact in the Euclidean space $\mathbb{R}$?

(i) $[0, 1)$ (ii) $[0, \infty)$ (iii) $\mathbb{Q} \cap [0, 1]$,

(iv) $\{\frac{1}{n} : n \in \mathbb{N}\}$,   (v) $\{\frac{1}{n} : n \in \mathbb{N}\} \cup \{0\}$.

9. Let A be a compact set in a metric space (S, d). Prove that for each $s \in S$ there is $a \in A$ such that $d(s, A) = d(s, a)$, where $d(s, A) = \inf \{d(s, u) : u \in A\}$ is the so called *lower distance* from s to A. Give an example showing that a need not be unique. Give another example showing that a need not exist if A is not compact.

10. Which of the following sets is compact in the Euclidean space $\mathbb{R}^2$?

(i) $\{(\xi_1, \xi_2) \in \mathbb{R}^2 : \xi_1 \geq 1 \text{ and } 0 \leq \xi_2 \leq 1/\xi_1\}$ (ii) $\{(\xi_1, \xi_2) \in \mathbb{R}^2 : \xi_1^2 + \xi_2^2 = 1\}$

(iii) $\{(\xi_1, \xi_2) \in \mathbb{R}^2 : |\xi_1| + |\xi_2| \leq 1\}$ (iv) $\{(\xi_1, \xi_2) \in \mathbb{R}^2 : \xi_1^2 + \xi_2^2 < 1\}$

11. If A is a bounded subset of the Euclidean space $\mathbb{R}^N$, show that the closure $\overline{A}$ is compact.

12. Let $f: X \to Y$ be a continuous mapping from a compact space X into a Hausdorff space Y such that $f(X)$ is dense in Y. Show that f is surjective.

13. Show that a compact subset of a topological space need not be closed by considering an infinite subset of a topological space X equipped with the finite complement topology.

14. Let $\mathcal{T}_1, \mathcal{T}_2$ be topologies on X, and let $\mathcal{T}_1$ be weaker than $\mathcal{T}_2$ (see Definition 4.9). If $(X, \mathcal{T}_1)$ is Hausdorff and $(X, \mathcal{T}_2)$ is a compact space, show that $\mathcal{T}_1 = \mathcal{T}_2$.

15. *Bolzano-Weierstrass property.* Prove the following facts:

(i) Every infinite subset of a compact topological space has a limit point.

(ii) If every infinite subset of a metric space has a limit point, the space is compact.

16. Show that the cartesian product of k metric spaces equipped with the product metric is totally bounded if and only if each of the component spaces is totally bounded.

17. (See Example 5.23.) Let $f: I \to \mathbb{R}$ be a continuous injective function on an arbitrary interval $I \subset \mathbb{R}$. Then the inverse function $g = f^{-1}$ defined on the interval $J = f(I)$ is also continuous.

18. If the sets A, B in a metric space are totally bounded, show that $A \cup B$ is totally bounded.

19. Show that a set A in a metric space is totally bounded if and only if $\overline{A}$ is totally bounded.

20. If A is a compact subset of a metric space, show that the set A' of all limit points of A is also compact.

21. *Hilbert cube.* Let Q be a subset of ℓ^2 consisting of $x = (\xi_1, \xi_2, \ldots)$ with $|\xi_k| \leq 1/k$ for all k. Show that Q is compact.

22. Let $f: S \to T$ be a uniformly continuous mappings between metric spaces. Show that the direct image $f(A)$ of a totally bounded set A is also totally bounded. Give an example showing that a merely continuous mapping f would not do.

23. *A fixed point theorem.* Let (S, d) be a compact metric space and $f: S \to S$ a self-map of S such that $d(f(x), f(y)) < d(x, y)$ for all $x, y \in S, \ x \neq y$. Prove that f has a unique fixed point on S. (Note that f need not be a contraction.)

24. *Dini's theorem.* Let (S, d) be a compact Hausdorff space, and (f_n) a sequence of continuous real valued functions on S such that $f_n \nearrow f$ for a continuous function f (monotonic pointwise convergence). Prove that (f_n) converge uniformly on S to f.

Suggestion. Given $\varepsilon > 0$, let $A_n = (f - f_n)^{-1}((-\infty, \varepsilon))$. Show that $\{A_n\}$ is an open cover for S, and use compactness.

25. Define $p_1(t) = 0$, $p_{n+1}(t) = p_n(t) + \frac{1}{2}(t - p_n(t)^2)$ for $t \in [0, 1]$ and $n = 1, 2, 3 \ldots$ Use Dini's theorem to show that (p_n) is a sequence of polynomials convergent uniformly on $[0, 1]$ to $p(t) = \sqrt{t}$. (Needed in the proof of the Stone–Weierstrass theorem.)

26. Let $\mathcal{F}$ be a subfamily of $C(S, \mathbb{R})$ which separates points of S and contains the constant function 1. If x, y are distinct points in S and α, β are distinct real numbers, then there exists a function $g \in \mathcal{F}$ such that $g(x) = \alpha$ and $g(y) = \beta$.

Suggestion. If h separates the points x and y, define g as a suitable linear combination of h and 1.

27. Prove that the family $\mathcal{P}$ consisting of all real (respectively complex) valued polynomials in k real variables on a compact set $K \subset \mathbb{R}^k$ is an algebra in $C(K, \mathbb{R})$ (respectively $C(K, \mathbb{C})$). Show that $\mathcal{P}$ contains the constant function 1 and separates points of K, and that the complex valued $\mathcal{P}$ is self-adjoint.

28. Prove the complex form of the Stone–Weierstrass theorem.

29. Give a detailed proof of the separability of $C(K, \mathbb{R})$ (respectively $C(K, \mathbb{C})$).

30. Let $K \subset \mathbb{C}$ be compact.

(i) Explain why the polynomials $p(\lambda) = \sum_{k=0}^n \alpha_k \lambda^k$ ($\alpha_k \in \mathbb{C}$, $\lambda \in K$) are not uniformly dense in $C(K, \mathbb{C})$ unless $K \subset \mathbb{R}$.

(ii) Show that the set of the so-called *bipolynomials* of the form $q(\lambda, \bar{\lambda}) = \sum_{k,j=0}^n \alpha_{k,j} \lambda^k \bar{\lambda}^j$ ($\alpha_k \in \mathbb{C}$, $\lambda \in K$) is dense in $C(K, \mathbb{C})$.

31. Let $\mathcal{T}$ be the subset of $C([0, 2\pi], \mathbb{C})$ consisting of all trigonometric polynomials of the form $\sigma(t) = \sum_{k=-n}^n a_k e^{ikt}$, where $a_k \in \mathbb{C}$ and n can vary. Prove the following:

(i) $\mathcal{T}$ is a self-adjoint algebra which contains the constant function 1.

(ii) $\mathcal{T}$ separates points of $[0, 2\pi)$.

(iii) $\mathcal{T}$ does not separate points of $[0, 2\pi]$.

32. (Harder.) Prove that a set A in a metric space (S, d) is totally bounded if and only if every sequence in A has a Cauchy subsequence.

33.* Let (S, d) be a compact metric space and let $f \colon S \to S$ be a mapping such that $d(f(x), f(y)) \geq d(x, y)$ for all $x, y \in S$ (a so called *expansive* mapping). Show that f is an isometry on S, that is, that $d(f(x), f(y)) = d(x, y)$ for all

$x, y \in S$ and that f is surjective.

Suggestion. Let a, b be any two points in S. Show that, for any $\varepsilon > 0$, there is an index k such that $d(f^k(a), a) < \varepsilon$ and $d(f^k(b), b) < \varepsilon$ (for this consider subsequences of $(f^n(a))$ and $(f^n(b))$). Conclude that $d(f(a), f(b)) \leq d(a, b)$. Then observe that $f(S)$ is dense in S.

34.* Show that $f(x) = 1 + 2x$ is an expansion on the space $S = (0, +\infty)$ equipped with the Euclidean metric, but that f is neither an isometry nor a surjection. Comment on the previous problem.

35.* Prove that the function h defined in Section 5.7 is a metric.

36.* Let (S, d) be a complete metric space. Let (B_m) be a contracting sequence of nonempty closed sets in S (that is, $B_m \supset B_{m+1}$), (δ_m) a sequence of positive reals convergent to 0, and (Q_m) a sequence of nonempty finite sets in S such that, for each m, $\text{dist}\,(x, Q_m) \leq \delta_m$ for all $x \in B_m$. Show that the set $A = \bigcap_{m=1}^{\infty} B_m$ is nonempty and compact.

37.* If (S, d) is complete, prove that the fractal space $(\mathcal{H}(S), h)$ is complete. For this assume that (A_n) is a Cauchy sequence in this space, and define

$$B_m = \text{cl}\left(\bigcup_{k=m}^{\infty} A_k \right) \quad \text{and} \quad A = \bigcap_{m=1}^{\infty} B_m.$$

Show that $h(A_n, A) \to 0$ in a series of steps.

(i) Prove that (B_m) is a contracting sequence of nonempty closed sets.

(ii) Select m and keep it fixed. For each n let K_n be a finite $(\frac{1}{2})^{m+1}$-net for A_n. Choose $p > m$ so that $h(A_n, A_p) < (\frac{1}{2})^{m+1}$ for all $n \geq p$ and set $Q_m = \bigcup_{i=m}^{p} K_i$. Show that $d(x, Q_m) \leq (\frac{1}{2})^m$ for all $x \in B_m$. From the preceding problem conclude that A is nonempty and compact.

(iii) Prove that $\rho(A, A_n) \to 0$ and $\rho(A_n, A) \to 0$.

38.* If (S, d) is a compact space, show that its Hausdorff hyperspace $(\mathcal{H}(S), h)$ is compact.

39.* Prove Theorem 5.38.

Chapter 6

Connectedness

Intuitively, a set is 'connected' if it is of one piece, if it cannot be decomposed into two or more 'separated' pieces. An interval in $\mathbb{R}$ can serve as an example of a connected set. Other examples are a disc in the plane or a strip consisting of all the points in the (x, y)-plane with the x coordinate between -1 and 1.

On the other hand, consider the set A in $\mathbb{R}$ obtained from the interval $(0, 2)$ by removing the point 1. We can view this as an example of a 'disconnected' set. We observe that A has a proper subset (distinct from $\emptyset$ and A), namely $(0, 1)$, which is simultaneously open and closed in the subspace A. If such a subset exists, we say that the set is disconnected, otherwise the set is connected. Observe that when we remove a point from a disc in the plane, it remains connected; however, it becomes disconnected when we remove a diameter.

In many application of mathematics we use yet another kind of connectedness. A set A can be said to be 'path connected' if any two points in A can be joined by a continuous curve lying entirely in A. This is an especially useful concept when we use contour integrals (for instance in the plane or the 3-space). It turns out that this 'path connectedness' differs from the concept of connectedness considered earlier.

Connectedness was first systematically studied by Hausdorff in [22].

6.1 Connected sets

We start with a definition of a disconnected set and a connected set in a topological space $(X, \mathcal{T})$ guided by the examples discussed in the introduction. W need the concept of the subspace topology $\mathcal{T}_A$ for a nonempty set $A \subset X$: It consists of all sets $A \cap G$, where G is open in X, that is, $G \in \mathcal{T}$.

Definition 6.1. A set A in a topological space X is *disconnected* if A has a proper subset (other that $\emptyset$ and A) which is both open and closed in the subspace topology of A. A set A is *connected* if it is not disconnected, that is, if the only subsets of A which are simultaneously open and closed in the subspace $(A, \mathcal{T}_A)$ are $\emptyset$ and A.

Example 6.2. Let A be the set in the Euclidean plane $\mathbb{R}^2$ obtained from the disc $\{(\xi_1, \xi_2) : \xi_1^2 + \xi_2^2 < 1\}$ by removing all the points (ξ_1, ξ_2) with $\xi_2 = 0$. The set A is disconnected as it contains a proper subset, which is both open and closed in A, namely the upper half-disc.

Example 6.3. A discrete space containing more than one point is disconnected as every subset is both open and closed.

Example 6.4. The set $\mathbb{Q}$ of all rational numbers is disconnected in $\mathbb{R}$: The set $U = \mathbb{Q} \cap (\sqrt{2}, +\infty)$ is open in $\mathbb{Q}$. It is also closed as its complement $\mathbb{Q} \cap (-\infty, \sqrt{2})$ is open in $\mathbb{Q}$.

Theorem 6.5. *Any interval is connected in the Euclidean line $\mathbb{R}$.*

Proof. Let I be an interval of any kind in $\mathbb{R}$. Then I is characterized by the property that, for any $u, v \in I$, the inequality $u < x < v$ implies $x \in I$.

Suppose that I is disconnected. Then there exists a proper subset E of I which is both open and closed in I. Choose a point $a \in E$ and a point $b \in F$, where $F = I \backslash E$ is the complement of E in I. These choices are possible since E is a proper subset of I. Note that F is also a proper subset of I both open and closed in I. Without loss of generality we may assume that $a < b$. We use the bisection method to reach a contradiction.

Define $[a_1, b_1] = [a, b]$. Bisect this interval. The half which has its left end point in E and its right end point in F will be denoted by $[a_2, b_2]$. Proceeding by induction we obtain a sequence of intervals $[a_n, b_n]$ contained in I whose end points form monotonic sequences convergent to the same limit $c \in I$. Since $c = \lim_{n \to \infty} a_n$ and since each a_n lies in the set E which is closed in I, we have $c \in E$. An analogous argument shows that $c \in F$, which is the desired contradiction. $\qquad\square$

Theorem 6.6. *Any connected subset of the Euclidean line $\mathbb{R}$ is an interval.*

Proof. Let A be a nonempty subset of $\mathbb{R}$. Let $a = \inf A$ if A is bounded below, and $a = -\infty$ otherwise; let $b = \sup A$ if A is bounded above, and $b = +\infty$ otherwise. If A is not an interval, there is a point ξ, $a < \xi < b$, with $\xi \notin A$. Set $E = A \cap (-\infty, \xi)$. Then E is a proper subset of A which is open in A and whose complement $A \cap (\xi, +\infty)$ is also open in A. So E is both open and closed in A and A is disconnected. $\square$

Theorem 6.7. *Let A be a connected subset of a topological space X, and let $f : X \to Y$ be a continuous mapping between topological spaces. Then the direct image $f(A)$ of A is a connected set in Y.*

Proof. Suppose the set $B = f(A)$ contains a subset V which is both open and closed in B. Let $g : A \to B$ be a restriction of the mapping f. Then g is continuous and the set $U = g^{-1}(V)$ is both open and closed in A. Since A is connected, either $U = \emptyset$ or $U = A$, which implies that either $V = \emptyset$ or $V = B$. So B is connected. $\square$

Theorem 6.8 (A general intermediate value theorem). *Let $f : X \to \mathbb{R}$ be a continuous mapping on a connected topological space X. If $a, b \in X$ and $f(a) < w < f(b)$, then there is $c \in X$ such that $f(c) = w$.*

Proof. The set $f(X)$ is connected in $\mathbb{R}$, and so it is an interval by Theorem 6.6. The result then follows. $\square$

Theorem 6.9. *Let $\{A_\alpha : \alpha \in D\}$ be a family of connected subsets of a topological space X such that any two sets have a nonempty intersection. Then*

$$A = \bigcup_{\alpha \in D} A_\alpha$$

is also a connected subset of X.

Proof. Let E be a subset of A which is both open and closed in A. For any index α, the set $E \cap A_\alpha$ is both open and closed in A_α; as A_α is connected, $E \cap A_\alpha$ equals either A_α or $\emptyset$. Since $A_\alpha \cap A_\beta \neq \emptyset$ for every $\beta \in D$, we have either $E \cap A_\alpha = A_\alpha$ for all $\alpha \in D$ or $E \cap A_\alpha = \emptyset$ for all $\alpha \in D$. This means that either $E = A$ or $E = \emptyset$, and A is connected. $\square$

Theorems 6.7 and 6.9 are very useful tools for establishing connectedness of various sets, as are the following results.

Theorem 6.10. *A topological space X is connected if and only if every continuous function on X to the discrete space $\{0, 1\}$ is constant.*

Proof. Problem 6.5. $\square$

Theorem 6.11. *If A is a connected subset of a topological space X and $A \subset B \subset \overline{A}$, then B is connected.*

Proof.   Problem 6.7.   □

Theorem 6.12. *If for any two points in a topological space X there exists a connected set which contains them, then X is connected.*

Proof.   Problem 6.6.   □

For the proof of the following theorem we need this construction: Let $X_1, \ldots, X_n$ be topological spaces and let $a = (a_1, \ldots, a_n)$ be a selected point in $X = X_1 \times \cdots \times X_n$. Then for each k, the mapping $\varphi_k(x) = (a_1, \ldots, a_{k-1}, x, a_{k+1}, \ldots, a_n)$ is a homeomorphism of X_k onto

$$Z_k(a) := \{a_1\} \times \ldots \{a_{k-1}\} \times X_k \times \{a_{k+1}\} \times \ldots \{a_n\}.$$

If X_k is a connected space, $Z_k(a)$ is a connected subset of X.

Theorem 6.13. *The product space $X = X_1 \times \cdots \times X_k$ of topological spaces X_i is connected if and only if each X_i is connected.*

Proof.   If X is connected, then each X_i is connected by Theorem 6.7 as $X_i = \pi_i(X)$, where π_i is a continuous projection onto X_i.

Conversely assume that each X_i is connected. We show that for any two points $p, q \in X$ there is a connected set which contains them. Then X will be connected in view of Theorem 6.12. For $p, q \in X$ we define points

$$a^{(k)} = (q_1, \ldots, q_k, p_{k+1}, \ldots, p_n), \quad k = 1, \ldots, n,$$

and construct the corresponding sets $Z_k := Z_k(a^{(k)})$ $(k = 1, \ldots, n)$. Then $a^{(k)} \in Z_k \cap Z_{k+1}$, that is, $Z_k \cap Z_{k+1} \neq \emptyset$ $(n = 1, \ldots, n-1)$. Then the set $Z = Z_1 \cup \cdots \cup Z_n$ is connected (Theorem 6.9), while $p \in Z_1 \subset Z$ and $q \in Z_n \subset Z$.   □

Theorem 6.14. *Every normed space is connected.*

Proof.   Let X be a real normed space with the induced metric

$$\rho(x, y) = \|x - y\|.$$

For each $x \neq 0$ in X define $A_x = \{\lambda x : \lambda \in \mathbb{R}\}$. We observe that each set A_x is connected since $A_x = f_x(\mathbb{R})$, where $f_x : (\mathbb{R}, d) \to (X, \rho)$ is a continuous mapping defined by $f_x(\lambda) = \lambda x$:

$$\rho(f_x(\lambda), f_x(\mu)) = \|f_x(\lambda) - f_x(\mu)\| = \|(\lambda - \mu)x\|$$
$$= |\lambda - \mu| \, \|x\| = \|x\| \cdot d(\lambda, \mu).$$

The family $\{A_x : x \in X\}$ has a point in common, namely $x = 0$. By Theorem 6.9, the union $\bigcup_{x \in X} A_x = X$ is connected.

The proof works also for complex normed spaces.   □

6.2 Path connected sets

Definition 6.15. Let a, b be two points in a topological space X. A *path* from a to b in X is a continuous mapping $\varphi \colon [0, 1] \to X$ with $\varphi(0) = a$ and $\varphi(1) = b$. We also say that φ *joins* a to b.

Definition 6.16. A set A in a topological space X is *path connected* if any points $a, b \in A$ can be joined by a path φ lying entirely in A (that is, $\varphi([0, 1]) \subset A$).

Theorem 6.17. *Any path connected set is connected.*

Proof. This follows from the definition of path connectedness and Problem 6.6. □

Example 6.18. There are sets which are connected but not path connected: Define

$$A = \{(1 - t^{-1}) \cos t, (1 - t^{-1}) \sin t) : t \geq 1\}, \quad B = \{(\cos t, \sin t) : t \in \mathbb{R}\},$$

and set $S = A \cup B = \overline{A}$ in the Euclidean plane. Then A is connected, being the continuous image of the connected space $[1, \infty)$, and hence $S = \overline{A}$ is connected by Theorem 6.11. However, S is not path connected (Problem 6.15).

However, there are certain classes of sets for which the two concepts of connectedness coincide.

Theorem 6.19. *Let A be an open subset of a normed space X. Then A is connected if and only if it is path connected.*

Proof. Suppose that A is connected. Choose $a \in A$ and keep it fixed for the rest of this argument. Let U be the set of all $x \in A$ which can be joined to a by a path in A. We show that U is both open and closed in A.

U is open in A: Let $x \in U$. There is a path φ in A joining a to x. Since A is open, there is a ball $B(x; r) \subset A$. We show that $B(x; r) \subset U$. Let $y \in B(x; r)$. Then

$$\psi(t) = \begin{cases} \varphi(2t) & \text{if } 0 \leq t \leq \frac{1}{2} \\ x + (2t - 1)(y - x) & \text{if } \frac{1}{2} \leq t \leq 1 \end{cases}$$

is a path from a to y contained entirely in A (verify).

U is closed in A: We show that the complement V of U in A is open in A. Given $y \in V$, there is a ball $B(y; r) \subset A$. If there were a point

$x \in B(y;r)$ which could be joined to a by a path in A, y itself could be joined to a by such a path by the construction in the first part of the proof. So $B(y;r) \subset V$. $\square$

6.3   Problems for Chapter 6

1. Show that the following subsets of the Euclidean plane are connected, using Theorems 6.7 and 6.9 whenever possible.

(i)   A ray $R_a = \{ta : t \geq 0\}$, where a is a nonzero point in $\mathbb{R}^2$.

(ii)   A line segment $L_{a,b} = \{a + t(b - a) : 0 \leq t \leq 1\}$, where a, b are distinct points in $\mathbb{R}^2$.

(iii)   A half-plane $\{(\xi_1, \xi_2) : \alpha_1 \xi_1 + \alpha_2 \xi_2 \geq \beta\}$.

(iv)   A disc $\{(\xi_1, \xi_2) : (\xi_1 - \alpha_1)^2 + (\xi_2 - \alpha_2)^2 \leq r^2\}$, where $r > 0$.

(v)   An annulus $\{(\xi_1, \xi_2) : r_1^2 \leq (\xi_1 - \alpha_1)^2 + (\xi_2 - \alpha_2)^2 \leq r_2^2\}$, where $0 < r_1 < r_2$.

2. Which of the following subsets of $\mathbb{R}^2$ are connected or path connected?

(i) $A = B((1, 0); 1) \cup B((-1, 0); 1)$

(ii) $A = \{\text{cl}\, B((1, 0); 1)\} \cup B((-1, 0); 1)$

(iii) $A = \{\text{cl}\, B((1, 0); 1)\} \cup \{\text{cl}\, B(-1, 0); 1)\}$

(iv) $A = \{(x, y) : x \in \mathbb{Q},\ y \in [0, 1]\} \cup \{(x, 1) : x \in \mathbb{R}\}$

(v)   the set of all points in $\mathbb{R}^2$ with at least one coordinate rational

3. Prove that a topological space X is disconnected if and only if $X = A \cup B$, where A, B are nonempty sets such that $\overline{A} \cap B = \emptyset = A \cap \overline{B}$.

4. Show that connectedness is a topological property.

5. Show that a set A is connected if and only if every continuous function f from A into the discrete space $\{0, 1\}$ is constant.

6. Let $A \subset X$ be such that, for each pair a, b of distinct points in A, there is a connected set $U_{a,b} \subset A$ containing a, b. Show that A is a connected set.

7. Let $A \subset B \subset \overline{A}$, where A is connected. Show that B is also connected.

8. Show that a topological space homeomorphic to a connected space is connected.

9. Let X be a Hausdorff topological space that has more than one point. If A is a connected subset of X, show that A is infinite. Give a counterexample showing that the result need not be true if X is not Hausdorff.

10. Prove that the product $X = X_1 \times \cdots \times X_n$ of path connected topological spaces is path connected.

11. Prove that a set A is connected if and only if every continuous function $f : A \to \mathbb{R}$ has the intermediate value property. (Problem 6.5.)

12. Let $\mathbb{R}$ be equipped with the topology $\mathcal{T} = \{\emptyset\} \cup \{U \subset \mathbb{R} : 0 \in U\}$. Is the space $(\mathbb{R}, \mathcal{T})$ connected? Is $\mathbb{R} \setminus \{0\}$ connected?

13. Consider $\mathbb{R}$ with the countable complement topology

$$\mathcal{T} = \{U \subset \mathbb{R} : U = \emptyset \text{ or } U^c \text{ is countable}\}.$$

Show that $(\mathbb{R}, \mathcal{T})$ is connected.

14. *The Cantor space.* For each $n \in \mathbb{N}$ let $X_n = \{0, 1\}$ with the discrete topology. The product space $C = \prod_n X_n$ is the *Cantor space*. Prove the following.

(i) C is compact.

(ii) C is *totally disconnected*, that is, nonempty connected subsets are singletons.

(iii) C does not have the discrete topology.

(iv) C is uncountable.

(v) Every point of C is a limit point of C.

15. Prove that the set S constructed in Example 6.18 is not path connected.

16. Let (S, d) be a pseudometric space containing points a, b with $d(a, b) > 0$. If there exists r, $0 < r < d(a, b)$, such that $d(a, x) \neq r$ for all $x \in S$, show that S is disconnected.

PART 2

Linear Analysis

Chapter 7

Normed Spaces

As we already know from metric space theory, a normed space is a special case of a metric space (X, d), in which the underlying set X is a real or complex vector space, and the metric d is induced by the formula $d(x, y) = \|x - y\|$. Vector spaces with a function $x \mapsto |x|$ similar to a norm were used from the beginning of the twentieth century, in particular by Fréchet. However, it was not assumed that $|\lambda x| = |\lambda||x|$. A definition of an abstract normed space in the form given below appeared only around 1922, given independently by Banach, Hahn and Wiener.

Normed spaces have a great deal more structure than metric spaces as they allow geometry of lines and linear subspaces. We will examine the metric properties of the closed unit ball and show that it is compact only in normed spaces of finite dimension. Another concept that is available in normed spaces but not in metric spaces is that of an infinite series.

7.1 Vector spaces and norms

In this section we explore some basic properties of norms and normed spaces. By X we denote a vector space over the field $\mathbb{K}$ of real or complex scalars.

First we survey some facts about vector spaces which do not involve a norm. A (finite or infinite) subset A of X is *linearly independent* if, for every finite set $F \subset A$ and any scalars $\lambda(u)$ $(u \in F)$, we have

$$\sum_{u \in F} \lambda(u)u = 0 \implies \lambda(u) = 0 \text{ for all } u \in F.$$

For any set $S \subset X$ we define the *span of S*, denoted by $\mathsf{sp}\, S$, as the least vector subspace of X containing S. It is known from Linear Algebra that sp S consists of all finite linear combinations of elements of S. A set B is called a *Hamel basis of X* if B is linearly independent and sp $B = X$ (see Appendix E, Section E.2). Then for any $x \in X$ there exists a unique finite subset $B(x)$ of B and uniquely determined scalars $\{\lambda(u) : u \in B(x)\}$ such that $x = \sum_{u \in B(x)} \lambda(u)u$.

In Appendix E we show that every nonzero vector space X possesses a Hamel basis, and that any two Hamel bases of X are bijective; this is described by saying that they have the same cardinal number (see Theorems E.18 and E.19). This cardinal number is called the *dimension of X*. A special case of this result is the well known theorem of Linear Algebra which states that any two bases of a finite dimensional vector space have the same number of elements. In Functional Analysis we are mostly interested in infinite dimensional vector spaces.

Definition 7.1. A *normed space* is a pair $(X, \|\cdot\|)$, where X is a vector space over the scalar field $\mathbb{K}$, and $\|\cdot\|$ is a norm on X. A *norm* on a vector space X is a real valued function $x \mapsto \|x\|$ satisfying the following axioms:

N1. $\|x\| \geq 0$ for all $x \in X$

N2. $\|x\| = 0 \implies x = 0$

N3. $\|\lambda x\| = |\lambda| \|x\|$ for all scalars λ and all $x \in X$

N4. $\|x + y\| \leq \|x\| + \|y\|$ for all $x, y \in X$

The axiom **N4** is called the *triangle inequality* (Figure 7.1). It says that the length of the side $x + y$ of a triangle with vertices 0, x and $x + y$ does not exceed the sum of lengths of the remaining two sides x, y.

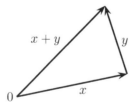

Figure 7.1

If we define d by

$$d(x, y) = \|x - y\| \text{ for all } x, y \in X, \tag{7.1}$$

then d is a metric on X; a verification of the Hausdorff axioms is left as an exercise. A *seminorm* (also called a pseudonorm) on a vector space

X satisfies the axioms for a norm with the exception of **N2**. If $\|\cdot\|$ is a seminorm, the function d described by Equation (7.1) is a pseudometric.

Definition 7.2. A set S in a vector space X is *convex* if for any two points $x, y \in S$, the line segment $\{z \in X : z = tx + (1 - t)y, \quad 0 \le t \le 1\}$ is contained in S.

Example 7.3. The *closed unit ball* $K = \{x \in X : \|x\| \le 1\}$ in a normed space $(X, \|\cdot\|)$ is a convex set: Indeed, if $x, y \in K$ and $0 \le t \le 1$, then $\|tx + (1 - t)y\| \le t\|x\| + (1-t)\|y\| \le t + (1-t) = 1$, that is, $tx + (1-t)y \in K$. Similarly, the open unit ball $\{x \in X : \|x\| < 1\}$ is convex.

We give several examples of normed or seminormed spaces, all discussed already in metric space theory.

Example 7.4. The space $\mathbb{R}^n$ with the Euclidean norm

$$\|x\| = (|\xi_1|^2 + \cdots + |\xi_n|^2)^{1/2}, \quad x = (\xi_1, \ldots, \xi_n),$$

is a real normed space. The space $\mathbb{C}^n$ with the Euclidean norm is a complex normed space. A check of the axioms **N1–N4** is left as an exercise. Note that the triangle inequality is a special case of Minkowski's inequality (C.7) for $p = 2$.

Example 7.5. For any $p \ge 1$, the space $\mathbb{R}^n$ (or $\mathbb{C}^n$) can be equipped with the ℓ^p norm:

$$\|x\|_p = \left(\sum_{k=1}^{n} |\xi_k|^p \right)^{1/p}.$$

A verification of **N1–N4** is left as an exercise; the triangle inequality coincides with Minkowski's inequality (C.7). Setting $\|x\| = \left(\sum_{k=2}^{n} |\xi_k|^p \right)^{1/p}$, we get a seminorm on $\mathbb{K}^n$.

Definition 7.6. A normed space $(X, \|\cdot\|)$ is said to be a *Banach space* if it is complete with respect to the metric induced by the norm on X.

Example 7.7. The normed spaces from Examples 7.4 and 7.5 are Banach spaces.

Example 7.8. Let $p \ge 1$ and let ℓ^p be the set of all vectors $x = (\xi_1, \xi_2, \xi_3, \ldots) \in \mathbb{K}^{\mathbb{N}}$ for which the series $\sum_{k=1}^{\infty} |\xi_k|^p$ converges (Example 1.14). Then ℓ^p is a real or complex Banach space under the norm

$$\|x\|_p = \left(\sum_{k=1}^{\infty} |\xi_k|^p \right)^{1/p} \qquad \text{(Problem 2.12)}.$$

To decide whether a metric d on a vector space X is induced by a norm, we consider the function

$$\varphi(x) = d(x, 0).$$

If $\|x\| = \varphi(x)$ fails any of the properties **N1** – **N4**, the answer is no. For d to be induced by a norm, we must have

$$d(x + z, y + z) = d(x, y)$$

for all $x, y, z \in X$, and $\|x\| := \varphi(x) = d(x, 0)$ must satisfy **N1** – **N4**.

Example 7.9. Let σ be the metric

$$\sigma(x, y) = \frac{|x - y|}{1 + |x - y|}$$

on $\mathbb{R}$. The associated function $\varphi(x) = \sigma(x, 0) = |x|/(1 + |x|)$ does not satisfy property **N3**; to see this, set for instance $\lambda = 2$ and $x = 1$ in

$$\varphi(\lambda x) = \frac{|\lambda x|}{1 + |\lambda x|}, \qquad |\lambda| \varphi(x) = \frac{|\lambda|\,|x|}{1 + |x|}.$$

This means that σ is not induced by a norm.

We observe that the addition of vectors in a normed space X with the scalar field $\mathbb{K}$ is continuous on the product normed space $X \times X$ to X; similarly, the multiplication of vector by scalars is a continuous function on the product space $\mathbb{K} \times X$ to X (Problem 7.10). This motivates the following definition.

Definition 7.10. A *topological vector space* is a pair (X, τ), where X is a vector space over the scalar field $\mathbb{K}$ and τ a topology on X such that the maps $(x, y) \mapsto x + y$ and $(\lambda, x) \mapsto \lambda x$ are continuous as functions on $X \times X$ and $\mathbb{K} \times X$, respectively, into X.

Two metrics d_1 and d_2 on a set X are equivalent if the identity mapping $I : (X, d_1) \to (X, d_2)$ is a homeomorphism, or, equivalently, if

$$x_n \xrightarrow{d_1} x \iff x_n \xrightarrow{d_2} x$$

(Theorem 2.12). We say that two norms defined on the same vector space X are *equivalent* if the metrics they induce are equivalent.

Theorem 7.11. *Norms $\|\cdot\|_1$ and $\|\cdot\|_2$ are equivalent if and only if there are positive constants α and β such that*

$$\alpha \|x\|_1 \le \|x\|_2 \le \beta \|x\|_1 \ \text{for all } x \in X.$$

Proof. Suppose that such constants exist. Then $\|x_n - x\|_1 \to 0$ implies $\|x_n - x\|_2 \to 0$ as $\|x_n - x\|_2 \leq \beta\|x_n - x\|_1$. A symmetrical argument shows that $\|x_n - x\|_2 \to 0$ implies $\|x_n - x\|_1 \to 0$.

Conversely, suppose that the norms are equivalent. For a proof by contradiction assume that there is no constant β such that $\|x\|_2 \leq \beta\|x\|_1$ for all $x \in X$. Choosing β to be equal to $1, 2, \ldots$ in succession, we obtain vectors $x_1, x_2, \ldots$ in X such that

$$\|x_n\|_2 > n\|x_n\|_1 > 0, \quad n = 1, 2, \ldots$$

Set $z_n = x_n/(n\|x_n\|_1)$. Then $\|z_n\|_1 = n^{-1} \to 0$, so that $\|z_n\|_2 \to 0$ by hypothesis. However,

$$\|z_n\|_2 = \frac{\|x_n\|_2}{n\|x_n\|_1} > 1 \text{ for all } n.$$

So $\|z_n\|_2 \not\to 0$, which is a contradiction. $\qquad\square$

Theorem 7.12. *Any two norms on $\mathbb{R}^n$ ($\mathbb{C}^n$) are equivalent.*

Proof. Let $\mathbb{K}$ be the field of real or complex scalars. For $x = (\xi_1, \ldots, \xi_n) \in \mathbb{K}^n$ let $\|x\|_2 = (|\xi_1|^2 + \cdots + |\xi_n|^2)^{1/2}$ be the Euclidean norm, and let $\|\cdot\|$ be an arbitrary norm in $\mathbb{K}^n$. Using properties of the norm and the Cauchy–Schwarz inequality, we get

$$\|x\| = \|\xi_1 e_1 + \cdots + \xi_n e_n\| \leq |\xi_1|\,\|e_1\| + \cdots + |\xi_n|\,\|e_n\| \leq \beta\|x\|_2,$$

where $e_1, \ldots, e_n$ is the standard Hamel basis in $\mathbb{K}^n$ and $\beta = (\|e_1\|^2 + \cdots + \|e_n\|^2)^{1/2}$.

The function $f(x) = \|x\|$ is continuous in the Euclidean space $\mathbb{K}^n$ as

$$|f(x) - f(x_0)| = |\,\|x\| - \|x_0\|\,| \leq \|x - x_0\| \leq \beta\|x - x_0\|_2.$$

Thus f attains its minimum α on the compact set $S = \{x \in \mathbb{K}^n : \|x\|_2 = 1\}$ (see Theorems 5.24 and 5.19). For any $x \in \mathbb{K}^n$,

$$\left\| \frac{x}{\|x\|_2} \right\| = f\left(\frac{x}{\|x\|_2} \right) \geq \alpha,$$

and $\|x\| \geq \alpha\|x\|_2$. (Check that $\alpha \neq 0$.) This proves that any norm on $\mathbb{K}^n$ is equivalent to the Euclidean norm. The result then follows from the fact that the equivalence of norms is a true equivalence relation. $\qquad\square$

Theorem 7.13. *Any two norms on a finite dimensional space X are equivalent.*

Proof. Let $\|\cdot\|_1, \|\cdot\|_2$ be norms on X. Select a basis $(a_1, \ldots, a_n)$ of X and define $\varphi(\xi_1, \ldots, \xi_n) = \xi_1 a_1 + \cdots + \xi_n a_n$; then φ is a linear isomorphism of $\mathbb{K}^n$ onto X. For each $u = (\xi_1, \ldots, \xi_n) \in \mathbb{K}^n$ set $p_i(u) = \|\varphi(u)\|_i$, $i = 1, 2$. Then p_1, p_2 are norms on $\mathbb{K}^n$, equivalent to each other by Theorem 7.12. The maps $\varphi^{-1} \colon (X, \|\cdot\|_1) \to (\mathbb{K}^n, p_1)$ and $\varphi \colon (\mathbb{K}^n, p_2) \to (X, \|\cdot\|_2)$ are isometries, and therefore homeomorphisms. Hence the identity mapping $I = \varphi \circ \varphi^{-1} \colon (X, \|\cdot\|_1) \to (X, \|\cdot\|_2)$ is a homeomorphism, and the norms $\|\cdot\|_1, \|\cdot\|_2$ are equivalent. $\qquad\square$

7.2   Compactness and finite dimension

In the Euclidean space $\mathbb{R}^n$ a set is compact if and only if it is closed and bounded. This characterization of compactness is not true in certain other normed spaces, notably in ℓ^1. The aim of this section is to show that the criterion is valid only in finite dimensional normed spaces.

Theorem 7.14. *Let $(X, \|\cdot\|)$ be a finite dimensional normed space. Then a set $A \subset X$ is compact if and only if A is closed and bounded.*

Proof. Let A be a subset of X. If A is compact, then it is closed and bounded by a general property of compact sets in metric spaces.

Suppose that A is closed and bounded. Let $a_1, \ldots, a_n$ be a basis of X. Then
$$\|x\|_2 = \|\xi_1 a_1 + \cdots + \xi_n a_n\|_2 = (|\xi_1|^2 + \cdots + |\xi_n|^2)^{1/2}$$
is a norm on X. The map $x \mapsto (\xi_1, \ldots, \xi_n)$ is an isometry of $(X, \|\ \|_2)$ onto the Euclidean space $\mathbb{K}^n$, and so the set A is compact in $(X, \|\ \|_2)$ by the Borel–Lebesgue theorem (Theorem 5.19). The result then follows from Theorem 7.13 and from the fact that compactness is preserved when passing to an equivalent norm. $\qquad\square$

In a normed space $(X, \|\cdot\|)$ we define the *closed unit ball* as the set
$$K = K(0; 1) = \{x \in X : \|x\| \leq 1\}.$$
Then K is a closed subset of X (verify). If X is finite dimensional, then K is compact by the preceding theorem. It is a remarkable fact that the converse is also true. For a proof of this theorem we need the following result which is a special case of the so called *Riesz's lemma*.

Lemma 7.15. *Let Y be a closed and proper subspace of a normed space X. Then there is a point x_0 in X such that*
$$\|x_0\| = 1 \quad and \quad \|x_0 - y\| \geq \tfrac{1}{2} \quad for\ all\ y \in Y.$$

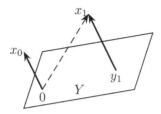

Figure 7.2

Proof. See Figure 7.2. Pick $x_1 \notin Y$, and let $m = \inf\{\|x_1 - y\| : y \in Y\}$. Since Y is closed and $x_1 \notin Y$, we have $m > 0$. By the definition of infimum there is $y_1 \in Y$ with $m \le \|x_1 - y_1\| < 2m$. Let $\alpha = \|x_1 - y_1\|$ and $x_0 = \alpha^{-1}(x_1 - y_1)$. Then $\|x_0\| = 1$, for any $y \in Y$ we have $y_1 + \alpha y \in Y$, $\alpha^{-1}(x_1 - y_1) - y = \alpha^{-1}(x_1 - (y_1 + \alpha y))$, and

$$\|x_0 - y\| = \alpha^{-1}\|x_1 - (y_1 + \alpha y)\| \ge \alpha^{-1}m > \tfrac{1}{2}.$$
□

Theorem 7.16. *Let $(X, \|\cdot\|)$ be a normed space whose closed unit ball K is compact. Then X is finite dimensional.*

Proof. Suppose that the closed unit ball K of X is compact. For a proof by contradiction we assume that X is infinite dimensional. First choose any x_1 of norm 1. This x_1 generates a one dimensional subspace X_1 of X, which is closed in X. Moreover, X_1 is a proper subspace of X since $\dim X$ is infinite. By the preceding lemma there is an element $x_2 \in X$ of norm 1 such that

$$\|x_2 - x_1\| \ge \tfrac{1}{2}.$$

The elements x_1, x_2 generate a two dimensional proper closed subspace X_2 of X. By Lemma 7.15 there is an element x_3 of norm 1 such that for all $y \in X_2$, $\|x_3 - y\| \ge 1/2$. In particular

$$\|x_3 - x_1\| \ge \tfrac{1}{2}, \qquad \|x_3 - x_2\| \ge \tfrac{1}{2}.$$

Proceeding by induction, we obtain a sequence (x_n) of elements $x_n \in K$ with

$$\|x_m - x_n\| \ge \tfrac{1}{2} \quad \text{if} \quad m \ne n.$$

This shows that (x_n) cannot have a convergent subsequence, and that is in contradiction with the compactness of K. Hence our assumption that $\dim X$ is infinite is false, and X is finite dimensional. □

A metric space is said to be *locally compact*, if every point has a compact neighbourhood. A normed space X is locally compact if and only if the closed unit ball in X is compact (Problem 7.14). Theorems 7.14 and 7.16 then combine to give the following result.

Theorem 7.17. *A normed space is locally compact if and only if it is finite dimensional.*

Let us give some examples of infinite dimensional normed spaces.

Example 7.18. For any $p \geq 1$, the space ℓ^p (real or complex) is infinite dimensional. Consider the vectors
$$e_1 = (1, 0, 0, \ldots), \qquad e_2 = (0, 1, 0, \ldots), \ldots$$
which are elements of ℓ^p. We show that, for every n, the set $\{e_1, \ldots, e_n\}$ is linearly independent:
$$\lambda_1 e_1 + \cdots + \lambda_n e_n = (\lambda_1, \ldots, \lambda_n, 0, 0, \ldots) = (0, \ldots, 0, 0, 0, \ldots)$$
implies $\lambda_i = 0$ for all i. Then the dimension of ℓ^p is greater than or equal to n. Since n is arbitrary, the space is infinite dimensional. Consequently, the closed unit ball of ℓ^p is closed and bounded, but not compact.

Example 7.19. The space $C[a, b]$ of all real valued continuous function on the interval $[a, b]$, equipped with the uniform norm
$$\|f\| = \sup_{t \in [a,b]} |f(t)|,$$
is an infinite dimensional Banach space. It is enough to consider the functions $f_k(t) = t^k$ for $k = 1, 2, \ldots$ Again, the closed unit ball in $C[a, b]$ is a closed bounded set which is not compact.

7.3   Series in normed spaces

A *series* $\sum_{n=1}^{\infty} x_n$ in a normed space $(X, \|\cdot\|)$ is defined by its terms x_n, which are elements of X. With every series $\sum_n x_n$ we associate the sequence of *partial sums*,
$$s_n = x_1 + \cdots + x_n.$$

Definition 7.20. The series $\sum_{n=1}^{\infty} x_n$ in a normed space $(X, \|\cdot\|)$ is said to *converge* to an element $x \in X$ if the sequence (s_n) of its partial sums converges to x in the norm of X:
$$\lim_{n \to \infty} \|s_n - x\| = \lim_{n \to \infty} \left\| \sum_{k=1}^{n} x_k - x \right\| = 0.$$

In this case we write $x = \sum_{n=1}^{\infty} x_n$. We say that the series $\sum_{n=1}^{\infty} x_n$ is *norm-absolutely convergent* if $\sum_{n=1}^{\infty} \|x_n\|$ is convergent in $\mathbb{R}$.

The preceding definition shows that the concept of a series can be successfully defined only in normed spaces, but not in general metric spaces, as it requires the addition of elements in the space. Note that the sum of a series is defined as a *limit*.

The following two theorems rephrase Theorem 2.33 given in §2.2.1 on the completion of normed spaces.

Theorem 7.21. *Let $(X, \|\cdot\|)$ be a Banach space. Then every norm-absolutely convergent series in X is convergent in X.*

Theorem 7.22. *Let $(X, \|\cdot\|)$ be a normed space in which every norm-absolutely convergent series is convergent in X. Then X is a Banach space.*

Definition 7.23. A sequence (a_n) of vectors in a normed space $(X, \|\cdot\|)$ is said to be a *Schauder basis* (or a *countable basis*) of X if every element x of X has a unique series expansion

$$x = \sum_{n=1}^{\infty} \xi_n a_n. \tag{7.2}$$

Example 7.24. Let $p \geq 1$. The sequence (e_n) defined by

$$e_1 = (1, 0, 0, \ldots), \qquad e_2 = (0, 1, 0, \ldots), \ldots$$

is a Schauder basis for the real (or complex) space ℓ^p (Problem 7.17).

Recall that any subset A of the vector space X generates a vector subspace M of X, which is the smallest vector subspace of X containing A. A typical element of M is a *finite* linear combination of vectors of A (whether A is finite or not). The subspace M is called the *span* of A, denoted by $\mathsf{sp}\, A$.

Definition 7.25. A set A in a normed space $(X, \|\cdot\|)$ is called *total* if the span of A is dense in X, that is, if $\overline{\mathsf{sp}\, A} = X$.

Let $A = \{a_1, a_2, \ldots\}$ be a Schauder basis of a normed space $(X, \|\cdot\|)$. Given $x \in X$, we have a unique expansion (7.2); so $x = \lim_n \sum_{k=1}^{n} \xi_k a_k = \lim_n x_n$, where each vector x_n lies in the span of A. This shows that each vector $x \in X$ lies in $\overline{\mathsf{sp}\, A}$, and so $X = \overline{\mathsf{sp}\, A}$. This proves the following result:

Theorem 7.26. *A Schauder basis is a total subset of a normed space X.*

Theorem 7.27. *A normed space which has a countable total subset is separable. In particular, any normed space with a Schauder basis is separable.*

Proof. Let $\{a_1, a_2, \ldots\}$ be a countable total subset of a complex normed space X. For a fixed n let A_n be the set of all vectors of the form

$$\rho_1 a_1 + \cdots + \rho_n a_n, \quad \text{where } \rho_k \in \mathbb{Q} + i\mathbb{Q}.$$

Then A_n is bijective to the set $(\mathbb{Q} + i\mathbb{Q})^n$, which is countable. Let A be the union of the A_n; then A is countable being the union of countably many countable sets (Theorem E.7). It remains to show that A is a dense subset of X. Given $x \in X$ and $\varepsilon > 0$, there is $y = \sum_{k=1}^{n} \xi_k a_k$ with $\xi_k \in \mathbb{C}$ such that $\|x - y\| < \frac{1}{2}\varepsilon$. Further, there is $z \in A_n$ such that $\|y - z\| < \frac{1}{2}\varepsilon$. Thus $\|x - z\| < \varepsilon$ with $z \in A$.   □

In 1932 Banach asked the question whether every separable Banach space had a Schauder basis. This remained an open problem until 1972, when the Swedish mathematician Per Enflo constructed a counterexample to Banach's conjecture.

7.4   Problems for Chapter 7

1. In a seminormed space $(X, \|\cdot\|)$ prove the following:

(i) $\|x - y\| \geq \big| \|x\| - \|y\| \big|$ for all $x, y \in X$.

(ii) $x_n \to x$ and $y_n \to y$ implies $x_n + y_n \to x + y$.

(iii) $x_n \to x$ and $|\lambda_n - \lambda| \to 0$ implies $\lambda_n x_n \to \lambda x$.

(iv) $x_n \to x$ implies $\|x_n\| \to \|x\|$, but not vice versa.

2. Show that every norm on the real vector space $\mathbb{R}$ is of the form $\|x\| = c|x|$ for some positive constant c. What is the situation in $\mathbb{C}$ over complex scalars? In $\mathbb{C}$ over real scalars?

3. Which of the following metrics on $\mathbb{K}$ are induced by a norm?

(i) $d(x, y) = |x - y|$ (ii) $\rho(x, y) = |x/(1 + |x|) - y/(1 + |y|)|$

(iii) $\sigma(x, y) = |e^{-x} - e^{-y}|$ (iv) $\omega(x, y) = |x - y|/(1 + |x - y|)$.

4. Which metric d on X is induced by a norm?

(i) $X = \mathbb{K}^{\mathbb{N}}$; $d(x, y) = \sum_{k=1}^{\infty} 2^{-k} |\xi_k - \eta_k|/(1 + |\xi_k - \eta_k|)$

(ii) $X = \mathbb{K}^n$; $d_p(x, y) = \left(\sum_{k=1}^{n} |\xi_k - \eta_k|^p\right)^{1/p}$, where $p \geq 1$

(iii) $X = \mathbb{K}^n$; $d_\infty(x, y) = \max\{|\xi_k - \eta_k| : k = 1, 2, \ldots, n\}$

(iv) X – any nontrivial vector space; d – the discrete metric on X

(v) $X = C[a,b]$; $d(f,g) = \int_a^b |f(t) - g(t)| \, dt$

(vi) $X = C[a,b]$; $d(f,g) = \int_a^b |f(t) - g(t)| / (1 + |f(t) - g(t)|) \, dt$

5. Show that $\|x\|_2 \leq \|x\|_1 \leq \sqrt{n}\|x\|_2$ for all $x \in \mathbb{K}^n$, where $\|x\|_p = (\sum_{i=1}^n |\xi_i|^p)^{1/p}$, $p \in \{1,2\}$.

6. Let d be a metric on a vector space X. Prove that d is induced by a norm if and only if it has the following properties: (i) $d(x + z, y + z) = d(x,y)$ for all $x, y, z \in X$, and (ii) $d(\alpha x, \alpha y) = |\alpha| d(x,y)$ for all $x, y \in X$ and all scalars α.

7. Given a real number $p \in [1, +\infty)$, find the best possible constants λ_n, μ_n such that $\lambda_n \|x\|_1 \leq \|x\|_p \leq \mu_n \|x\|_1$, for all $x \in \mathbb{K}^n$ ($\|\cdot\|_p$ is the ℓ^p norm).

8. Show that, for each fixed $x = (\xi_1, \ldots, \xi_n) \in \mathbb{K}^n$, $\|x\|_\infty = \lim_{p \to \infty} \|x\|_p$.

9. Let M be a vector subspace of a normed space X. Show that the closure $\overline{M}$ of M (in the norm of X) is also a vector space.

10. Prove that in a normed space X with the scalar field $\mathbb{K}$ the function $(x, y) \mapsto x + y$ is continuous on $X \times X$ to X, and that the function $(\lambda, x) \mapsto \lambda x$ is continuous on $\mathbb{K} \times X$ to X.

11. Verify that each of the following is a (real or complex) normed space. Which is infinite dimensional? Which is complete? Here $p \geq 1$.

(i) The space ℓ^p with the ℓ^p norm.

(ii) The space ℓ^∞ with the ℓ^∞ norm.

(iii) The space **c** with the ℓ^∞ norm.

(iv) The space $\mathbf{c}_0$ with the ℓ^∞ norm.

(v) The space $B(E)$ with the uniform norm; E is a nonempty set.

(vi) The space $L^1[a,b]$ equipped with the L^1 norm $\|f\|_1 = \int_a^b |f(t)| \, dt$.

(vii) The space $C[a,b]$ equipped with the L^1 norm $\|f\|_1 = \int_a^b |f(t)| \, dt$.

(viii) The space $L^p[a,b]$, equipped with the L^p norm $\|f\|_p = (\int_a^b |f(t)|^p \, dt)^{1/p}$.

12. Show $p(x) = \sum_{n=1}^\infty \alpha_n |\xi_n|$ is a seminorm on ℓ^1 if (α_n) is bounded, $\alpha_n \geq 0$.

13. Prove that any *finite* dimensional normed space is complete. Hence conclude that any finite dimensional subspace of a normed space is closed.

14. Prove that a normed space X is locally compact if and only if the closed unit ball in X is compact.

15. Let $p_0 \geq 1$. Show that, for a fixed $x \in \ell^{p_0}$, the function $\varphi(p) = \|x\|_p$ is decreasing in the interval $[p_0, \infty)$. Hence prove that $\lim_{p \to \infty} \|x\|_p = \|x\|_\infty$.

16. Give an example of a normed space X and a norm-absolutely convergent series in X which is not convergent in X.

17. Let $e_1 = (1, 0, 0, \ldots)$, $e_2 = (0, 1, 0, \ldots)$, $\ldots$ in $\mathbb{C}^{\mathbb{N}}$.

(i) Show that the sequence $(e_1, e_2, \ldots)$ is a Schauder basis in any complex space ℓ^p for $p \geq 1$; that is, show that, for any vector $x = (\xi_1, \xi_2, \ldots) \in \ell^p$, we have $x = \sum_{k=1}^{\infty} \xi_k e_k$.

(ii) Show that $(e_1, e_2, \ldots)$ is *not* a Schauder basis for the complex space ℓ^∞ or **c**, but is a Schauder basis for **c**$_0$.

18. *Quotient spaces.* Let X be a normed space, and M a subspace of X. We define the *quotient space* X/M as the space of all cosets $x + M = \{x + y : y \in M\}$, where $x \in X$. Note that $x + M = y + M \iff x - y \in M$. Addition is defined by $(x + M) + (y + M) = (x + y) + M$, multiplication by scalars by $\lambda(x + M) = \lambda x + M$. (Check that these definitions are independent of the choice of representatives.) Define $\|x + M\| = \operatorname{dist}(x, M) = \inf\{\|x - y\| : y \in M\}$.

(i) Show that the quotient space X/M becomes a seminormed space with these definitions. Show that X/M is a normed space if and only if M is closed.

(ii) If X is a Banach space and M is closed, prove that the quotient space X/M is also a Banach space. (Use norm-absolutely convergent series.)

19. A set A in a complex vector space X is *balanced* if $\lambda A \subset A$ for any scalar λ with $|\lambda| \leq 1$, and *absorbing* if for each $x \in X$ there exists a scalar α such that $x \in \alpha A$. Let A be absorbing, balanced and convex. Show that $p(x) = \inf\{\alpha > 0 : x \in \alpha A\}$ defines a seminorm.

20. Let $q \geq 1$. A function $p \colon X \to [0, \infty)$ is a *q-seminorm* on a complex vector space X if $p(\lambda x) = |\lambda| p(x)$ and $p(x + y) \leq 2^{1 - 1/q} (p(x)^q + p(y)^q)^{1/q}$ for all $\lambda \in \mathbb{C}$ and all $x, y \in X$. Prove that (i) every seminorm is a q-seminorm, and (ii) every q-seminorm p is a seminorm on X.

Suggestion. (ii) Show that $A = \{x \in X : p(x) \leq 1\}$ is absorbing, balanced and convex. For convexity show first that if $a, b \in A$, then A contains the midpoint of a, b, and then all points of the form $\lambda a + (1 - \lambda) b$, where $\lambda = k/2^n$, $k, n \in \mathbb{N}$ and $0 \leq k \leq n$. Extend to all $\lambda \in [0, 1]$.

21. (*Tychonoff's TVS theorem.*) All n-dimensional Hausdorff topological vector spaces (TVS) over the same scalar field $\mathbb{K}$ are linearly homeomorphic.

22. Show that a finite dimensional vector space X possesses exactly one topology $\mathcal{T}$ which makes X into a Hausdorff TVS.

Suggestion. Select a basis $a_1, \ldots, a_n$ of X; show that $\|x\| = \|\xi_1 a_1 + \cdots + \xi_n a_n\| = |\xi_1| + \cdots + |\xi_n|$ is a norm on X which makes X into a Hausdorff TVS. If X is a Hausdorff TVS under another topology $\mathcal{T}$, show that the identity map from $(X, \|\cdot\|)$ to $(X, \mathcal{T})$ is a linear homeomorphism.

Chapter 8

Inner Product Spaces

An inner product space is a special case of a normed space whose geometry is much richer than that of a normed space mainly because of the presence of orthogonality which is not available in a general normed space. The theory of inner product spaces was initiated around 1912 by the German mathematician David Hilbert (1862–1943) in his work on integral equations. A complete inner product space is called a Hilbert space in his honour. The geometric terminology, so successfully used in the theory of inner product spaces, is mainly due to Hilbert's student Erhart Schmidt. The first abstract axiomatic study of inner product spaces was carried out by von Neumann in 1928.

8.1 Inner products

Real inner products are subsumed in complex inner products, so we restrict ourselves to complex spaces. However, there are some results valid in complex, but not real, inner product spaces. These differences will be pointed out.

Definition 8.1. An *inner product space* consists of a complex vector space X and an *inner product* on X, which associates with every pair of vectors x, y in X a complex number $\langle x, y \rangle$ with the following properties:

I1. $\langle x, x \rangle \geq 0$ for all $x \in X$

I2. $\langle x, x \rangle = 0 \implies x = 0$

I3. $\langle x, y \rangle = \overline{\langle y, x \rangle}$ for all $x, y \in X$

I4. $\langle \alpha u + \beta v, y \rangle = \alpha \langle u, y \rangle + \beta \langle v, y \rangle$ for all $u, v, y \in X$ and scalars α, β

A *semi-inner product* on X satisfies only **I1**, **I3** and **I4**.

14 means that, for each fixed $y \in X$, the function $x \mapsto \langle x, y \rangle$ is linear. If X is a real vector space, then the inner product takes real values, and $\langle x, y \rangle = \langle y, x \rangle$ for all $x, y \in X$. From the axioms we can deduce (Problem 8.1)

$$\langle x, 0 \rangle = 0 \text{ for all } x \in X, \qquad \langle x, \lambda y \rangle = \overline{\lambda} \langle x, y \rangle \text{ for all } x, y \in X.$$

Example 8.2. *Euclidean space.* The space $\mathbb{C}^n$ becomes an inner product space when equipped with the so called standard inner product

$$\langle x, y \rangle = \xi_1 \overline{\eta}_1 + \cdots + \xi_n \overline{\eta}_n,$$

where $x = (\xi_1, \dots, \xi_n)$, $y = (\eta_1, \dots, \eta_n)$.

Example 8.3. *Hilbert sequence space.* The complex space ℓ^2 becomes an inner product space when we define

$$\langle x, y \rangle = \sum_{n=1}^{\infty} \xi_n \overline{\eta}_n,$$

where $x = (\xi_1, \xi_2, \dots)$, $y = (\eta_1, \eta_2, \dots)$. The series converges since $|\xi_n \eta_n| \leq \frac{1}{2}(|\xi_n|^2 + |\eta_n|^2)$. Check the axioms for an inner product. The space ℓ^2 was investigated extensively by David Hilbert at the beginning of the twentieth century, and is often called the Hilbert sequence space.

Example 8.4. The space $L^2(a, b)$ (see Section 2 of Chapter 2 and Section 3 of Appendix D) consists of complex valued functions f on (a, b) which are pointwise limits of step functions and such that $|f|^2$ is Lebesgue integrable on (a, b). The inner product is defined by

$$\langle f, g \rangle = \int_a^b f(t) \overline{g(t)} \, dt.$$

Definition 8.5. Two vectors x, y in an inner product space X are said to be *orthogonal*, written $x \perp y$, if

$$\langle x, y \rangle = 0.$$

If $A \subset X$ and $x \in X$, we say that x is *orthogonal to A* if $x \perp u$ for all $u \in A$; we write $x \perp A$. The set $A^\perp$ of all $x \in H$ for which $x \perp A$ is called the *orthogonal complement* of A in H:

$$A^\perp = \{x \in H : x \perp A\}.$$

Some of the properties of orthogonal complements are summarized in Problem 8.22. Orthogonality is one of the most important concepts in inner product space theory.

In an inner product space X we define

$$\|x\| = \langle x, x \rangle^{1/2},$$

and verify presently that this function satisfies the axioms for a norm. For any pair $x \perp y$ we have *Pythagoras's theorem* (Problem 8.2):

$$\|x + y\|^2 = \|x\|^2 + \|y\|^2.$$

Theorem 8.6 (Schwarz inequality). *If X is an inner product space, then*

$$|\langle x, y \rangle| \leq \|x\| \, \|y\| \text{ for all } x, y \in X. \tag{8.1}$$

The equality holds if and only if x and y are linearly dependent.

Proof. Problem 8.3. □

The Schwarz inequality (8.1) holds also in a semi-inner product space, without the conclusion about equality.

Theorem 8.7. *If X is an inner product space, then $\|x\| = \langle x, x \rangle^{1/2}$ has the properties of a norm.*

Proof. We leave the proof of **N1** through **N3** as an exercise, and establish only the triangle inequality. For this we note that $\mathrm{Re}\langle x, y \rangle \leq |\langle x, y \rangle| \leq \|x\| \|y\|$ in view of the Schwarz inequality. Thus

$$
\begin{aligned}
\|x + y\|^2 &= \langle x + y, x + y \rangle = \langle x, x \rangle + \langle x, y \rangle + \langle y, x \rangle + \langle y, y \rangle \\
&= \|x\|^2 + 2\,\mathrm{Re}\langle x, y \rangle + \|y\|^2 \\
&\leq \|x\|^2 + 2\|x\| \, \|y\| + \|y\|^2 \\
&= (\|x\| + \|y\|)^2,
\end{aligned}
$$

and $\|x + y\| \leq \|x\| + \|y\|$. □

The norm $\|x\| = \langle x, x \rangle^{1/2}$ in an inner product space X is referred to as the norm *induced by the inner product* on X. A semi-inner product induces a seminorm.

Definition 8.8. An inner product space H is called a *Hilbert space* if it is complete with respect to the norm induced by the inner product on H. (The name is chosen in honour of David Hilbert.)

Example 8.9. The inner product spaces introduced in Examples 8.2–8.4 are Hilbert spaces.

A norm induced by an inner product satisfies the *parallelogram identity*

$$\|x + y\|^2 + \|x - y\|^2 = 2\|x\|^2 + 2\|y\|^2$$

illustrated in Figure 8.1. See Problem 8.8.

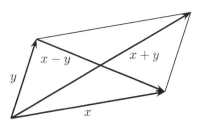

Figure 8.1

Jordan and von Neumann proved in [25] that any norm which satisfies the parallelogram identity is induced by an inner product:

$$\langle x, y \rangle = \tfrac{1}{4}(\|x + y\|^2 - \|x - y\|^2 + i\|x + iy\|^2 - i\|x - iy\|^2)$$

(Problem 8.29).

The parallelogram identity gives us the means for testing whether a given norm is induced by some inner product.

Example 8.10. The norm in ℓ^1 is not induced by any inner product: It is enough to show that the norm does not satisfy the paralellogram identity. For this choose for instance $x = (1, 1, 0, 0, \ldots)$, $y = (1, -1, 0, 0, \ldots)$.

Example 8.11. The uniform norm in the space $C[0, 1]$ is not induced by any inner product: It is enough to show that the elements

$$f(t) = 1, \qquad g(t) = t$$

do not satisfy the parallelogram identity.

8.2   Orthonormal sets

Definition 8.12. A subset A of an inner product space X is *orthonormal* if the elements of A are pairwise orthogonal and have norm 1, that is,

$$\langle x, y \rangle = \begin{cases} 0 & \text{if } x \neq y, \\ 1 & \text{if } x = y. \end{cases}$$

An orthonomal set is linearly independent (exercise). In a finite dimensional inner product space X we can always choose an orthonormal basis. There is an efficient construction for such a choice, known as the *Gram–Schmidt process*, which is usually discussed in linear algebra: If $v_1, \ldots, v_q$ is a spanning set for X, we define $u_1 = v_1$. For any $i \in \{2, \ldots, q\}$ we set

$$u_i = v_i - \sum_{j=1}^{i-1} \lambda_{ji} u_j,$$

where $\lambda_{ji} = \langle u_j, v_i \rangle / \langle u_j, u_j \rangle$ if $u_j \neq 0$ and $\lambda_{ji} = 0$ otherwise.

Theorem 8.13. *Let M be a finite dimensional subspace of an inner product space X with an orthonormal basis $a_1, \ldots, a_n$. For each $x \in X$ define a vector Px by*

$$Px = \sum_{k=1}^{n} \langle x, a_k \rangle a_k. \qquad (8.2)$$

Then $Px \in M$ and $x - Px \in M^{\perp}$.

Proof. Start by showing that $\langle Px, a_n \rangle = \langle x, a_n \rangle$ for all n. See Problem 8.13. □

Definition 8.14. Let M be a subspace of an inner product space X. A mapping $P \colon X \to X$ is an *orthogonal projection of X onto M* if

$$Px \in M \text{ and } x - Px \in M^{\perp} \text{ for all } x \in X. \qquad (8.3)$$

If such a mapping exists, we say that M *admits an orthogonal projection*.

A subspace M admits *at most one* orthogonal projection. Indeed, let $Q \colon X \to X$ be another orthogonal projection onto M. Then $Qx - Px \in M$ and $Qx - Px = (x - Px) - (x - Qx) \in M^{\perp}$. Since $M \cap M^{\perp} = \{0\}$ (check), we have $Qx - Px = 0$, and $Qx = Px$ for all $x \in X$. We often write $P = P_M$.

Finite dimensional subspaces of X admit orthogonal projections by Theorem 8.13. In a Hilbert space, every *closed* subspace M admits an orthogonal projection P_M (Theorem 8.23). Problem 8.20 shows that nonclosed subspaces do not admit orthogonal projections.

The French mathematician Joseph Fourier (1768–1830) studied expansions of functions in terms of trigonometric functions (the so called classical Fourier series), and his work motivated much of the research into orthogonal expansions in Hilbert spaces. An orthonormal basis for M makes it possible to express orthogonal projections in an economical way.

Definition 8.15. Let A be an orthonormal set in an inner product space X. For any element $x \in X$, the complex numbers $\langle x, a \rangle$, where $a \in A$, are called the *Fourier coefficients* of x with respect to A.

Theorem 8.16 (Fourier series theorem). *Let (a_n) be an orthonormal sequence in a Hilbert space H, and let $N = \mathsf{sp}(a_n)$ be the span of the a_n. Then the space $M = \overline{N}$ admits an orthogonal projection $P = P_M$ given by the so-called* Fourier series

$$Px = \sum_{k=1}^{\infty} \langle x, a_k \rangle a_k. \tag{8.4}$$

The sum is independent of the rearrangement of terms in the series (8.4).

Proof.   For any $x \in H$, the real term series $\sum_{k=1}^{\infty} |\langle x, a_k \rangle|^2$ converges by Bessel's inequality (Problem 8.14). Keep x fixed. The sequence

$$s_n = s_n(x) = \sum_{k=1}^{n} \langle x, a_k \rangle a_k$$

is Cauchy as $\|s_m - s_n\|^2 = \|\sum_{k=m+1}^{n} \langle x, a_k \rangle a_k\|^2 = \sum_{k=m+1}^{n} |\langle x, a_k \rangle|^2$ (Pythagoras's identity). Since H is complete, the sequence (s_n) converges; define

$$Px = \lim_{n \to \infty} s_n(x).$$

Then $Px \in M$. To show that $x - Px \perp M$, we show first that $x - Px \perp a_k$ for all k (exercise). Each $u \in N$ is a finite linear combination of some of the a_n; so $x - Px \perp u$. If $y \in M$, then $y_n \to y$ for some sequence $y_n \in N$, and $x - Px \perp y$ by Problem 8.11.

Let (b_n) be a rearrangement of (a_n). By the preceding part of the proof, $Qx = \sum_{k=1}^{\infty} \langle x, b_k \rangle b_k$ defines an orthogonal projection onto M. Since such a projection is unique, $Qx = Px$ for all $x \in X$.   □

Total orthonormal sets play an important role in the theory of Fourier series. Recall that a set in a normed space X is *total* if its span is dense in X (Definition 7.25). If an inner product space X possesses a total countable orthonormal set, then X is separable. In fact, such an orthonormal set forms a Schauder basis for X. In contrast to the situation in a general Banach space, a separable Hilbert space always has a Schauder basis. A total orthonormal set in a Hilbert space H is called an *orthonormal basis* of H.

Theorem 8.17. *Any countable orthonormal basis $\{a_n\}$ in a Hilbert space H forms a Schauder basis for H.*

Proof.   Let $M = \mathsf{sp}\{a_1, a_2, \dots\}$ so then $\overline{M} = H$. For each $x \in H$ define Px by (8.4). Then $x - Px \perp H$ by Theorem 8.16, which means that $x - Px = 0$. Therefore

$$x = \sum_{k=1}^{\infty} \langle x, a_k \rangle a_k \text{ for all } x \in H.$$

The uniqueness of the expansion follows from the uniqueness of the orthogonal projection onto a subspace.   □

Theorem 8.18. *An infinite dimensional Hilbert space H is separable if and only if H possesses a countable orthonormal basis.*

Proof.   Let $A = \{a_1, a_2, \dots\}$ be a total orthonormal sequence in H. By the preceding theorem, this sequence is a Schauder basis for H, and so H is separable by Theorem 7.27.

Conversely, let H be separable, and let $S = \{s_1, s_2, \dots\}$ be a countable dense subset of H. Let z_1 be the first nonzero vector in the sequence (s_n). Given $z_1, \dots, z_k$, choose z_{k+1} to be the first s_n which is not in the vector subspace generated by $z_1, \dots, z_k$. Then the sequences (s_n) and (z_k) generate the same vector subspace M of H, where $\overline{M} = H$. Apply the Gram-Schmidt process to the linearly independent sequence (z_k) to obtain a total orthonormal sequence (a_k).   □

The preceding theorem guarantees that every separable Hilbert space contains a (countable) total orthonormal set. The next theorem extends this result to nonseparable Hilbert spaces.

Theorem 8.19. *Every nonzero Hilbert space H contains an orthonormal basis S (which may be indexed by some index set D as $S = \{u_\alpha : \alpha \in D\}$). Then every $x \in H$ has a unique expansion independent of the indexing of S,*

$$x = \sum_{\alpha \in D} \langle x, u_\alpha \rangle u_\alpha, \tag{8.5}$$

with only countably many of the Fourier coefficients $\langle x, u_\alpha \rangle$ nonzero.

In view of the independence of the sum on the indexing of S we usually write

$$\sum_{\alpha \in D} \langle x, u_\alpha \rangle u_\alpha = \sum_{u \in S} \langle x, u \rangle u.$$

To prove this theorem, first we need a result named after Hausdorff which is equivalent to the Axiom of choice or Zorn's lemma. (See Theorem E.17 in Appendix E.)

The Hausdorff maximality principle. *Let $\mathcal{A}$ be a nonempty family of sets in which every chain $\mathcal{C}$ has an upper bound. Then $\mathcal{A}$ has a maximal element.*

(A *chain* $\mathcal{C}$ is a family of sets with the property that for any two sets $A, B \in \mathcal{C}$ we have either $A \subset B$ or $B \subset A$. A set U is an *upper bound* for $\mathcal{C}$ if $A \subset U$ for all $A \in \mathcal{C}$. A *maximal element* of $\mathcal{A}$ is a set S with the property that, for any $A \in \mathcal{A}$, S is not a proper subset of A.)

Proof.   (Theorem 8.19.) Let $\mathcal{A}$ be the family of all orthonormal subsets of H. Since H is nonzero, the family $\mathcal{A}$ is nonempty. If $\mathcal{C}$ is a chain in $\mathcal{A}$, then the union $U = \bigcup_{A \in \mathcal{C}} A$ is an orthonormal set (verify). Also, U is an upper bound of $\mathcal{C}$. By the Hausdorff maximality principle, there is a maximal element S in the family $\mathcal{A}$. Let M be the span of S. If $\overline{M} \neq H$, then there is $x \notin \overline{M}$. By Problem 8.19, there are only countably many elements $u \in S$ for which $\langle x, u \rangle \neq 0$; let S_x be the set of these elements. According to Theorem 8.16, we can define v (independently of the enumeration of S_x) by

$$v = x - \sum_{u \in S_x} \langle x, u \rangle u.$$

Then $v \neq 0$ (otherwise x would be in $\overline{M}$), and $v \perp S$. But the set $T = S \cup \{v/\|v\|\}$ is then orthonormal, and S is a proper subset of T contrary to the assumption of maximality of S.

The existence and uniqueness of the expansion for x can be deduced from Theorem 8.16. We define $\sum_{\alpha \in D} \langle x, u_\alpha \rangle u_\alpha := \sum_{u \in S_x} \langle x, u \rangle u$, where the countable sum over S_x is independent of the enumeration of S_x, and therefore of the indexing of S. $\square$

From Linear Algebra we know that any two bases of a nonzero finite dimensional vector space have the same number of elements. If H is a nonzero Hilbert space, then any two orthonormal bases are bijective (have the same cardinal number—see Appendix E).

Theorem 8.20. *Let S_1 and S_2 be orthonormal bases of a Hilbert space H. Then S_1 and S_2 have the same cardinal number, that is, are bijective.*

Proof.   We assume that H does not possess a finite orthonormal basis, as this case is covered in Linear Algebra. For each $x \in S_1$ let $S_2(x)$ be the set of all $u \in S_2$ for which $\langle x, u \rangle \neq 0$. The set $S_2(x)$ is countable, that is, card $S_2(x) \leq \aleph_0$. Let $y \in S_2$. Since S_1 is total, there exists $x \in S_1$ such that $y \in S_2(x)$ (otherwise $S_1 \cup \{y\}$ would be orthonormal). Therefore

$$S_2 = \bigcup_{x \in S_1} S_2(x).$$

This implies card $S_2 \leq \aleph_0$ card $S_1 = $ card S_1. A symmetrical argument shows that card $S_1 \leq$ card S_2. Then, by the Schröder–Bernstein theorem (Theorem E.1 (iii)), card $S_1 = $ card S_2. $\qquad\square$

Definition 8.21. The cardinal number of any orthonormal basis of a Hilbert space H is called the *Hilbert dimension* of H.

It is an interesting fact that the Hilbert dimension of a Hilbert space H determines the space uniquely up to an isometric isomorphism (see Problem 8.26). No such characterization exists for Banach spaces.

The existence of orthonormal bases is very important in approximation theory. The following theorem gives an example of an orthonormal basis essential in the theory of classical Fourier series.

Theorem 8.22. *The functions*

$$e_m(t) = \frac{1}{\sqrt{2\pi}}e^{imt}, \quad m = 0, \pm 1, \pm 2, \dots$$

form an orthonormal basis of the complex space $L^2[0, 2\pi]$.

Proof. The given sequence is orthonormal by Problem 8.15.

For brevity write $\Omega = [0, 2\pi]$, $C(\Omega)$ for the set of all continuous complex valued functions on Ω and $C_0(\Omega)$ for the subset of $C(\Omega)$ consisting of h satisfying $h(0) = h(2\pi)$.

We show that the given sequence is total in $L^2(\Omega)$. Let $f \in L^2(\Omega)$ and let $\varepsilon > 0$ be given. We use the fact that continuous functions are dense in the space $(L^2(\Omega), \|\cdot\|_{L^2})$, which will be proved in the third part of this book (see Theorem 16.8). This enables us to select $g \in C(\Omega)$ such that $\|f - g\|_{L^2} < \frac{1}{3}\varepsilon$. By modifying the function g on a suitably small interval $[0, \delta]$ we obtain a function $h \overset{\bullet}{\in} C_0(\Omega)$ satisfying $\|g - h\|_{L^2} < \frac{1}{3}\varepsilon$; the details are left as an exercise. Finally, by Theorem 5.37 there exists a trigonometric polynomial $\sigma(t) = \sum_{k=-n}^{n} a_k e^{ikt}$ such that $\|h - \sigma\|_\infty < (2\pi)^{-1/2}\frac{1}{3}\varepsilon$, where $\|\cdot\|_\infty$ denotes the uniform norm. Then

$$\|f - \sigma\|_{L^2} \leq \|f - g\|_{L^2} + \|g - h\|_{L^2} + \|h - \sigma\|_{L^2}$$
$$\leq \|f - g\|_{L^2} + \|g - h\|_{L^2} + \sqrt{2\pi}\|h - \sigma\|_\infty$$
$$< \frac{1}{3}\varepsilon + \frac{1}{3}\varepsilon + \sqrt{2\pi}\frac{1}{\sqrt{2\pi}}\frac{1}{3}\varepsilon = \varepsilon.$$

(Note that $\|u\|_{L^2} = (\int_0^{2\pi}|u|^2\,dm)^{1/2} \leq \sqrt{2\pi}\|u\|_\infty$ for any $u \in C(\Omega)$.) $\qquad\square$

Let X be a vector space. We say that X is the *sum of subspaces* $M, N \subset X$, written $X = M + N$, if each vector $x \in X$ can be expressed as the sum $x = m + n$, where $m \in M$ and $n \in N$ (m, n need not be unique). If, in addition, $M \cap N = \{0\}$, we say that X is the *direct sum of subspaces* M, N, and write $X = M \oplus N$. If X is a normed space which is a direct sum of *closed* subspaces M, N, we say that $X = M \oplus N$ is a *topological direct sum*.

Theorem 8.23. *Let M be a closed subspace of a nonzero Hilbert space H. Then M admits an orthogonal projection P_M, and H is the topological direct sum $H = M \oplus M^\perp$.*

Proof.   Problem 8.21.   □

8.2.1 *Pointwise convergence of classical Fourier series*

Joseph Fourier studied pointwise expansions of the series

$$(\mathcal{F}f)(t) = a_0 + \sum_{k=1}^{\infty} (a_k \cos kt + b_k \sin kt),$$

where f is a 2π-periodic function defined on the real line, and where the coefficients a_k, b_k are those given in Problem 8.16. Fourier wrote a memoir in 1807 which infuriated many mathematicians of the day, as it claimed—erroneously— that any 2π-periodic function f can be expressed as the series $\mathcal{F}f$. Some twenty years later Dirichlet published a proof that the Fourier series of every 2π-periodic continuous function converges to that function. There the matter rested until 1873 when Paul Du Bois-Reymond constructed a 2π-periodic continuous function whose Fourier series diverged. So much for Dirichlet's proof! In the early 1900s Luzin made a conjecture that the Fourier series of every L^2 function f on $(0, 2\pi)$ converges to f almost everywhere in $(0, 2\pi)$. Finally in 1966 Lennart Carleson proved Luzin's conjecture. For this and other achievements Carleson won the prestigious Abel prize for mathematics in 2006.

8.3   Problems for Chapter 8

Unless otherwise specified, H stands for a Hilbert space.

1. If X is an inner product space prove that for all $x, y \in X$ and all complex numbers λ: $\langle x, 0 \rangle = 0 = \langle 0, x \rangle$ and $\langle x, \lambda y \rangle = \overline{\lambda}\langle x, y \rangle$.

2. *Pythagoras's theorem.* If $x \perp y$ in an inner product space X, show that

$$\|x + y\|^2 = \|x\|^2 + \|y\|^2.$$

Draw a diagram. Extend the formula to m mutually orthogonal vectors.

3. *Schwarz inequality.* For any $x \in H$ define $\|x\| = \sqrt{\langle x, x \rangle}$. Prove that
$$|\langle x, y \rangle| \le \|x\| \, \|y\|.$$
Show that the equality holds if and only if x and y are linearly dependent.

4. If $\langle a, x \rangle = \langle b, x \rangle$ for all $x \in H$, show that $a = b$.

5. Show that the following are inner product spaces:

(i) $\mathbb{R}^n$ with $\langle x, y \rangle = y^\mathsf{T} x = \xi_1 \eta_1 + \cdots + \xi_n \eta_n$ (real inner product)

(ii) $\mathbb{C}^n$ with $\langle x, y \rangle = y^\mathsf{H} x = \xi_1 \overline{\eta}_1 + \cdots + \xi_n \overline{\eta}_n$ (complex inner product)

(iii) real space $L^2[a, b]$ with $\langle f, g \rangle = \int_a^b f(t) g(t) \, dt$

(iv) complex space $L^2[a, b]$ with $\langle f, g \rangle = \int_a^b f(t) \overline{g(t)} \, dt$

(v) complex space $C[a, b]$ with $\langle f, g \rangle = \int_a^b f(t) \overline{g(t)} \, dt$

(vi) real space ℓ^2 with $\langle x, y \rangle = \sum_{i=1}^\infty \xi_i \eta_i$

(vii) complex space ℓ^2 with $\langle x, y \rangle = \sum_{i=1}^\infty \xi_i \overline{\eta}_i$

(viii) the set of all complex $x = (\xi_1, \xi_2, \xi_3, \dots)$ with only finitely many ξ_k nonzero, equipped with the ℓ^2 inner product

(ix) $\mathbb{C}^n$ with $\langle x, y \rangle = y^\mathsf{H} A x$, where A is a Hermitian matrix with positive eigenvalues

For each of these inner products write down the induced norm. Which of these spaces are complete?

6. (Semi-inner products.) Let $\alpha_i \ge 0$ for $i = 1, \dots, n$. Show that $\langle x, y \rangle = \sum_{i=1}^n \alpha_i \xi_i \overline{\eta}_i$ defines a semi-inner product on $\mathbb{C}^n$ which becomes an inner product if and only if $\alpha_i > 0$ for all i. More generally, if A is a hermitian matrix with nonnegative eigenvalues, show that $\langle x, y \rangle = y^\mathsf{H} A x$ defines a semi-inner product on $\mathbb{C}^n$ which becomes an inner product if and only if all the eigenvalues of A are nonzero.

7. Let (α_i) be a bounded sequence of nonnegative real numbers. Show that $\langle x, y \rangle = \sum_{i=1}^\infty \alpha_i \xi_i \overline{\eta}_i$ defines a semi-inner product on ℓ^2 which is an inner product if and only if $\alpha_i > 0$ for all i.

8. Prove the *parallelogram identity* for any inner product space.

9. Which of the following norms on complex spaces are induced by an inner product?

(i) ℓ^p norm on $\mathbb{C}^n$ for $p \ge 1$;

(ii) uniform norm on $B(E)$, $E \ne \emptyset$;

(iii)   integral norm $\|f\| = \int_a^b |f(t)| \, dt$ on $C[a, b]$.

Suggestion. Parallelogram identity.

10.   Let $\langle a_n, x \rangle \to \langle a, x \rangle$ for all $x \in H$ and $\|a_n\| \to \|a\|$. Show that $\|a_n - a\| \to 0$.

11.   Show that an inner product is a continuous function on $X \times X$ to $\mathbb{C}$.

12.   Show that $\|x\| = \sup_{y \neq 0} |\langle x, y \rangle| / \|y\|$.

13.   Let $a_1, \ldots, a_n$ be an orthonormal set in an inner product space X. Show that

$$Px = \sum_{i=1}^{n} \langle x, a_i \rangle a_i$$

defines the orthogonal projection P of X onto the space M generated by $a_1, \ldots, a_n$, that is, $Px \in M$ and $x - Px \in M^{\perp}$ for each $x \in H$.

14.   Let $a_1, a_2, a_3, \ldots$ be an orthonormal sequence in an inner product space X. Prove *Bessel's inequality*

$$\sum_{i=1}^{\infty} |\langle x, a_i \rangle|^2 \leq \|x\|^2.$$

Suggestion. Prove the corresponding inequality for partial sums using the result of the preceding problem and Pythagoras's theorem.

15.   Let H be the complex space $L^2[0, 2\pi]$ with the complex L^2 inner product. Prove that the functions

$$e_m(t) = \frac{1}{\sqrt{2\pi}} e^{imt}, \quad m = 0, \pm 1, \pm 2, \ldots$$

form a total orthonormal sequence in H. Write down the Fourier expansion for an L^2 function f on $[0, 2\pi]$. (For the totality in H see Example 8.22.)

16.   Let H be the real space $L^2[0, 2\pi]$ with the real L^2 inner product.

(i)   Prove that the functions $u_0, u_1, u_2, \ldots, v_1, v_2 \ldots$ defined by

$$\frac{1}{\sqrt{2\pi}}, \quad \frac{1}{\sqrt{\pi}} \cos t, \quad \frac{1}{\sqrt{\pi}} \cos 2t, \quad \ldots, \quad \frac{1}{\sqrt{\pi}} \sin t, \quad \frac{1}{\sqrt{\pi}} \sin 2t, \quad \ldots$$

form a total orthonormal sequence in H. (Use the totality result of the preceding problem.)

(ii)   Find the Fourier coefficients $\langle f, u_k \rangle$ and $\langle f, v_k \rangle$. Use this result to prove that the *classical* Fourier series for f,

$$a_0 + \sum_{k=1}^{\infty} (a_k \cos kt + b_k \sin kt),$$

converges to f with respect to the norm of H; the coefficients are given by

$$a_0 = \frac{1}{2\pi} \int_0^{2\pi} f(t)\, dt, \quad a_k = \frac{1}{\pi} \int_0^{2\pi} f(t) \cos kt\, dt, \quad b_k = \frac{1}{\pi} \int_0^{2\pi} f(t) \sin kt\, dt.$$

17. Let H be the real space $L^2[0, \pi]$.

(i) Show that the sequence $(f_n(t)) = (\cos nt : n = 0, 1, 2, \ldots)$ is a total orthogonal sequence in H. Find the norms of the elements f_n. (Use the totality result of Problem 8.15.)

(ii) Show that the sequence $(g_n(t)) = (\sin nt : n = 1, 2, 3, \ldots)$ is a total orthogonal sequence in H. Find the norms of the elements g_n. (Use the totality result of Problem 8.15.)

18. Let H be the real space $L^2[-1, 1]$. Show that the orthonormalization of the sequence
$$1, t, t^2, t^3, \ldots$$
yields the sequence of *Legendre polynomials* given by
$$L_k(t) = c_k (d/dt)^k (t^2 - 1)^k, \quad k = 0, 1, 2, 3, \ldots,$$
where c_k are obtained by normalization. Prove that this sequence is total in H. (Example 8.22.)

19. Let A be an orthonormal set in an inner product space X. Show that, for each $x \in X$, the set of all $a \in A$ with nonzero Fourier coefficients $\langle x, a \rangle$ is countable.

20. Let M be a subspace of an inner product space X which admits an orthogonal projection $P = P_M$. Show that:

(i) P is a linear mapping.

(ii) $\|Px\| \leq \|x\|$ for all $x \in X$.

(iii) $P^2 = P$.

(iv) M is closed.

(v) $\text{dist}\,(x, M) = \|x - Px\|$ (Px is the closest point to x of all points in M).

Suggestion. (i) Show that $(Px + Py) \in M$ and $(x + y) - (Px + Py) \in M^\perp$ and use the uniqueness of the orthogonal projection to show additivity of P. Proceed similarly to show $P(\lambda x) = \lambda Px$.

21. *Projection theorem.* Accepting as a fact that every nonzero Hilbert space has an orthonormal basis, prove that every *closed* nonzero vector subspace M of H admits an orthogonal projection $P = P_M$. Hence show that $H = M \oplus M^\perp$.
Suggestion. Let S be an orthonormal basis for M and $Px = \sum_{u \in S} \langle x, u \rangle u$. Consider the decomposition $x = Px + (x - Px)$.

22. *Orthogonal complements.* Let A, B be subsets of H. If $A^\perp$ is the orthog-

onal complement of A defined by $A^\perp = \{x \in H : x \perp A\}$, prove the following facts:

(i) $A^\perp$ is a closed subspace of H.

(ii) $A \cap A^\perp \subset \{0\}$.

(iii) $A \subset B \implies A^\perp \supset B^\perp$.

(iv) $A \subset A^{\perp\perp}$.

(v) $M^{\perp\perp} = M$ if and only if M is a closed subspace of H.

23. Let S be an orthonormal basis for H. Prove that, for any $x, y \in H$,

(i) $x = \sum_{u \in S} \langle x, u \rangle u$ (Fourier series)

(ii) $\|x\|^2 = \sum_{u \in S} |\langle x, u \rangle|^2$ (Parseval's identity for norms)

(iii) $\langle x, y \rangle = \sum_{u \in S} \langle x, u \rangle \overline{\langle y, u \rangle}$ (Parseval's identity for inner products)

24. If $M \perp N$ with M, N closed subspaces, show that $M + N$ is closed in H.

25. Let S be a subset of H such that $S^\perp = \{0\}$. Show that S is a total set in H, that is, $\overline{\mathrm{sp}\, S} = H$.

26. Let D be a nonempty set of any cardinality. By $\ell^2(D)$ we denote the set of all functions $x \colon D \to \mathbb{C}$ such that $x(\alpha) \neq 0$ for only countably many $\alpha \in D$, and such that the series $\sum_{\alpha \in D} |x(\alpha)|^2$ converges. Define

$$\langle x, y \rangle = \sum_{\alpha \in D} x(\alpha)\overline{y(\alpha)}, \quad x, y \in \ell^2(D).$$

Prove that:

(i) $\ell^2(D)$ is a Hilbert space.

(ii) $\ell^2 = \ell^2(\mathbb{N})$.

(iii) Every Hilbert space H is isometrically isomorphic to the space $\ell^2(D)$, where the cardinality of D is equal to the Hilbert dimension of H. (See Definition 8.21.)

27. *Existence of the closest point.* Let C be a closed convex subset of H. Prove that for any point $x \in H$ there exists a unique $x_0 \in C$ such that $\|x - x_0\| = \mathrm{dist}\,(x, C)$. (See Section 7.1 for the definition of convexity.)

28. Let M be a closed proper subspace of H. Show that M is convex, and that for any $x \in H$, the closest point $x_0 \in M$ to x satisfies $x_0 = Px$, where P is the orthogonal projection of H onto M.

29. *Jordan–von Neumann characterization of inner products.* If a given norm $\|\cdot\|$ satisfies the parallelogram identity, prove that it is induced by an inner product. See the article [25] by these two authors.

30. *Another characterization of inner products.* In addition to the Jordan–von Neumann characterization there are many other such conditions (see [1]). In this problem, which may be regarded as a miniproject, the reader is asked to prove that a norm $\|\cdot\|$ on a vector space X is induced by an inner product if and only if for any finite set of vectors $x_1, \ldots, x_n$ in X, we have

$$\sum_{k=1}^{n} \left\| x_k - \frac{1}{n} \sum_{j=1}^{n} x_j \right\|^2 = \sum_{k=1}^{n} \|x_k\|^2 - n \left\| \frac{1}{n} \sum_{j=1}^{n} x_j \right\|^2.$$

See the Rassias' article [39] and Amir's book [1]. Look up other characterizations of inner products in [1].

Chapter 9

Linear Operators and Functionals

In linear algebra we study matrices and the linear transformation they induce. A $m \times n$ real matrix A defines a linear transformation $T \colon \mathbb{R}^n \to \mathbb{R}^m$ by $T(x) = Ax$, where $x \in \mathbb{R}^n$ is regarded as a column vector or $n \times 1$ matrix. We recall that a linear transformation T is characterized by the equations

$$T(x + y) = T(x) + T(y) \ \text{ and } \ T(\alpha x) = \alpha T(x)$$

valid for any two vectors x, y and any scalar α. The set of all solutions to the homogeneous system $Tx = 0$ is often called the kernel or the nullspace of T; the set of all Tx is often called the image or the range of T. Theorems about these spaces studied in linear algebra are elementary results of operator theory.

More generally, linear transformations arise in the solution of differential and integral equations; a common requirement for all these examples is that those transformations operate from one normed spaces into another (or into the same space). Following a traditional terminology, we will call these transformations *linear operators*. The name suggests that in some cases we operate on functions (points of a normed space) to produce other functions (again points of perhaps another normed space).

The importance of linear operators in analysis was demonstrated at the end of the 19th century by the Swedish mathematician Ivar Fredholm (1866–1927) who studied integral operators and laid foundations to spectral theory. David Hilbert (1862–1943) then studied 'equations with infinitely many variables', working with linear operators acting on what we now call the Hilbert sequence space. Hungarian mathematician Frigyes Riesz (1880–1956) contributed much to the progress of operator theory, and became one of the principal founders of functional analysis. Much of the geometric concepts and notation in inner product spaces we use today is due to Erhard Schmidt (1876–1959).

9.1   Linear operators

One of the most important concepts studied in functional analysis is that of a bounded linear operator and functional.

Definition 9.1. A linear transformation $T\colon X \to Y$, where X, Y are normed spaces, is called *a bounded linear operator* if there exists a positive constant c such that

$$\|Tx\| \le c\|x\| \text{ for all } x \in X. \tag{9.1}$$

(Note that $\|Tx\|$ refers to the norm in Y, and $\|x\|$ to the norm in X.)

For a bounded linear operator $T\colon X \to Y$ we define the *operator norm* $\|T\|$ to be the smallest constant c satisfying (9.1). Explicitly,

$$\|T\| = \sup_{x \ne 0} \frac{\|Tx\|}{\|x\|}. \tag{9.2}$$

Observe that

$$\|Tx\| \le \|T\| \, \|x\| \text{ for all } x \in X. \tag{9.3}$$

Since $\|Tx - Ty\| \le \|T\|\|x - y\|$ for all $x, y \in X$, we have the following result.

Theorem 9.2. *A bounded linear operator* $T\colon X \to Y$ *is uniformly continuous on* X.

Example 9.3. The identity operator $I\colon X \to X$ is a bounded linear operator with norm $\|I\| = 1$. The zero operator $0\colon X \to Y$ is a bounded linear operator with norm $\|0\| = 0$.

Example 9.4. *A diagonal operator on* ℓ^1. Let (α_n) be a bounded sequence of real numbers, so that $c = \sup_{n \in \mathbb{N}} |\alpha_n|$ is finite. The operator T defined by

$$Tx = T(\xi_1, \xi_2, \xi_3, \ldots) = (\alpha_1 \xi_1, \alpha_2 \xi_2, \alpha_3 \xi_3, \ldots) \text{ for all } x \in \ell^1$$

is a bounded linear operator on ℓ^1 with norm $\|T\| = c$. Indeed, the inequality $\sum_n |\alpha_n \xi_n| \le c \sum_n |\xi_n|$ shows that for any $x \in \ell^1$, Tx is in ℓ^1, and $\|Tx\| \le c\|x\|$. The linearity of T is easy to check. Finally we show that $\|T\| = c$. From $\|Tx\| \le c\|x\|$ we conclude that $\|T\| \le c$. By the definition of supremum, for each $\varepsilon > 0$ there exists α_m such that $|\alpha_m| > c - \varepsilon$. Then for $x = e_m$, where $e_1, e_2, e_3, \ldots$ is the standard Schauder basis for ℓ^1, we have $Tx = \alpha_m x$, and

$$\frac{\|Tx\|}{\|x\|} = |\alpha_m| > c - \varepsilon.$$

This proves that $\|T\| = c$.

It turns out that the boundedness of a linear operator is equivalent to its continuity.

Theorem 9.5. *Let $T: X \to Y$ be a linear operator between normed spaces X, Y. Then T is continuous if and only if T is bounded.*

Proof. Let T be bounded. Then the continuity of T follows from Theorem 9.2.

Conversely, assume that T is continuous at $x_0 = 0$. If $\varepsilon = 1$, then by the definition of continuity, there exists $\delta > 0$ such that

$$\|x\| \leq \delta \implies \|Tx\| \leq \varepsilon = 1.$$

Given any $u \neq 0$, set $x = (\delta/\|u\|)u$. Then $u = (\|u\|/\delta)x$, $\|x\| = \delta$, and from the linearity of T and from the continuity condition it follows that

$$\|Tu\| = \frac{\|u\|}{\delta}\|Tx\| \leq \frac{\|u\|}{\delta} = \frac{1}{\delta}\|u\|.$$

Hence T satisfies the boundedness condition with a constant $c = 1/\delta$ (the inequality works also for $u = 0$). $\qquad\square$

Note. Observe that the continuity of a linear operator at one point of X is sufficient for its (uniform) continuity on all of X.

Example 9.6. If X, Y are normed spaces and if X is finite dimensional, then any linear operator $T: X \to Y$ is bounded.

Let $a_1, \ldots, a_n$ be a basis for X. For any $x \in X$ define the ℓ^1-norm $\|x\|_1$ by

$$\|x\|_1 = \left\|\sum_{k=1}^n \xi_k a_k\right\|_1 = \sum_{k=1}^n |\xi_k|.$$

Recall that any two norms on a finite dimensional vector space are equivalent (Theorem 7.13). So there is a constant $c > 0$ such that $\|x\|_1 \leq c\|x\|$ for all $x \in X$. Set also $\alpha = \max(\|Ta_1\|, \ldots, \|Ta_n\|)$. Then

$$\|Tx\| = \left\|\sum_{k=1}^n \xi_k Ta_k\right\| \leq \sum_{k=1}^n |\xi_k|\,\|Ta_k\| \leq \alpha\|x\|_1 \leq \alpha c\|x\|.$$

Example 9.7. *Fredholm integral operator.* Let $J = [a, b]$ and let $k: J \times J \to \mathbb{C}$ be a continuous function of two variables. The space $X = C(J)$ of all continuous functions $f: J \to \mathbb{C}$ becomes a complex Banach space when equipped with the uniform norm. For any $x \in X$ we define a function Tx on J by

$$(Tx)(t) = \int_a^b k(t,s)x(s)\, ds, \quad t \in J.$$

The function is well defined since, for each fixed $t \in J$, the function $s \mapsto k(t,s)x(s)$ is continuous.

First we show that, for any $x \in X$, the function Tx is continuous. For this recall that the set $J \times J$ is compact in the Euclidean plane $\mathbb{R}^2$. Then the function $k \colon J \times J \to \mathbb{C}$ is uniformly continuous on $J \times J$. Given any $\varepsilon > 0$, there is $\delta > 0$ such that, for any points $(t_1, s_1), (t_2, s_2) \in J \times J$,

$$(|t_1 - t_2|^2 + |s_1 - s_2|^2)^{1/2} < \delta \implies |k(t_1, s_1) - k(t_2, s_2)| < \varepsilon.$$

Let $t_1, t_2 \in J$ satisfy $|t_1 - t_2| < \delta$. Then the Euclidean distance between (t_1, s) and (t_2, s) is less than δ, and so

$$|k(t_1, s) - k(t_2, s)| < \varepsilon$$

for any $s \in J$. Then we have

$$|(Tx)(t_1) - (Tx)(t_2)| \le \int_a^b |k(t_1, s) - k(t_2, s)|\, |x(s)|\, ds \le (b-a)\varepsilon\|x\|.$$

This shows that Tx is (uniformly) continuous on J. Hence

$$x \in X \implies Tx \in X.$$

A proof that T is linear is left as an exercise.

Finally we show that T is bounded. Since k is continuous on the compact set $J \times J$, k is bounded, say $|k(t,s)| \le c$ for all $(t,s) \in J \times J$. Then

$$|(Tx)(t)| \le \int_a^b |k(t,s)|\, |x(s)|\, ds \le (b-a)c\|x\|.$$

This means that

$$\|Tx\| \le (b-a)c\|x\| \text{ for all } x \in X,$$

and $\|T\| \le (b-a)c$.

Definition 9.8. A *functional* is an operator which maps a normed space X into its scalar field $\mathbb{R}$ or $\mathbb{C}$. A linear functional f on X is *bounded* if there is $c > 0$ such that $|f(x)| \le c\|x\|$ for all $x \in X$. (So f is a special case of a bounded linear operator.) We define the functional norm of f by

$$\|f\| = \sup_{x \neq 0} \frac{|f(x)|}{\|x\|}.$$

Example 9.9. For each continuous function $x\colon [a,b] \to \mathbb{R}$ define $f(x)$ by

$$f(x) = \int_a^b x(t)\, dt.$$

Then f maps the Banach space $X = C[a,b]$ of all continuous functions equipped with the uniform norm linearly into $\mathbb{R}$, while

$$|f(x)| \le \int_a^b |x(t)|\, dt \le (b-a)\|x\|.$$

So f is a bounded linear functional on X with $\|f\| \le b - a$. Choosing $x(t) = 1$, we conclude that $\|f\| = b - a$.

Let X, Y be normed spaces. The set of all bounded linear operators $T\colon X \to Y$ will be denoted by $\mathcal{B}(X,Y)$. We write $\mathcal{B}(X)$ for $\mathcal{B}(X,X)$. (An alternative notation in the literature is $\mathcal{L}(X,Y)$ for $\mathcal{B}(X,Y)$.) We observe that $\mathcal{B}(X,Y)$ is a vector space under pointwise operations of addition of operators and multiplication of operators by scalars. The operator norm

$$\|T\| = \sup_{x \ne 0} \frac{\|Tx\|}{\|x\|}$$

has the properties **N1**–**N4** of the norm (Problem 9.1). If Y is a Banach space, then $\mathcal{B}(X,Y)$ is also a Banach space under the operator norm (Problem 9.1). If $T \in \mathcal{B}(X,Y)$ and $S \in \mathcal{B}(Y,Z)$, we define the *product* ST as the composition

$$ST = S \circ T;$$

we observe that $ST \in \mathcal{B}(X,Z)$.

Definition 9.10. An operator $T \in \mathcal{B}(X,Y)$ is said to be *invertible in* $\mathcal{B}(X,Y)$ if there is an operator $S \in \mathcal{B}(Y,X)$ such that $TS = I_Y$ and $ST = I_X$. The operator S is the *inverse* of T, and is usually denoted by T^{-1}. If $T \in \mathcal{B}(X)$, the inverse S satisfies $TS = I = ST$.

A bijective linear operator $T\colon X \to Y$ has an *algebraic* inverse $S\colon Y \to X$. It is known from elementary linear algebra that S is linear. This does not mean that a bijective operator $T \in \mathcal{B}(X,Y)$ is always invertible in $\mathcal{B}(X,Y)$. Note that for the invertibility of T the inverse S, in addition to being linear, must be bounded. The preceding definition can be rephrased by saying that $T \in \mathcal{B}(X,Y)$ is invertible in $\mathcal{B}(X,Y)$ if and only if T is a linear homeomorphism from X to Y. If T and S are composable invertible operators, we have

$$(TS)^{-1} = S^{-1}T^{-1}$$

(exercise). Suppose that T is a linear operator on X. From Theorem 9.5 and Problem 9.47 it follows that T is bounded and invertible in $\mathcal{B}(X)$ if and only if T is surjective and there are positive constants a, b such that

$$a\|x\| \leq \|Tx\| \leq b\|x\| \quad \text{for all } x \in X.$$

The space $\mathcal{B}(X)$ is an example of a (noncommutative) normed algebra.

Definition 9.11. A *normed algebra* is a set $\mathcal{A}$ of elements $A, B, C, \dots$ satisfying the following conditions:

(i) $\mathcal{A}$ is a normed space.

(ii)   There is a product on $\mathcal{A}$ subject to the following laws:

$$(AB)C = A(BC), \qquad\qquad A(B + C) = AB + AC,$$
$$(B + C)A = BA + CA, \qquad\qquad \lambda(AB) = (\lambda A)B = A(\lambda B).$$

(iii) $\|AB\| \leq \|A\|\,\|B\|$ for all $A, B \in \mathcal{A}$.

A *unital normed algebra* is a normed algebra with a multiplicative unit I satisfying $\|I\| = 1$. A *Banach algebra* is a complete normed algebra. If X is a Banach space, then $\mathcal{B}(X)$ is a unital Banach algebra under the product $TS = T \circ S$ with the unit the operator identity $I = I_X$.

9.2   Linear operators on Hilbert spaces

Let H_1, H_2 be two Hilbert spaces with inner products $\langle x_1, x_2 \rangle_1$ and $\langle y_1, y_2 \rangle_2$. We usually supress the subscripts as the inner products are usually understood from the context.

Definition 9.12. Let $T\colon H_1 \to H_2$ be a bounded linear operator between Hilbert spaces H_1, H_2. A bounded linear operator $T^*\colon H_2 \to H_1$ is the *Hilbert space adjoint* of T if

$$\langle Tx, y \rangle = \langle x, T^*y \rangle \quad \text{for all } x \in H_1, \ y \in H_2. \tag{9.4}$$

We show that every bounded linear operator between Hilbert spaces has a Hilbert space adjoint. For this we need the Riesz representation of functionals given later in this chapter (Theorem 9.32).

Theorem 9.13. *For each $T \in \mathcal{B}(H_1, H_2)$ the Hilbert space adjoint $T^* \in \mathcal{B}(H_2, H_1)$ of T exists and is unique. Moreover,*

$$\|T^*\| = \|T\|.$$

Proof. Let $T: H_1 \to H_2$ be a bounded linear operator. With each fixed element $y \in H_2$ we associate a function $f_y: H_1 \to \mathbb{C}$ defined by

$$f_y(x) = \langle Tx, y \rangle \text{ for all } x \in H_1.$$

Then f_y is a bounded linear functional on H_1 (exercise). By the Riesz representation theorem for Hilbert space (Theorem 9.32), there is a unique element y^* of H_1 such that

$$f_y(x) = \langle x, y^* \rangle \text{ for all } x \in H_1.$$

Define $Sy = y^*$ for each $y \in H_2$. Then, for each $x \in H_1$ and each $y \in H_2$,

$$\langle Tx, y \rangle = f_y(x) = \langle x, y^* \rangle = \langle x, Sy \rangle.$$

The proof of the linearity of S is left as an exercise. Using the equality $\langle Tx, y \rangle = \langle x, Sy \rangle$ and the Schwarz inequality, we get

$$\|Sy\|^2 = \langle Sy, Sy \rangle = \langle TSy, y \rangle \le \|TSy\| \, \|y\| \le \|T\| \, \|Sy\| \, \|y\|.$$

Then $\|Sy\| \le \|T\| \, \|y\|$ (whether or not $Sy = 0$). This shows that S is bounded with $\|S\| \le \|T\|$. A proof of the uniqueness of an operator $S \in \mathcal{B}(H_2, H_1)$ satisfying $\langle Tx, y \rangle = \langle x, Sy \rangle$ for all $x \in H_1$ and all $y \in H_2$ is left as an exercise. Hence we can write $T^* = S$.

We have already proved that $\|T^*\| \le \|T\|$. To show that $\|T^*\| = \|T\|$, we observe that $T^{**} = T$ by Problem 9.20; then $\|T\| = \|(T^*)^*\| \le \|T^*\|$. $\square$

Example 9.14. Let $T: \mathbb{C}^n \to \mathbb{C}^m$ be a linear operator with the standard matrix $A = [\alpha_{jk}]$; this means that $T(x) = Ax$, where A is an $m \times n$ matrix and x an $n \times 1$ matrix. The *hermitian transpose* A^{H} of the matrix A is defined by

$$A = \begin{bmatrix} \alpha_{11} & \alpha_{12} & \cdots & \alpha_{1n} \\ \alpha_{21} & \alpha_{22} & \cdots & \alpha_{2n} \\ \cdots & \cdots & \cdots & \cdots \\ \alpha_{m1} & \alpha_{m2} & \cdots & \alpha_{mn} \end{bmatrix}, \qquad A^{\mathsf{H}} = \begin{bmatrix} \overline{\alpha}_{11} & \overline{\alpha}_{21} & \cdots & \overline{\alpha}_{m1} \\ \overline{\alpha}_{12} & \overline{\alpha}_{22} & \cdots & \overline{\alpha}_{m2} \\ \cdots & \cdots & \cdots & \cdots \\ \overline{\alpha}_{1n} & \overline{\alpha}_{2n} & \cdots & \overline{\alpha}_{mn} \end{bmatrix}.$$

We observe that $(AB)^{\mathsf{H}} = B^{\mathsf{H}} A^{\mathsf{H}}$ and $(A^{\mathsf{H}})^{\mathsf{H}} = A$. Further, the standard inner product on $\mathbb{C}^n$ can be written in terms of the hermitian transpose as

$$\langle x, y \rangle = \xi_1 \overline{\eta}_1 + \cdots + \xi_n \overline{\eta}_n = y^{\mathsf{H}} x.$$

To find the adjoint of T:

$$\langle Tx, y \rangle = \langle Ax, y \rangle = y^{\mathsf{H}} A x = (A^{\mathsf{H}} y)^{\mathsf{H}} x = \langle x, A^{\mathsf{H}} y \rangle.$$

So

$$\langle x, T^* y \rangle = \langle Tx, y \rangle = \langle x, A^{\mathsf{H}} y \rangle$$

for all $x \in \mathbb{C}^n$ and all $y \in \mathbb{C}^m$. This shows that the Hilbert space adjoint of a linear operator $T \colon \mathbb{C}^n \to \mathbb{C}^m$ with the standard matrix A is the linear operator $T^* \colon \mathbb{C}^m \to \mathbb{C}^n$ with the standard matrix A^H (the hermitian transpose of A).

Example 9.15. *Fredholm integral operator.* Let $J = [a, b]$, and let $k \in L^2(J \times J)$, that is, let k be a complex valued measurable function on $J \times J$ such that $|k|^2$ is Lebesgue integrable. Such a function is called an L^2 *integral kernel.* For any $x \in L^2(J)$ define $y = Tx$ a.e. by

$$y(t) = \int_J k(t, s)\, x(s)\, ds.$$

To show that $y \in L^2(J)$, we use the Schwarz inequality:

$$|y(t)|^2 \leq \int_a^b |k(t, s)|^2\, ds \int_a^b |x(s)|^2\, ds,$$

$$\int_a^b |y(t)|^2\, dt \leq \int_a^b \int_a^b |k(t, s)|^2\, dt\, ds \int_a^b |x(s)|^2\, ds.$$

Hence $\|y\| \leq \|k\| \|x\|$, where

$$\|k\| = \left(\int_{J \times J} |k(t, s)|^2\, dt\, ds \right)^{1/2}.$$

T is linear, and the above calculation shows that $\|T\| \leq \|k\|$. So T is a bounded linear operator on the Hilbert space $L^2(J)$ into itself. We find the Hilbert space adjoint of T:

$$\langle Tx, y \rangle = \int_J Tx(t)\overline{y(t)}\, dt = \int_J \left(\int_J k(t, s)\, x(s)\, ds \right) \overline{y(t)}\, dt$$

$$= \int_J x(s) \left(\int_J k(t, s)\, \overline{y(t)}\, dt \right) ds = \int_J x(s) \overline{\left(\int_J \overline{k(t, s)}\, y(t)\, dt \right)} ds,$$

where the bar denotes complex conjugation. So $T^* y(s) = \int_J \overline{k(t, s)}\, y(t)\, dt$; on interchanging s and t,

$$T^* y(t) = \int_J \overline{k(s, t)}\, y(s)\, ds = \int_J \widetilde{k}(t, s)\, y(s)\, ds,$$

where the kernel $\widetilde{k}$ defined by

$$\widetilde{k}(t, s) = \overline{k(s, t)}$$

is called the *conjugate transpose* of the kernel k.

General properties of Hilbert space adjoints are summarized in Problem 9.20. The equation

$$\|T\|^2 = \|T^*T\|$$

is called the C^*-*identity*, and was used by I. Gelfand as the departure point for his theory of C^*-algebras.

Definition 9.16. An operator $A \in \mathcal{B}(H)$ is called *self-adjoint* if $A^* = A$; this is equivalent to

$$\langle Ax, y \rangle = \langle x, Ay \rangle \text{ for all } x, y \in H.$$

The set of all self-adjoint operators on H will be denoted by $S(H)$.

If A is self-adjoint, then $\langle Ax, x \rangle$ is real for all $x \in H$ as $\overline{\langle Ax, x \rangle} = \langle x, Ax \rangle = \langle Ax, x \rangle$.

Example 9.17. Every diagonal operator in the real space ℓ^2 is self-adjoint. In the complex ℓ^2, a diagonal operator is self-adjoint if it has real coefficients. Real symmetric matrices represent self-adjoint operators in real Euclidean spaces, and Hermitian matrices represent self-adjoint operators in complex Euclidean spaces. A verification of these facts is left as an exercise.

Theorem 9.18. *For any self-adjoint operator $A \in \mathcal{B}(H)$ we have*

$$\|A\| = \sup_{\|x\|=1} |\langle Ax, x \rangle|.$$

Proof. By the Schwarz inequality,

$$|\langle Ax, x \rangle| \leq \|Ax\| \, \|x\| \leq \|A\| \, \|x\|^2 \text{ for all } x \in H.$$

Therefore $\beta = \sup \{|\langle Ax, x \rangle| : \|x\| = 1\}$ is finite, and

$$\beta \leq \|A\|.$$

Observe that

$$|\langle Ax, x \rangle| \leq \beta \|x\|^2 \text{ for all } x \in H. \tag{9.5}$$

For a proof of the reverse inequality $\|A\| \leq \beta$ we need the following formula obtained by a direct calculation for any self-adjoint operator A:

$$4 \operatorname{Re}\langle Ax, y \rangle = \langle A(x+y), x+y \rangle - \langle A(x-y), x-y \rangle \text{ for all } x, y \in H. \tag{9.6}$$

Let $\|x\| = 1$. If $Ax \neq 0$, put $y = \|Ax\|^{-1}Ax$. Then $\|y\| = 1$, and

$$\|Ax\| = \langle Ax, y \rangle = \operatorname{Re}\langle Ax, y \rangle$$

$$\begin{aligned}
&= \tfrac{1}{4}\{\langle A(x+y), x+y \rangle - \langle A(x-y), x-y \rangle\} \\
&\leq \tfrac{1}{4}\{|\langle A(x+y), x+y \rangle| + |\langle A(x-y), x-y \rangle|\} \\
&\leq \tfrac{1}{4}\beta\{\|x+y\|^2 + \|x-y\|^2\} = \tfrac{1}{2}\beta(\|x\|^2 + \|y\|^2) \\
&= \beta
\end{aligned}$$

by (9.6), (9.5) and the parallelogram identity. So $\|A\| \leq \beta$, and the result follows. $\qquad\square$

In many ways self-adjoint operators play a role in the set $\mathcal{B}(H)$ similar to that of the real numbers in the set $\mathbb{C}$ of complex numbers. This analogy is explored in the rest of this section.

Theorem 9.19 (Cartesian decomposition). *Every operator $T \in \mathcal{B}(H)$ can be uniquely written in the form*

$$T = A + iB$$

where A, B are self-adjoint operators.

Proof. We solve the equation $T = A + iB$ for unknown self-adjoint operators A, B. From $T = A + iB$ we get $T^* = A - iB$, and hence

$$A = \tfrac{1}{2}(T + T^*), \qquad B = -i\tfrac{1}{2}(T - T^*)$$

are the unique solutions to $T = A + iB$. (Check that A, B are indeed self-adjoint.) $\qquad\square$

Definition 9.20. An operator A is called *positive* if $A \in S(H)$ and $\langle Ax, x \rangle \geq 0$ for all $x \in H$. We define a relation $\leq$ on $S(H)$ by setting

$$A \leq B \iff B - A \geq 0.$$

Note that, for any $A \in S(H)$,

$$-\|A\|\, I \leq A \leq \|A\|\, I.$$

We observe that $\leq$ is a partial order on $S(H)$, that is, the relation is reflexive, antisymmetric and transitive (see Appendix E). In addition,

$$A \leq B \implies A + C \leq B + C \text{ for all } C \in S(H)$$

and

$$A \leq B \implies cA \leq cB \text{ for all } c \geq 0.$$

In general, the product of positive operators is not positive unless the operators commute (this will be discussed later). Positive operators correspond to positive reals in our analogy.

Theorem 9.21 (Reid's inequality). *If $A \geq 0$, then*

$$\|Ax\|^2 \leq \|A\|\langle Ax, x \rangle \ \text{ for all } \ x \in H.$$

Proof. Problem 9.24. □

Theorem 9.22 (Vigier's theorem). *Every bounded monotonic sequence (A_n) in $S(H)$ converges pointwise to an operator $A \in S(H)$.*

Proof. Assume that the sequence is increasing, that is, $A_n \leq A_{n+1}$ for all n. By assumption there are constants a, b such that $aI \leq A_n \leq bI$ for all n. Replacing each A_n by $A_n - aI$, we may assume that

$$0 \leq A_n \leq cI \text{ for all } n \text{ and some } c > 0.$$

By Theorem 9.18, $\|A_n\| \leq c$ for all n. If $m > n$, then $A_m - A_n \geq 0$ and, by Reid's inequality,

$$\begin{aligned}
\|(A_m - A_n)x\|^2 &\leq \|A_m - A_n\|\langle (A_m - A_n)x, x \rangle \\
&\leq 2c(\langle A_m x, x \rangle - \langle A_n x, x \rangle).
\end{aligned}$$

The numerical sequence $(\langle A_n x, x \rangle)$ is bounded and monotonic, and hence, Cauchy. Then $(A_n x)$ is Cauchy by the above inequality, and so $\lim_{n \to \infty} A_n x = Ax$ exists for each x in the Hilbert space H. The operator A is linear; since the norms $\|A_n\|$ are uniformly bounded, A is a bounded operator. In addition, $A \in S(H)$ (Problem 9.25). □

Definition 9.23. If $A, B \in \mathcal{B}(H)$ and $B^2 = A$, then B is called a *square root* of A.

Theorem 9.24 (The positive square root theorem). *Every positive operator A on a Hilbert space H has a unique positive square root $A^{1/2}$. The square root double commutes with A, that is,*

$$TA = AT \implies TA^{1/2} = A^{1/2}T.$$

Proof. First assume that $0 \leq A \leq I$. The equation $B^2 = A$ is equivalent to

$$Q = \tfrac{1}{2}(I - A + Q^2), \tag{9.7}$$

where $Q = I - B$. We find a solution to (9.7) by iteration

$$Q_0 = 0, \qquad Q_{n+1} = \tfrac{1}{2}(I - A + Q_n^2), \quad n = 0, 1, 2, \ldots \tag{9.8}$$

By induction we show that

(i)   each Q_n is a polynomial in $(I - A)$ with positive coefficients,

(ii) each $Q_n - Q_{n-1}$ is a polynomial in $(I - A)$ with positive coefficients,

(iii) $Q_n \leq I$.

(i) and (iii) follow from (9.8). Then (ii) is derived from (i) and the identity

$$Q_{n+1} - Q_n = \tfrac{1}{2}(Q_n^2 - Q_{n-1}^2) = \tfrac{1}{2}(Q_n - Q_{n-1})(Q_n + Q_{n-1})$$

(the last factorization is possible as Q_n, Q_{n-1} commute being polynomials in $(I - A)$). Since $C = I - A \geq 0$, also $C^n \geq 0$ for all n (Problem 9.26). So $Q_n \geq 0$ and $Q_n - Q_{n-1} \geq 0$ for all n (linear combinations of positive operators with positive coefficients); this means that (Q_n) is an increasing operator sequence with $0 \leq Q_n \leq I$. By Vigier's theorem, (Q_n) converges pointwise to an operator Q with $0 \leq Q \leq I$. Taking the limit in (9.8) we get (9.7), so that $B^2 = (I - Q)^2 = A$ and $B = I - Q \geq 0$.

If A is positive but $A \leq I$ does not hold, set $c = \|A\|^{-1}$ and $T = cA$. Then $0 \leq T \leq I$ and T has a positive square root S by the preceding result. Let $B = c^{-1/2}S$; then $B^2 = A$ and $B \geq 0$.

The proof of the uniqueness of a positive square root is left as an exercise. From the construction of $A^{1/2}$ as the pointwise limit of the operators $I - Q_n$ we deduce that $AT = TA \implies A^{1/2}T = TA^{1/2}$.   □

Example 9.25. As an application of the square root theorem we show that any operator $T \in \mathcal{B}(H)$ for which $\langle Tx, x \rangle$ is always real is self-adjoint. Indeed, if T has this property, then $T = A + iB$ with A, B self-adjoint, and $\langle Tx, x \rangle = \langle Ax, x \rangle + i\langle Bx, x \rangle$ for all $x \in H$; hence $\langle Bx, x \rangle = 0$ for all $x \in H$. Thus B is positive, and has a positive square root $B^{1/2}$. Then $\|B^{1/2}x\|^2 = \langle B^{1/2}x, B^{1/2}x \rangle = \langle Bx, x \rangle = 0$, that is, $B^{1/2} = 0$. Then also $B = (B^{1/2})^2 = 0$, and $T = A$ is self-adjoint.

Definition 9.26. An operator $U \in \mathcal{B}(H)$ is *unitary* if

$$U^*U = I = UU^*.$$

Unitary operators correspond to complex numbers z with $|z| = 1$.

Theorem 9.27 (Properties of unitary operators). *Let U, V be unitary. Then:*

(i)   *U is an isometry.*

(ii) *$\|U\| = 1$.*

(iii) *UV is unitary.*

Proof. Exercise. □

There is an analogue of polar form for operators. We give a special case for invertible operators.

Theorem 9.28 (Polar form for operators on a Hilbert space). *Let $T \in \mathcal{B}(H)$ be an invertible operator. Then there exists a positive operator P and a unitary operator U such that*

$$T = UP. \tag{9.9}$$

Proof. For any $T \in \mathcal{B}(H)$ the operator T^*T is positive, so that we can define $P = (T^*T)^{1/2}$; then $P^2 = T^*T$. Set $U = TP^{-1}$. Then $T = UP$ where P is positive and U unitary:

$$U^*U = P^{-1}T^*TP^{-1} = P^{-1}P^2P^{-1} = I$$

and

$$UU^* = TP^{-1}P^{-1}T^* = T(T^*T)^{-1}T^* = TT^{-1}(T^*)^{-1}T^* = I. \qquad □$$

It can be shown that the polar decomposition (9.9) is unique (if T is invertible). There is also a reverse polar decomposition $T = P_1U_1$ with P_1 positive and U_1 unitary (exercise).

9.3 Bounded linear functionals

A linear functional is a special case of a linear operator. A linear functional maps a normed space X linearly into the scalar field $\mathbb{R}$ or $\mathbb{C}$; the scalar field is a one dimensional normed space with modulus playing the role of norm. Boundedness and norm of a bounded linear fuctional are defined as for operators. This means that a linear functional f on a normed space X is bounded if there is a constant $c > 0$ such that

$$|f(x)| \le c \, \|x\| \text{ for all } x \in X.$$

The *norm* of a bounded linear functional f is defined by

$$\|f\| = \sup_{x \neq 0} \frac{|f(x)|}{\|x\|}. \tag{9.10}$$

Observe that, for any $x \in X$,

$$|f(x)| \le \|f\| \, \|x\|.$$

Example 9.29. *Definite integral as a functional* (Example 9.9). Let $X = C[a, b]$ be the space of all real valued continuous functions on the interval $[a, b]$, with functions written as $x(t)$, $y(t)$, $z(t)$, etc., equipped with the uniform norm

$$\|x\| = \sup_{t \in [a,b]} |x(t)|.$$

Define a functional f on the space X by setting

$$f(x) = \int_a^b x(t)\, dt \text{ for all } x \in X.$$

Then f is a bounded linear functional on X with $\|f\| = b - a$.

Example 9.30. *Evaluation functional.* Let $X = C[a, b]$ be the space from the previous example with the uniform norm, and let t_0 be a point in the interval $[a, b]$, which is kept fixed throughout. Define f by

$$f(x) = x(t_0) \text{ for all } x \in X.$$

Then f is a bounded linear functional whose norm is equal to 1. Details are left as an exercise.

Example 9.31. *Representation of functionals on ℓ^p.* Given $p > 1$, we consider the complex space ℓ^p and investigate what form bounded linear functionals on ℓ^p take.

Let f be a bounded linear functional on ℓ^p, and let

$$e_1 = (1, 0, 0, \dots), \quad e_2 = (0, 1, 0, \dots), \quad e_3 = (0, 0, 1, \dots), \dots$$

be the standard Schauder basis for ℓ^p. Define $\alpha_k = f(e_k)$ for all k. Using the continuity of f, we get

$$f(x) = f\left(\sum_{n=1}^{\infty} \xi_n e_n\right) = f\left(\lim_{n \to \infty} \sum_{k=1}^{n} \xi_k e_k\right)$$

$$= \lim_{n \to \infty} f\left(\sum_{k=1}^{n} \xi_k e_k\right) = \lim_{n \to \infty} \sum_{k=1}^{n} f(e_k)\xi_k$$

$$= \sum_{n=1}^{\infty} \alpha_n \xi_n.$$

This gives a representation for f in the form

$$f(x) = \sum_{n=1}^{\infty} \alpha_n \xi_n.$$

We want to show that the vector $a = (\alpha_1, \alpha_2, \ldots)$ is in ℓ^q, where $q = p/(p-1)$ is the conjugate index of p, and that

$$\|f\| = \|a\|_q.$$

Define

$$\xi_k = \begin{cases} |\alpha_k|^q \alpha_k^{-1} & \text{if } \alpha_k \neq 0, \\ 0 & \text{if } \alpha_k = 0. \end{cases}$$

Then

$$|\xi_k|^p = |\alpha_k|^q = \alpha_k \xi_k.$$

So, for any $n \in \mathbb{N}$,

$$\sum_{k=1}^{n} |\alpha_k|^q = \sum_{k=1}^{n} \alpha_k \xi_k = f\left(\sum_{k=1}^{n} \xi_k e_k\right)$$

$$\leq \|f\| \left(\sum_{k=1}^{n} |\xi_k|^p\right)^{1/p} = \|f\| \left(\sum_{k=1}^{n} |\alpha_k|^q\right)^{1/p}.$$

Dividing by $(\sum_{k=1}^{n} |\alpha_k|^q)^{1/p}$ and noting that $1 - 1/p = 1/q$, we have

$$\left(\sum_{k=1}^{n} |\alpha_k|^q\right)^{1/q} \leq \|f\| \text{ for all } n \in \mathbb{N},$$

which shows that $a \in \ell^q$ and that $\|a\|_q \leq \|f\|$. To prove the reverse inequality for the norms, note that, by Hölder's inequality,

$$|f(x)| = \left|\sum_{k=1}^{\infty} \alpha_k \xi_k\right| \leq \|a\|_q \|x\|_p \text{ for all } x \in \ell^q.$$

Conversely, if $a \in \ell^q$, then according to Problem 9.14, $f(x) = \sum_{k=1}^{\infty} \alpha_k \xi_k$ defines a bounded linear functional on ℓ^p.

F. Riesz gave several representation theorems for bounded linear functionals on various normed spaces. The following result is one of them.

Theorem 9.32 (Riesz representation theorem for Hilbert space).
Let H be a Hilbert space and f a bounded linear functional on H. Then there exists a unique vector $a \in H$ such that $f(x) = \langle x, a \rangle$ for all $x \in H$, and $\|f\| = \|a\|$.

Proof. Problem 9.17. □

Definition 9.33. If X is a normed space, then the set of all bounded linear functionals on X equipped with the functional norm (9.10) is called the *dual* of X and is denoted by X'.

Theorem 9.34. *For every normed space X, its dual space X' is a Banach space.*

Proof. Note that $X' = \mathcal{B}(X, \mathbb{K})$, where $\mathbb{K}$ is the scalar field. We recall that the space $\mathcal{B}(X, Y)$ of bounded linear operators from X to Y is complete provided Y is complete (Problem 9.1). The result follows as both $\mathbb{R}$ and $\mathbb{C}$ are complete. $\qquad\square$

Example 9.31 shows that, for $p > 1$, the dual of ℓ^p can be identified with ℓ^q, where q is the conjugate index of p. Let T be the mapping that carries every bounded linear functional f on ℓ^p onto the ℓ^q vector $a = (f(e_1), f(e_2), \ldots)$. From Example 9.31 it follows that T is a bijection and an isometry. It can be verified that T is also a linear operator from $(\ell^p)'$ onto ℓ^q. Recall that a linear bijection T is called an isomorphism between normed spaces. The foregoing results can be summarized as follows.

Theorem 9.35. *For any $p > 1$, the dual $(\ell^p)'$ of ℓ^p is isometrically isomorphic to ℓ^q, where q is the conjugate index of p.*

Example 9.36. The dual of ℓ^1 is isometrically isomorphic to ℓ^∞. (Problem 9.38.)

Let $p > 1$ and let q be the conjugate index of p. Then the dual of ℓ^p is isometrically isomorphic to ℓ^q, and the dual of ℓ^q isometrically isomorphic to ℓ^p. This pleasing symmetry is not present in the case that $p = 1$. While the dual of ℓ^1 is isometrically isomorphic to ℓ^∞, the dual of ℓ^∞ is not isometrically isomorphic to ℓ^1. A proof of this fact requires Theorem 9.44 given in the next section. Indeed, ℓ^1 is separable, while ℓ^∞ is not.

Example 9.37. Let H be a Hilbert space and let $R\colon H' \to H$ be the mapping satisfying $y'(x) = \langle x, Ry' \rangle$, $x \in H, y' \in H'$, whose existence follows from Theorem 9.32. It is not difficult to check that R is a conjugate linear isometric bijection of H' onto H: $R(y' + z') = Ry' + Rz'$ and $R(\lambda y') = \overline{\lambda} R(y')$, $y', z' \in H, \lambda \in \mathbb{C}$. R is usually called the *Riesz mapping*. This can be expressed by saying that H' is isometrically conjugate-isomorphic to H.

9.4 The Hahn-Banach theorem

The Hahn-Banach theorem deals with extension of bounded linear functionals defined on subspaces of normed spaces. It can be used to show that every normed space has many nonzero bounded linear functionals. The theorem is a very important tool of functional analysis, especially in the form of its corollaries. The theorem was discovered by Hahn in 1927 (for bounded linear functionals on normed spaces), and extended by Banach in 1929 to a more general setting. Bohnenblust and Sobczyk generalized the theorem to complex vector spaces. We give a proof of Hahn's original version of the theorem for real spaces, while the complex case is left as an exercise (Problem 9.53).

Theorem 9.38 (The Hahn-Banach theorem). *Let X be a normed space, and Y a vector subspace of X. Every bounded linear functional f defined on Y can be extended to a linear functional F on X with the same norm $\|F\| = \|f\|$.*

Proof. Assume that X is a real normed space. Without loss of generality we assume that $\|f\| = 1$. Let $\mathcal{A}$ be the family consisting of every linear extension g of f to a subspace $D(g)$ containing Y with $\|g\| = 1$. We identify each functional g with its graph, a subset of $X \times \mathbb{R}$. Note that g_2 is an extension of g_1 if and only if $g_2 \supset g_1$ in terms of graphs.

We verify that $\mathcal{A}$ satisfies the conditions of the Hausdorff maximality principle (Theorem E.17). Let $\mathcal{C}$ be a chain in $\mathcal{A}$. Define h by

$$h = \bigcup_{g \in \mathcal{C}} g.$$

Then $D = \bigcup_{g \in \mathcal{C}} D(g)$ is a subspace of X (as $\{D(g) : g \in \mathcal{C}\}$ is a chain of subspaces), and h is a linear extension of f with $\|h\| = 1$. Then $h \in \mathcal{A}$, and h is an upper bound for the chain $\mathcal{C}$. The Hausdorff maximality principle then applies to ensure the existence of a maximal element F of $\mathcal{A}$. Let F be defined on a subspace $Z = D(F)$. We show that $Z = X$.

Assume on the contrary that there is a vector $x_0 \in X \backslash Z$. Put $N = \mathsf{sp}(x_0) \oplus Z$, and define $G \colon N \to \mathbb{R}$ by

$$G(\lambda x_0 + z) = \lambda c + F(z),$$

where c is a real constant. We show that c can be chosen so that $\|G\| = 1$.

If $z \in Z$, set $m(z) = F(z) - \|z - x_0\|$, and $M(z) = F(z) + \|x_0 - z\|$. For any $u, v \in Z$,

$$F(u - v) \leq \|u - v\| \leq \|u - x_0\| + \|x_0 - v\|,$$

so that

$$m(u) = F(u) - \|u - x_0\| \le F(v) + \|x_0 - v\| = M(v).$$

Choose any c with

$$\sup_{u \in Z} m(u) \le c \le \inf_{v \in Z} M(v).$$

Then

$$|c - F(y)| \le \|x_0 - y\| \text{ for all } y \in Z.$$

So, for any $\lambda \ne 0$,

$$\begin{aligned}
|G(\lambda x_0 + z)| = |\lambda c + F(z)| &= |\lambda| \, |c - F(y)| \\
&\le |\lambda| \, \|x_0 - y\| = \|\lambda x_0 - \lambda y\| \\
&= \|\lambda x_0 + z\|
\end{aligned}$$

where $y = -z/\lambda$. So $\|G\| \le 1$; since G is an extension of f and $\|f\| = 1$, we have $\|G\| = 1$.

For an extension to complex spaces see Problem 9.53.   □

Theorem 9.39 (Corollary 1 to the Hahn-Banach theorem). *Let M be a vector subspace of a normed space X, and let $x_0 \in X$ be such that*

$$d(x_0, M) = \inf_{u \in M} \|x_0 - u\| > 0.$$

Then there is $f \in X'$ such that

$$f(M) = 0, \quad f(x_0) = d(x_0, M), \quad \|f\| = 1.$$

Proof.   Let $Y = \mathsf{sp}(x_0) \oplus M$, and define $g : Y \to \mathbb{R}$ by

$$g(\lambda x_0 + z) = \lambda d,$$

where $d = d(x_0, M)$. Then g is linear. It is left as a nontrivial exercise to verify that $\|g\| = 1$. By the Hahn-Banach theorem there is a functional $f \in X'$ which is an extension of g and has norm 1. Then f satisfies the requirements of the Corollary (check).   □

Example 9.40. *Banach limits.* There exists a functional $\mathrm{LIM} : \ell^\infty \to \mathbb{C}$ such that

(i) $\mathrm{LIM}\,(\alpha \xi_n + \beta \eta_n) = \alpha \, \mathrm{LIM}\, \xi_n + \beta \, \mathrm{LIM}\, \eta_n.$

(ii) $\mathrm{LIM}\, \xi_{n+1} = \mathrm{LIM}\, \xi_n.$

(iii)   If (ξ_n) is a real nonnegative sequence, then $\mathrm{LIM}\, \xi_n \ge 0.$

(iv) If (ξ_n) is a convergent sequence, then LIM $\xi_n = \lim\limits_{n \to \infty} \xi_n$.

(v) If (ξ_n) is a real sequence, then $\liminf\limits_{n \to \infty} \xi_n \leq$ LIM $\xi_n \leq \limsup\limits_{n \to \infty} \xi_n$.

First we show the existence of LIM for the real space ℓ^∞. Let $S: \ell^\infty \to \ell^\infty$ be the left shift operator $(Sx)_n = (x)_{n+1}$ for all n, that is,

$$S(\xi_1, \xi_2, \xi_3, \dots) = (\xi_2, \xi_3, \xi_4, \dots), \quad x = (\xi_n) \in \ell^\infty.$$

Let M be the space $M = \{x - Sx: x \in \ell^\infty\}$. We show that the element $e = (1, 1, 1, \dots)$ has distance $\text{dist}(e, M) = \inf_{u \in M} \|e - u\| = 1$ from M. Suppose that $\|e - (x - Sx)\| < 1$ for some $x \in \ell^\infty$. Then there exists $\varepsilon > 0$ such that $1 - \xi_n + \xi_{n+1} \leq |1 - \xi_n + \xi_{n+1}| \leq 1 - \varepsilon$ for all n. Hence $\xi_{n+1} \leq \xi_n - \varepsilon$ for all n, which contradicts the boundedness of (ξ_n). This proves that $\text{dist}(e, M) \geq 1$; since also $\|e - (e - Se)\| = 1$, we have $\text{dist}(e, M) = 1$.

By Corollary 2 to the Hahn-Banach theorem, there exists a bounded linear functional f on the real space ℓ^∞ such that

$$\|f\| = 1, \quad f(M) = 0, \quad f(e) = \text{dist}(e, M) = 1.$$

When we set

$$\text{LIM } \xi_n = f(x) \text{ for each } x = (\xi_n) \in \ell^\infty,$$

we can verify the properties (i)–(iv):

(i) This holds since f is linear.

(ii) From $f(M) = 0$ we get $f(x - Sx) = 0$ for all $x \in \ell^\infty$. If $x \in \ell^\infty$, then LIM $\xi_{n+1} = f(Sx) = f(x) = $ LIM ξ_n.

(iii) We write $x \geq 0$ if $\xi_n = (x)_n \geq 0$ for all n. Let $x \in \ell^\infty$ and $x \geq 0$. For some $\alpha > 0$, $0 \leq \alpha\|x\| \leq 1$, so $\|\alpha x - e\| \leq 1$; since $\|f\| = 1$, we have $f(\alpha x - e) \geq -1$, that is, $\alpha f(x) = f(e) + f(\alpha(x) - e) = 1 + f(\alpha x - e) \geq 0$. Thus $x \geq 0$ implies $f(x) \geq 0$.

(iv) Suppose that $\lim\limits_{n \to \infty} \xi_n = \xi$. From $f(x - Sx) = 0$ we deduce by induction that $f(S^n x) = f(x)$ for all n. Let $\varepsilon > 0$. There exists n such that $|\xi_i - \xi| \leq \varepsilon$ for all $i \geq n$. Then

$$|f(x) - \xi| = |f(S^n x - \xi e)| \leq \|S^n x - \xi e\| = \sup_{i \geq n} |\xi_i - \xi| \leq \varepsilon.$$

Since ε was arbitrary, $f(x) = \xi$, that is, LIM $\xi_n = \lim_{n \to \infty} \xi_n$.

(v) Let $\alpha_n = \inf_{i \geq n} \xi_i$ and $\beta_n = \sup_{i \geq n} \xi_i$. Then $\alpha_n \leq \xi_n \leq \beta_n$ for all n. From property (iii) and the linearity of LIM we deduce that

$$\text{LIM } \alpha_n \leq \text{LIM } \xi_n \leq \text{LIM } \beta_n.$$

The sequences (α_n) and (β_n) converge with limits equal to $\liminf_{n\to\infty} \xi_n$ and $\limsup_{n\to\infty} \xi_n$, respectively. The result then follows from property (iv).

Extension to complex sequences is done by setting $\mathrm{LIM}\,(\xi_n + i\eta_n) = f(x) + if(y)$.

The following result is a special case of Corollary 1.

Theorem 9.41 (Corollary 2 to the Hahn-Banach theorem). *Let M be a closed subspace of a normed space X and let $x_0 \in X \backslash M$. Then there is $f \in X'$ such that*

$$f(M) = 0, \quad f(x_0) \neq 0, \quad \|f\| = 1.$$

Theorem 9.42 (Corollary 3 to the Hahn-Banach theorem). *For each $x_0 \neq 0$ in a normed space X there is $f \in X'$ such that $\|f\| = 1$ and $f(x_0) = \|x_0\|$.*

Proof. Set $M = \{0\}$ in Corollary 2. $\qquad\qquad\qquad\qquad\qquad\qquad\square$

Corollary 3 is probably the most useful and the most often used of all the preceding results. As an example of its application we prove two important facts.

Theorem 9.43. *For any element x of a normed space X we have*

$$\|x\| = \max_{0 \neq f \in X'} \frac{|f(x)|}{\|f\|}.$$

Proof. Let $f \in X'$ be nonzero. Then

$$|f(x)| \leq \|f\|\,\|x\|,$$

so that $\sup\{|f(x)|/\|f\| : 0 \neq f \in X'\} \leq \|x\|$. By Corollary 3, there is $f \in X'$ with norm 1 such that $f(x) = \|x\|$; so the supremum is attained.$\square$

Theorem 9.44. *If X is a normed space whose dual X' is separable, then X itself is separable.*

Proof. X' contains a countable dense set, say $\{f_n : n \in \mathbb{N}\}$. By the definition of $\|f_n\|$ as a supremum we conclude that, for each n, there is $x_n \in X$ such that $\|x_n\| = 1$ and $|f_n(x_n)| \geq \frac{1}{2}\|f_n\|$. Let Y be the closed subspace of X generated by the x_n. For a proof by contradiction assume that $Y \neq X$. By Corollary 2 to the Hahn–Banach theorem there is $f \in X'$

such that $f(Y) = 0$ and $f \neq 0$. Let $\varepsilon > 0$. Since $\{f_n\}$ is dense in X', there is k with $\|f - f_k\| < \varepsilon$. Then

$$\tfrac{1}{2}\|f_k\| \leq |f_k(x_k)| = |f_k(x_k) - f(x_k)| \leq \|f_k - f\|\,\|x_k\| = \|f_k - f\| < \varepsilon,$$

and

$$\|f\| \leq \|f - f_k\| + \|f_k\| < \varepsilon + 2\varepsilon = 3\varepsilon.$$

Since $\varepsilon > 0$ was arbitrary, $\|f\| = 0$, which is a contradiction. This proves that $\{x_1, x_2, \dots\}$ is a countable total subset of X, and hence X is separable by Theorem 7.27. $\qquad\square$

9.5 Adjoint operator in normed spaces

First we introduce a notation that will bring out a certain symmetry between a normed space X and its dual X'. We will write $x, y, z \dots$ for elements of X, and $x', y', z', \dots$ for elements of X'. If $x \in X$ and $x' \in X'$, we write

$$(x, x') = x'(x).$$

The scalar valued function $F(x, x') = (x, x')$ is *bilinear* on $X \times X'$, and satisfies

$$|(x, x')| \leq \|x\|\,\|x'\|.$$

This notation shows that the action of x' on x generates a scalar function on $X \times X'$ in some ways similar to an inner product; note, however, that inner product operates on ordered pairs from the same space, while (x, x') operates on ordered pairs from two different spaces.

Definition 9.45. Let X, Y be normed spaces and let $T \in \mathcal{B}(X, Y)$. An operator $T' \in \mathcal{B}(Y', X')$ is an *adjoint* of T if

$$(Tx, y') = (x, T'y'), \quad \text{for all } x \in X, \quad \text{for all } y' \in Y'. \tag{9.11}$$

Note that in (Tx, y') the bilinear function $(\,,\,)$ operates on $Y \times Y'$, while in $(x, T'y')$ it operates on $X \times X'$. Note also that if the spaces X, Y are inner product spaces, we can also define the Hilbert space adjoint T^* of T, which does not coincide with the adjoint T' of T. The next result is concerned with the existence of adjoint operators.

Theorem 9.46. *Let X, Y be normed spaces. Every operator $T \in \mathcal{B}(X, Y)$ has a unique adjoint T'.*

Proof. For any $y' \in Y'$ define g to be the composition operator $g = y'T$; since $T \in \mathcal{B}(X,Y)$ and $y' \in \mathcal{B}(Y,\mathbb{K})$ ($\mathbb{K}$ is the scalar field), $g \in \mathcal{B}(X,\mathbb{K})$, that is, $g \in X'$. Define $V \colon Y' \to X'$ by

$$Vy' = y'T \text{ for all } y' \in Y'.$$

Then V is linear, and $\|Vy'\| = \|y'T\| \leq \|y'\| \, \|T\|$; so V is bounded with

$$\|V\| \leq \|T\|.$$

Moreover, for any $x \in X$ and any $y' \in Y'$ we have

$$(Tx, y') = y'(Tx) = (y'T)(x) = (Vy')(x) = (x, Vy').$$

So V satisfies the defining equation (9.11) for an adjoint of T.

Assume that U is another adjoint of T. Then

$$(Uy')(x) = (x, Uy') = (Tx, y') = (x, Vy') = (Vy')(x) \text{ for all } x \in X.$$

So $Uy' = Vy'$ for all $y' \in Y'$, which means that $U = V$. Hence the adjoint of T is unique, and can be denoted by T'. Note that we proved the equation

$$\|T'\| \leq \|T\|, \tag{9.12}$$

which is a half of $\|T'\| = \|T\|$. $\qquad \square$

Theorem 9.47. *For any $T \in \mathcal{B}(X,Y)$, $\|T'\| = \|T\|$.*

Proof. In view of (9.12) it is enough to show that $\|T\| \leq \|T'\|$. Given $x \in X$, Corollary 3 to the Hahn-Banach theorem ensures that there is $y' \in Y'$ such that $\|y'\| = 1$ and $y'(Tx) = \|Tx\|$. Then, for each $x \in X$,

$$\|Tx\| = y'(Tx) = (Tx, y') = (x, T'y') \leq \|x\| \, \|T'y'\|$$
$$\leq \|x\| \, \|T'\| \, \|y'\| = \|x\| \, \|T'\|.$$

This proves that $\|T\| \leq \|T'\|$. $\qquad \square$

In connection with dual spaces we introduce a new type of convergence of sequences, called the weak convergence: A sequence (x_n) in a normed space X *converges weakly* to $x \in X$ if

$$(x_n, x') \to (x, x') \text{ for all } x' \in X'.$$

The convergence in the norm of X is then often called the *strong convergence*. We write $x_n \rightharpoonup x$ if (x_n) converges weakly to x, and $x_n \to x$ if it converges strongly to x. Clearly,

$$x_n \to x \implies x_n \rightharpoonup x.$$

We can then prove that for any $T \in \mathcal{B}(X,Y)$,

$$x_n \rightharpoonup x \implies Tx_n \rightharpoonup Tx \tag{9.13}$$

(Problem 9.52).

9.6 Problems for Chapter 9

1. Let X, Y be normed spaces. Show that the set $\mathcal{B}(X, Y)$ of all bounded linear operators $T \colon X \to Y$ is a normed space under the operator norm defined by $\|T\| = \sup \{\|Tx\| : \|x\| = 1\}$. If Y is a Banach space, show that $\mathcal{B}(X, Y)$ is also a Banach space.

2. Let X, Y be normed spaces and $T \colon X \to Y$ a linear operator. Show that T is bounded if and only if T maps every bounded subset of X onto a bounded subset of Y.

3. Let $T \in \mathcal{B}(X, Y)$ and let $N = N(T) = T^{-1}(\{0\})$ be the nullspace of T. Look up the definition of the quotient space X/N (Problem 7.18). Define $\widehat{T}$ on X/N to Y by

$$\widehat{T}(x + N) = Tx, \quad x \in X.$$

Verify that $\widehat{T}$ is well defined (remember that $x + N = y + N \iff y - x \in N$). Prove that $\widehat{T}$ is a bounded linear operator and that $\|\widehat{T}\| = \|T\|$.

4. For $T, S \in \mathcal{B}(X)$ show that $\|TS\| \leq \|T\| \|S\|$. (Remember that $(TS)x = T(Sx)$.)

5. Let $p \geq 1$ and let R and L be the right and left shift operators respectively on the normed space ℓ^p; that is,

$$R(\xi_1, \xi_2, \xi_3, \ldots) = (0, \xi_1, \xi_2, \ldots), \quad L(\xi_1, \xi_2, \xi_3, \ldots) = (\xi_2, \xi_3, \ldots)$$

for all $x = (\xi_1, \xi_2, \xi_3, \ldots) \in \ell^p$. Show that:

(i) R and L are linear and bounded on ℓ^p.

(ii) R is injective but not surjective, and L is surjective but not injective. Contrast with the situation in finite dimensional spaces.

(iii) $LR = I$, but $RL \neq I$.

(iv) $\|L^n x\| \to 0$ for all $x \in \ell^p$, but $\|L^n\| \not\to 0$.

Find the norms $\|R\|$ and $\|L\|$.

6. For $T \in \mathcal{B}(X, Y)$ show that $\|T\| = \sup_{\|x\|=1} \|Tx\| = \sup_{\|x\| \leq 1} \|Tx\| = \sup_{\|x\| < 1} \|Tx\|$.

7. Let $X = C^1[0, 1]$, the space of all real continuously differentiable functions on $[0, 1]$, and let $Y = C[0, 1]$, the space of all real continuous functions on $[0, 1]$. Define norms

$$\|f\| = \sup_{t \in [0,1]} |f(t)| \quad \text{on both} \quad X, \; Y,$$

$$\|f\|_1 = \|f\| + \|f'\| \quad \text{only on} \quad X,$$

where $f'(t) = df(t)/dt$. Let D be the differential operator $Df = f'$.

(i) Check that $D: (X, \| \cdot \|_1) \to (Y, \| \cdot \|)$ is a bounded linear operator with $\|D\| = 1$.

(ii) Show that $D: (X, \| \, \|) \to (Y, \| \, \|)$ is unbounded.

8. Let X be the real space $C[a, b]$ equipped with the L^3 norm

$$\|x\|_3 = \left(\int_a^b |x(t)|^3 \, dt \right)^{1/3} \quad \text{for all } x \in X.$$

Define functionals f and g on X by $f(x) = \int_a^b x(t) \, dt$, and $g(x) = x(c)$ where $c \in (a, b)$ is kept fixed. Determine whether the functionals are (i) linear and (ii) bounded. Find their functional norm when it exists.

9. Let (α_n) be a bounded sequence of complex numbers and let T be the diagonal operator on the complex space ℓ^1 defined by

$$T(\xi_1, \xi_2, \xi_3, \ldots) = (\alpha_1 \xi_1, \alpha_2 \xi_2, \alpha_3 \xi_3 \ldots).$$

Find $\|T\|$ if

$$\text{(i) } \alpha_n = \frac{2n - i}{n + 2i}, \qquad \text{(ii) } \alpha_n = \frac{n + 1 + i}{n - 1 - i}.$$

10. Let the space $X = C[a, b]$ be equipped with the uniform norm. On X we define the *Volterra integral operator* T with the kernel k by

$$(Tx)(t) = \int_a^t k(t, s) x(s) \, ds \quad \text{for all } t \in [a, b],$$

where k is a function continuous in both variables in the triangle

$$\Delta = \{(t, s) \in \mathbb{R}^2 : a \le s \le t \le b\}.$$

Prove that, for each $x \in X$, $Tx \in X$, that $T \in \mathcal{B}(X)$ and that $\|T\| \le M(b - a)$, where $M = \max\{|f(t, s)| : (t, s) \in \Delta\}$. More generally, prove that

$$\|T^n\| \le M^n \frac{(b - a)^n}{n!}, \quad n = 1, 2, \ldots$$

Suggestion. Define the iterated kernels by $k_1 = k$,

$$k_n(t, s) = \int_a^t k(t, u) k_{n-1}(u, s) \, du, \quad n = 2, 3, \ldots$$

and show that

$$(T^n x)(t) = \int_a^t k_n(t, s) x(s) \, ds, \quad n = 1, 2, \ldots$$

Then prove by induction that $|k_n(t, s)| \le M^n (t - s)^{n-1}/(n - 1)!$.

11. Let $T\colon \mathbb{C}^n \to \mathbb{C}^m$ be a linear operator whose matrix with respect to the standard basis is $[\alpha_{jk}]$ – an $m \times n$ matrix. Show that $\|T\|$ is given by the formula in the following table when $\mathbb{C}^n$ and $\mathbb{C}^m$ are equipped with the norms specified below. (In (v), $p > 1$.)

	norm in $\mathbb{C}^n$	norm in $\mathbb{C}^m$	operator norm $\|T\|$
(i)	ℓ^∞	ℓ^∞	$\displaystyle\max_{1\le j\le m} \sum_{k=1}^{n} |\alpha_{jk}|$
(ii)	ℓ^1	ℓ^1	$\displaystyle\max_{1\le k\le n} \sum_{j=1}^{m} |\alpha_{jk}|$
(iii)	ℓ^2	ℓ^2	$\displaystyle\max_{1\le k\le n} \sqrt{\lambda_k};\ \lambda_k$ are the eigenvalues of A^*A
(iv)	ℓ^1	ℓ^∞	$\displaystyle\max_{1\le j\le m}\ \max_{1\le k\le n} |\alpha_{jk}|$
(v)	ℓ^1	ℓ^p	$\displaystyle\max_{1\le k\le n} \left(\sum_{j=1}^{m} |\alpha_{jk}|^p\right)^{1/p}$

Suggestion. Case (iii) requires a diagonalization of the Hermitian matrix A^*A. In (v) write $\|Tx\|_p = \|\sum_k \xi_k Te_k\|_p \le \sum_k |\xi_k| \|Te_k\|_p$.)

12. Let $p > 1$. Show that

$$f(x) = \int_{[a,b]} x \, dm$$

defines a bounded linear functional on $L^p[a,b]$, and find its norm.

13. Let $[\alpha_{ij}]$ be an infinite complex matrix $(i, j \in \mathbb{N})$ such that $c_j = \sum_{i=1}^{\infty} |\alpha_{ij}|$ converges for each $j \in \mathbb{N}$, and such that $c = \sup_j c_j < \infty$.

(i) Show that the equation

$$Tx = \begin{bmatrix} \alpha_{11} & \alpha_{12} & \alpha_{13} & \cdots \\ \alpha_{21} & \alpha_{22} & \alpha_{23} & \cdots \\ \alpha_{31} & \alpha_{32} & \alpha_{33} & \cdots \\ \cdots & \cdots & \cdots & \end{bmatrix} \begin{bmatrix} \xi_1 \\ \xi_2 \\ \xi_3 \\ \cdots \end{bmatrix} = \begin{bmatrix} \sum_{j=1}^{\infty} \alpha_{1j}\xi_j \\ \sum_{j=1}^{\infty} \alpha_{2j}\xi_j \\ \sum_{j=1}^{\infty} \alpha_{3j}\xi_j \\ \cdots \end{bmatrix}, \quad x = (\xi_1, \xi_2, \xi_3, \dots) \in \ell^1,$$

defines a bounded linear operator $T\colon \ell^1 \to \ell^1$ with $\|T\| \le c$.

(ii) Show that $\|T\| = c$.

14. Let $p, q > 1$ be conjugate indices $(1/p + 1/q = 1)$, and let $a = (\alpha_1, \alpha_2, \alpha_3, \dots)$ be an element of ℓ^q. Prove that, for each $x \in \ell^p$, the series

$$f(x) = \sum_{k=1}^{\infty} \alpha_k \xi_k$$

converges absolutely. Show that f is a bounded linear functional on ℓ^p with $\|f\| \le \|a\|_q$.

15. Let (α_n) be a bounded sequence of complex numbers, and let

$$A(\xi_1, \xi_2, \xi_3, \ldots) = (0, \alpha_1\xi_1, \alpha_2\xi_2, \ldots)$$

for all $x = (\xi_1, \xi_2, \ldots)$ in the complex space ℓ^2. Show that A is a bounded linear operator on ℓ^2 and find its norm $\|A\|$. Find the adjoint operator A^* of A. Show that $A^*A \neq AA^*$ provided $A \neq 0$. Find the eigenvalues of A^*.

16. Let a be a fixed element of a Hilbert space H. Show that the function $f(x) = \langle x, a \rangle$ defined for all $x \in H$ is a bounded linear functional on H, and that $\|f\| = \|a\|$.

17. Prove Theorem 9.32: If $f \neq 0$, show that $N = f^{-1}(\{0\})$ is a closed subspace of H, and $N^\perp$ is a one-dimensional subspace of H. (Such a subspace N is called a *hyperplane* in H.) Choose $a = \lambda b$ for a suitable λ.

18. Let R and L be the right and left shift operators on the complex Hilbert space ℓ^2 (Problem 9.5). Find the adjoint operators R^* and L^*.

19. Let S, T be self-adjoint operators on a Hilbert space H. Prove that the operator ST is self-adjoint if and only if $ST = TS$.

20. Prove the following facts about bounded linear operators on a Hilbert space:

$$(T + S)^* = T^* + S^*, \qquad\qquad (\alpha T)^* = \overline{\alpha}T^*,$$
$$(TS)^* = S^*T^*, \qquad\qquad (T^*)^* = T,$$
$$\|T^*\| = \|T\|, \qquad\qquad \|T^*T\| = \|T\|^2.$$

21. For each operator $A \in \mathcal{B}(H)$ on a Hilbert space H prove that

$$H = N(A) \oplus^\perp \overline{R(A^*)}.$$

22. Use the C^*-identity $\|U\|^2 = \|U^*U\|$ to prove the *law of $*$-cancellation* for Hilbert space operators: $T^*TA = 0 \implies TA = 0$. Deduce that $T^*TA = T^*TB \implies TA = TB$.

23. Show that each of the operators given below is self-adjoint on the specified Hilbert space:

(i) any diagonal operator on the real ℓ^2

(ii) any diagonal operator with real diagonal entries on the complex ℓ^2 space

(iii) any linear operator on $\mathbb{R}^n$ to $\mathbb{R}^n$ with a symmetric matrix

(iv) any linear operator on $\mathbb{C}^n$ to $\mathbb{C}^n$ with a Hermitian matrix

24. *Reid's inequality.* If A is a positive operator on a Hilbert space H, show that

$$\|Ax\|^2 \leq \|A\|\langle Ax, x \rangle \text{ for all } x \in H.$$

Suggestion. Set $[x, y] = \langle Ax, y \rangle$ for all $x, y \in H$. Show that $[x, y]$ has all the properties of an inner product except $[x, x] = 0$ need not imply $x = 0$. Apply the Schwarz inequality to $[x, y]$, and then set $y = Ax$.

25. Let (S_n) be a sequence of self-adjoint operators on a Hilbert space H which converges pointwise to an operator $S \in \mathcal{B}(H)$. Show that S is also self-adjoint.

26. If $A \geq 0$, show that $A^n \geq 0$ for all n.
Suggestion. Consider separately A^{2n} and A^{2n+1}.

27. (i) Show that the inverse of a positive invertible operator is positive.

(ii) Prove that the product of two commuting positive operators A, B is positive.
Suggestion. (ii) Use the double commutativity of square roots to show that $A^{1/2}B^{1/2} = B^{1/2}A^{1/2}$. Then write $AB = A^{1/2}B^{1/2}B^{1/2}A^{1/2}$.

28. If $A, B \in S(H)$ and C is a positive operator that commutes with both A and B, show that $A \leq B$ implies $AC \leq BC$.

29. If $A \geq 0$, show that $\|A^{1/2}\| = \|A\|^{1/2}$.

30. Show that a bounded linear operator P on a Hilbert space H is an orthogonal projection onto a closed subspace M of H if and only if $P^2 = P$ and $P^* = P$.

31. Let P be an orthogonal projection P on a Hilbert space H. Show that:

(i) $0 \leq P \leq I$.

(ii) $\langle Px, x \rangle = \|Px\|^2$ for all $x \in H$.

(iii) If $P \neq 0$, then $\|P\| = 1$.

(iv) $x \in R(P) \iff Px = x$.

(v) $\|Px\| = \|x\| \implies Px = x$.

32. Let (P_n) be an increasing sequence of orthogonal projections on a Hilbert space H. Show that (P_n) converges pointwise to an orthogonal projection P.

33. Let P, Q be orthogonal projection on a Hilbert space H. Show that:

(i) $P \leq Q \iff R(P) \subset R(Q) \iff QP = P \iff PQ = P$.

(ii) PQ is an orthogonal projection if and only if $PQ = QP$.

(iii) $Q - P$ is an orthogonal projection if and only if $PQ = P$.

(iv) $P + Q$ is an orthogonal projection if and only if $PQ = 0$.

34. Find the eigenvalues and eigenvectors of the following bounded linear operators on the space ℓ^1:
(i) a diagonal operator; (ii) the left shift; (iii) the right shift.

35. Let the space $X = C[a, b]$ be equipped with the uniform norm, and let T be the Fredholm integral operator on X with kernel k (see Example 9.7). Assume that the kernel k is *degenerate*, that is, that it can be written in the form

$$k(t, s) = \sum_{j=1}^{n} a_j(t) b_j(s),$$

where $\{a_1, \ldots, a_n\}$ and $\{b_1, \ldots, b_n\}$ are linearly independent elements of X. Show that each eigenfunction f of T corresponding to a nonzero eigenvalue is a linear combination $f = \xi_1 a_1 + \cdots + \xi_n a_n$.

36. Find the nonzero eigenvalues and corresponding eigenfunctions of a Fredholm integral operator T with kernel k if

  (i) $a = 0,\ b = 1,\ k(t, s) = t(1 + s)$

  (ii) $a = 0,\ b = \pi,\ k(t, s) = \sin t \cos s$

  (iii) $a = 0,\ b = 1,\ k(t, s) = e^{t+s}$

Suggestion. Note that the kernels are degenerate and use preceding problem.

37. Find the nonzero eigenfunctions and corresponding eigenvalues of the Fredholm integral operator

$$(Tf)(t) = \int_0^{\pi} (\cos^2 t \cos 2s + \cos 3t \cos^3 s) f(s)\, ds.$$

Suggestion. Problem 9.35.

38. Show that the dual of ℓ^1 is isometrically isomorphic to ℓ^∞.

39. Show that the dual of ℓ^p $(p > 1)$ is isometrically isomorphic to ℓ^q $(q = p/(p-1))$.

40. Show that, for each $f \in (c_0)'$, there is $a = (\alpha_n) \in \ell^1$ such that $f(x) = \sum_{n=1}^{\infty} \alpha_n \xi_n$. Show that the assignment $f \mapsto a$ is an isometric isomorphism of $(c_0)'$ onto ℓ_1.

41. For each $a = (\alpha_0, \alpha_1, \alpha_2, \ldots) \in \ell^1$ and $x = (\xi_1, \xi_2, \xi_3, \ldots) \in c$ define

$$f(x) = \alpha_0 \xi_0 + \sum_{n=1}^{\infty} \alpha_n \xi_n,$$

where $\xi_0 = \lim_{n \to \infty} \xi_n$. Show that f is a bounded linear functional on c. Show also that each bounded linear functional on c can be represented in this form for some $a \in \ell^1$, and that the assignment $a \mapsto f$ is an isometric isomorphism. *Suggestion.* Express each $x \in c$ as $x = \xi_0 e_0 + \sum_{n=1}^{\infty} (\xi_n - \xi_0) e_n$.

42. Show that each bounded linear operator A on ℓ^p $(p > 1)$ into ℓ^∞ is representable by an infinite matrix (α_{ij}) such that $c = \sup_i \left(\sum_{j=1}^{\infty} |\alpha_{ij}|^q \right)^{1/q} < \infty$, where $q = p/(p-1)$, with $y = Ax$ expressed as in the preceding problem. Show that $\|A\| = c$.

43. Show that every bounded linear operator A on $\mathbf{c}_0$ into ℓ^∞ determines and is determined by an infinite matrix (α_{ij}) such that $c = \sup_i \sum_{j=1}^\infty |\alpha_{ij}| < \infty$, and that $y = Ax$ is expressed by $\eta_i = \sum_{j=1}^\infty \alpha_{ij}\xi_j$, $i = 1, 2, \ldots$ Show that $\|A\| = c$.

44. (Harder.) Discuss representation of bounded linear operators in terms of infinite matrices:

(i)	on $\mathbf{c}_0$ into $\mathbf{c}$,	(ii)	on $\mathbf{c}_0$ into $\mathbf{c}_0$,
(iii)	on $\mathbf{c}$ into ℓ^∞,	(iv)	on $\mathbf{c}$ into $\mathbf{c}$,
(v)	on $\mathbf{c}$ into $\mathbf{c}_0$,	(vi)	on ℓ^1 into ℓ^∞,
(vii)	on ℓ^p into $\mathbf{c}$,	(viii)	on ℓ^p into $\mathbf{c}_0$.

(For details see the functional analysis textbook [43] by Taylor and Lay.)

45. If $B \in \mathcal{B}(X, Y)$ and $A \in \mathcal{B}(Y, Z)$, show that $(AB)' = B'A'$.

46. Show that if $A \in \mathcal{B}(X, Y)$ is invertible, then A' is also invertible.
Suggestion. Preceding problem.

47. *Invertible operator.* For any $A \in \mathcal{B}(X, Y)$ define the so-called *minimum modulus* of A by $g(A) = \inf_{x \neq 0} \|Ax\|/\|x\|$. Show that A is invertible if and only if A is surjective and $g(A) > 0$.
Suggestion. Show that $\|Ax\| \geq g(A)\|x\|$ for all $x \in X$. (Recall that a linear bijection has a linear inverse, which in general is not bounded.)

48. Let $A \in \mathcal{B}(X, Y)$. Show that if X is complete and $g(A) > 0$, then the range of A is closed. ($g(A)$ is defined in the preceding problem.)

49. Let X be a Banach space and let $\|A\| < 1$ for some $A \in \mathcal{B}(X)$. Show that the operator series $\sum_{n=0}^\infty A^n$ converges in the operator norm to $(I - A)^{-1}$.
Suggestion. Show that the series is norm-absolutely convergent and use the fact that $\mathcal{B}(X)$ is complete. Consider $(I - A)\sum_{n=0}^N A^n$.

50. *Canonical (James) embedding.* Given $x \in X$, define $\hat{x}$ to be the scalar valued function $\hat{x}(x') = x'(x)$ for all $x' \in X'$.

(i) Show that $\hat{x}$ is a bounded linear functional on X' with $\|\hat{x}\| = \|x\|$.

(ii) Let $\pi \colon X \to X''$ be the mapping $\pi(x) = \hat{x}$. Show that π is linear and isometric. Show that π is injective, but not necessarily surjective.

51. Let X be a Banach space and let $A \in \mathcal{B}(X)$. Prove that if A' is invertible, then A is also invertible.

52. If $T \in \mathcal{B}(X, Y)$, show that $x_n \rightharpoonup x$ in X implies $Tx_n \rightharpoonup Tx$ in Y. (Here $\rightharpoonup$ denotes the weak convergence.)
Suggestion. You need the adjoint operator.

53. Prove the Hahn–Banach theorem for the case of a complex normed space X.
Suggestion. (Sobzcyk and Bohnenblust.) The idea is that X can be also regarded

as a real normed space. Let f be a $\mathbb{C}$-linear functional on the complex space X. Show that $f(x) = g(x) - ig(ix)$, where $g = \operatorname{Re} f$ is an $\mathbb{R}$-linear functional on the real space X. Prove that $\|f\| = \|g\|$ by observing that $\|g\| \leq \|f\|$, and then showing that $|f(x)| = e^{-i\theta} f(x) = f(e^{-i\theta}x) = \operatorname{Re} f(e^{-i\theta}x) = g(e^{-i\theta}x) \leq \|g\| \, \|x\|$ (with a suitable θ).

54.   Let H be a Hilbert space, H' the dual of H, and $R\colon H' \to H$ the conjugate linear *Riesz mapping* of Example 9.37. Let A' and A^* be the (normed space) adjoint and Hilbert space adjoint of $A \in \mathcal{B}(H)$, respectively. Prove that

$$A' = R^{-1}A^*R.$$

Chapter 10

Self-adjoint Compact Operators

⋆ *This chapter contains more specialized material and can be omitted on first reading.* ⋆

An investigation of linear integral equations was one of the most important factors that stimulated the development of functional analysis. In particular, the work of Ivar Fredholm and David Hilbert at the beginning of the twentieth century helped to shape the face of modern analysis. A linear integral operator T acting on the Hilbert space $L^2(a, b)$ has a special kind of continuity: The weak convergence of a sequence (f_n) of L^2 functions implies the norm convergence of (Tf_n) in L^2, that is, $f_n \rightharpoonup f \implies Tf_n \to Tf$. Such an operator was called *completely continuous* (or 'vollstetig' in German). In Hilbert spaces this is equivalent to T being compact in the sense that T transforms every bounded set into a set with compact closure. In 1918 F. Riesz published a paper on compact operators that was to influence operator theory for decades. A highly readable account of the theory of integral equations and of compact operators is given by Riesz and Sz.-Nagy [40]. In discussion of the Sturm-Liouville systems we follow Pryce [38].

10.1   General theory

Throughout, H is a complex Hilbert space. Recall that a bounded linear operator $A : H \to H$ is called *self-adjoint* if

$$\langle Ax, y \rangle = \langle x, Ay \rangle \text{ for all } x \in H.$$

Recall that all the values $\langle Ax, x \rangle$ are real.

Let X be a normed space, and let $K = \{x \in X : \|x\| = 1\}$ be the closed unit sphere in X. An operator $T \in \mathcal{B}(X)$ is said to be *compact* if the set $\overline{T(K)}$ is compact. Compact operators are very important in applications, especially in the solution of differential equations, and in statistical and quantum mechanics. For instance, linear operators on finite dimensional spaces are compact, as are Fredholm integral operators discussed earlier, both on $C[a, b]$ and $L^2[a, b]$.

Theorem 10.1.   *Let A be a nonzero self-adjoint compact operator on H. Then* *there is an eigenvalue λ of A with $|\lambda| = \|A\|$. Any corresponding eigenvector x with $\|x\| = 1$ is a solution to the extremal problem*

$$\max_{\|u\|=1} |\langle Au, u\rangle|. \tag{10.1}$$

Proof. By Theorem 9.18, $\|A\| = \sup_{\|u\|=1} |\langle Au, u\rangle|$. By the definition of supremum, there is a sequence (x_n) in H with $\|x_n\| = 1$ and $|\langle Ax_n, x_n\rangle| \to \|A\|$. The real sequence $(\langle Ax_n, x_n\rangle)$ does not converge in general, but we can extract a monotonic subsequence $(\langle Ay_n, y_n\rangle)$ which converges to λ, where $|\lambda| = \|A\|$. The operator A is compact by hypothesis, and the vectors Ay_n lie in $\overline{A(K)}$; so there is a subsequence (Az_n) of (Ay_n) convergent to some element $w \in \overline{A(K)}$, that is, $Az_n \to w$. We show that $(\lambda I - A)z_n \to 0$:

$$\begin{aligned} 0 \leq \|(\lambda I - A)z_n\|^2 &= \|\lambda z_n - Az_n\|^2 = \lambda^2 - 2\lambda\langle Az_n, z_n\rangle + \|Az_n\|^2 \\ &\leq \lambda^2 - 2\lambda\langle Az_n, z_n\rangle + \|A\|^2 \\ &\to \lambda^2 - 2\lambda^2 + \|A\|^2 = \|A\|^2 - \lambda^2 = 0. \end{aligned}$$

So $\lambda z_n = (\lambda I - A)z_n + Az_n \to w$, and $Aw = \lim_{n \to \infty} A(\lambda z_n) = \lambda \lim_{n \to \infty} Az_n = \lambda w$. Setting $x = \|w\|^{-1}w$ completes the proof. $\qquad\square$

Theorem 10.2. *Let A be a nonzero self-adjoint compact operator on a Hilbert space H. Then H has an orthonormal basis consisting of eigenvectors of A.*

Proof. Let Λ be the set of all eigenvalues of A. By Theorem 10.1, Λ is nonempty. For each $\lambda \in \Lambda$ define $X_\lambda = N(\lambda I - A)$, that is, X_λ is the set consisting of the zero vector and of all eigenvectors corresponding to λ. For each $\lambda \in \Lambda$ choose an orthonormal basis B_λ for X_λ. Since $X_\lambda \perp X_\mu$ for distinct eigenvalues λ, μ of A, the set $B = \bigcup_{\lambda \in \Lambda} B_\lambda$ is orthonormal. Set $X = \overline{\mathrm{sp}\, B}$. We show that $X = H$.

From the definition of X and the self-adjointness of A we can deduce that $A(X) \subset X$ and $A(X^\perp) \subset X^\perp$. Let S be the restriction of A to $X^\perp$. Then S is a compact self-adjoint operator on $X^\perp$. Suppose that $S \neq 0$. By Theorem 10.1, there exists $\lambda \in \mathbb{C}$ and a nonzero vector $v \in X^\perp$ such that $Sv = \lambda v$ and $|\lambda| = \|S\| \neq 0$. Hence $Av = Sv = \lambda v$, which means that $\lambda \in \Lambda$, and $v \in X_\lambda \subset X$. From $v \in X^\perp \cap X$ we obtain a contradictory conclusion $v = 0$. This proves that $S = 0$.

Suppose that $X^\perp \neq \{0\}$. For each unit vector $w \in X^\perp$ we get $Aw = Sw = 0$. Again, $w \in X^\perp \cap X = \{0\}$, which contradicts the fact that w is a unit vector. This proves $X^\perp = \{0\}$, and $X = H$. $\qquad\square$

Theorem 10.3 (Spectral expansion). *Let A be a self-adjoint compact operator on a Hilbert space H. Let Λ be the set of all eigenvalues of A, and let $P(\mu)$ be the orthogonal projection of H onto the nullspace $N(\mu I - A)$ of $\mu I - A$ for each $\mu \in \Lambda$. Then*

$$Ax = \sum_{\mu \in \Lambda} \mu P(\mu)x, \quad x \in H. \tag{10.2}$$

Proof. By the preceding theorem there exists an orthonormal basis B of H consisting of eigenvectors of A. For each eigenvalue μ of A let $B(\mu)$ be the set of all $u \in B$ such that $u \in N(\mu I - A)$. Then for each $x \in H$,

$$
Ax = A\left(\sum_{u \in B} \langle x, u \rangle u\right) = \sum_{u \in B} \langle x, u \rangle Au = \sum_{\mu \in \Lambda} \sum_{u \in B(\mu)} \langle x, u \rangle Au
$$
$$
= \sum_{\mu \in \Lambda} \mu \sum_{u \in B(\mu)} \langle x, u \rangle u = \sum_{\mu \in \Lambda} \mu P(\mu) x. \qquad \square
$$

$$\square$$

Theorem 10.3 is one of the most important results about a self-adjoint compact operator A. It can be shown that the set Λ of all eigenvalues of A is countable, and that 0 is an accumulation point of the eigenvalues if the space H has infinite dimension. In general, 0 may or may not be an eigenvalue of a self-adjoint compact operator A.

10.2 Sturm-Liouville systems

The application of the general theory of self-adjoint compact operators to Sturm-Liouville systems presented in this section is based on treatment given in Pryce [38]. We are going to study the so-called Sturm-Liouville system, which consists of a second order differential equation

$$(py')' + (\lambda - q)y = 0 \qquad (10.3)$$

for $y \in C^2[a, b]$, with $p \in C^1[a, b]$ strictly positive, $q \in C[a, b]$ real, and λ a scalar parameter (possibly complex), together with the boundary conditions

$$a_1 y(a) + a_2 y'(a) = 0, \qquad (10.4)$$
$$b_1 y(b) + b_2 y'(b) = 0, \qquad (10.5)$$

where $(a_1, a_2) \neq (0, 0) \neq (b_1, b_2)$ (real constants).

We describe a technique which transforms the Sturm-Liouville system into an operator equation with a self-adjoint compact operator.

We define vector spaces that are going to be used in the sequel; $C^2[a, b]$ is the space of all twice continuously differentiable functions on $[a, b]$.

$$
\begin{aligned}
&Y = C^2[a, b], &\qquad &Y_0 = Y_1 \cap Y_2, \\
&Y_1 = \{y \in Y : y \text{ satisfies } (10.4)\}, &\qquad &X = C[a, b], \\
&Y_2 = \{y \in Y : y \text{ satisfies } (10.5)\}, &\qquad &U = L^2[a, b].
\end{aligned}
$$

Observe that $Y_0 \subset Y \subset X \subset U$.

Define a linear operator $L \colon Y \to X$ by

$$Ly = (-py')' + qy.$$

Then the Sturm-Liouville system (10.3)–(10.5) is equivalent to the equation

$$Ly = \lambda y, \qquad y \in Y_0. \tag{10.6}$$

We are going to construct the Green's function for the given Sturm-Liouville system. Let $u, v \in Y$ be such that

$$Lu = 0, \ u \in Y_1, \qquad\qquad Lv = 0, \ v \in Y_2.$$

The existence of such u, v is a standard result of the theory of ordinary differential equations, and can be proved by means of the contraction mapping theorem. The Green's function G of the Sturm-Liouville system is now defined by

$$G(t, s) = \begin{cases} v(t)u(s) & (s \leq t) \\ u(t)v(s) & (s > t) \end{cases}$$

for $(t, s) \in [a, b] \times [a, b]$. We can show that G is continuous on $[a, b] \times [a, b]$, and that

$$G(t, s) = G(s, t). \tag{10.7}$$

Let T be the Fredholm integral operator defined on U by

$$Tf(t) = \int_a^b G(t, s) f(s) \, ds, \quad t \in [a, b]. \tag{10.8}$$

Since G is continuous on the compact set $K = [a, b] \times [a, b]$ in the Euclidean plane, G is also an L^2 function on K, and T is a bounded linear operator $T \colon U \to U$ by general theory of the Fredholm integral operator. This operator is self–adjoint in view of equation (10.7). Moreover, it can be proved that T is compact. We summarize some facts about the Fredholm integral operator T in the following theorem:

Theorem 10.4. *Let T be the Fredholm integral operator defined by (10.8), where G is the Green's function of the Sturm-Liouville system (10.3)–(10.5). Then*

(i) *T is a compact self–adjoint linear operator on U to U.*

(ii) *The eigenvalues of T are nonzero, and form a sequence which converges to 0.*

(iii) *Each eigenvector of T lies in Y_0.*

(iv) *$f \in Y_0$ is an eigenvector of L with eigenvalue λ if and only if f is an eigenvector of T with eigenvalue $\mu = \lambda^{-1}$.*

Proof. (i) The compactness follows from general theory of Fredholm integral operators. T is self–adjoint in view of (10.7).

(ii) The statement about nonzero eigenvalues follows from the construction of T, the rest from general theory of self-adjoint compact operators described in Section 10.1: First we check that $g \in U$ implies $Tg \in X$. Then, for any $f \in X$, we prove that $y = Tf \in Y_0$ and $Ly = f$; this requires rewriting Tf in the form

$$Tf(t) = \int_a^t u(s)v(t)f(s)\,ds + \int_t^b v(s)u(t)f(s)\,ds.$$

We can in fact prove that $T_0 = T|_X$ is the inverse operator to $L_0 = L|_{Y_0}$.

Let f be an eigenvector of T with eigenvalue μ. Then $Tf = \mu f$. By the previous result, $y = Tf \in Y_0$ and $Ly = f = \mu^{-1}Tf = \lambda y$. This, together with $T_0 = L_0^{-1}$, completes the proof. $\square$

In the preceding argument we saw that $T_0 = T|_X$ is the inverse operator to $L_0 = L|_{Y_0}$. The differential operator L_0 is linear but unbounded, and its range is not contained in its domain. The operator T, which is an extension of the inverse operator T_0, is a much more amenable operator. This is the main reason for constructing the Green's function of the Sturm-Liouville system and the operator T.

Theorem 10.5 (Sturm-Liouville theorem). *Let S be the Sturm-Liouville system defined in Section* 10.1. *Then*

(i) *S has an infinite number of eigenvalues forming a sequence (λ_n) satisfying $|\lambda_n| \to \infty$.*

(ii) *Each eigenspace of S is one–dimensional.*

(iii) *The normalized eigenfunctions of S form a total orthonormal sequence in $L^2[a,b]$.*

Proof. Follows from the preceding theorem. $\square$

Example 10.6. Let us consider the Sturm-Liouville system on the interval $[0, \pi]$ induced by the differential equation

$$y'' + \lambda y = 0 \text{ for } t \in [0, \pi]$$

with the boundary conditions

$$y(0) = y(\pi) = 0.$$

The Green's function for this system is given by $G(t,s) = t - st/\pi$ if $s \leq t$ and $G(t,s) = G(s,t)$ otherwise. We can find the eigenvalues and eigenvectors of the Fredholm integral operator T with kernel G: The eigenvalues are $\lambda_n = n^2$ with the corresponding eigenvectors $s_n(t) = \sqrt{2/\pi} \sin nt$. These vectors form a total orthonormal sequence in $L^2[0, \pi]$.

10.3   Problems for Chapter 10

1. Let A be a self-adjoint compact operator on a Hilbert space H. If λ is a nonzero complex number such that $\lambda I - A$ is injective, prove that $\lambda I - A$ is invertible in $\mathcal{B}(H)$.

2. Let A be a self-adjoint operator on a Hilbert space H. If λ is a complex number such that $\lambda I - A$ is surjective, prove that $\lambda I - A$ is invertible in $\mathcal{B}(H)$. *Suggestion.* $H = N(\lambda I - A) \oplus^{\perp} \overline{R(\lambda I - A)}$.

3. Let A be a self-adjoint compact operator on a Hilbert space H. If λ is a nonzero eigenvalue of A, show that the nullspace $N(\lambda I - A)$ is finite dimensional.

Chapter 11

Introductory Functional Analysis

Functional analysis could be broadly described as a topological study of normed spaces. It originated at the end of the 19th century with the work of Ivar Fredholm and Vito Volterra on integral equations, and continued into the twentieth century with the research of David Hilbert, Frigyes Riesz and many others. It is characterized by abstract view point which encompasses ideas of linear algebra, linear differential and integral equations, approximation theory, etc. In effect, a study of bounded linear operators and functionals presented in previous chapters is a part of functional analysis. In the 1920s a group of mathematicians loosely centred about the prestigious journal *Studia Mathematica* which was founded by the famous Polish mathematician Stefan Banach concentrated their efforts on deep properties of complete normed spaces, which were later named Banach spaces after the leading investigator in the field. This period of intensive and rapid development culminated in 1932 in the publication of Banach's celebrated treatise [3], which could be said to define the field of functional analysis.

Banach's book presents the three basic principles of functional analysis, the Hahn-Banach theorem (Hahn 1927, Banach 1929), the uniform boundedness principle (Hahn 1922, Hildebrandt 1923, Banach and Steinhaus 1927), and the open mapping theorem (Banach 1929 and Schauder 1930). Consequences of these principles include the Banach-Steinhaus theorem, the Banach inverse mapping theorem and the closed graph theorem. We have already encountered the Hahn-Banach theorem, and so in this chapter we discuss the uniform boundedness principle and the open mapping theorem. Proofs of both of these results require the so called Baire's theorem.

René-Louis Baire (1874–1932)
Ivar Fredholm (1866–1927)

Vito Volterra (1860–1940)

David Hilbert (1862–1943)

Frigyes Riesz (1880–1956)

Stefan Banach (1892–1945)

Hans Hahn (1874–1932)

Hugo Steinhaus (1887–1972)

Juliusz Schauder (1899–1943)

11.1   Uniform boundedness principle

Theorem 11.1 (Baire's theorem). *Let a complete metric space (S, d) be a countable union $S = \bigcup_{n=1}^{\infty} A_n$ of closed sets A_n. Then one of the sets A_n has nonempty interior.*

Proof.   Assume that each set A_n has empty interior. This means that any ball in S meets the complement A_n^c of A_n; A_1^c is then nonempty and open. Hence there is a ball $B_1 = B(x_1; r_1) \subset A_1^c$ with $r_1 < \frac{1}{2}$.

The set $B(x_1; \frac{1}{2}r_1) \cap A_2^c$ is nonempty and open. There exists a ball

$$B_2 = B(x_2; r_2) \subset B(x_1; \tfrac{1}{2}r_1) \cap A_2^c, \quad r_2 < \tfrac{1}{2}r_1.$$

Proceeding by induction, we obtain a sequence of balls $B_n = B(x_n; r_n)$ such that for all n

$$B_{n+1} \subset B(x_n; \tfrac{1}{2}r_n) \cap A_{n+1}^c \subset B_n, \quad r_{n+1} < \tfrac{1}{2}r_n.$$

Since $r_n < (\frac{1}{2})^n$, the sequence (x_n) is Cauchy, and therefore convergent to some point x in the complete space S. Let $n \in \mathbb{N}$ be fixed. There exists $k \in \mathbb{N}$ such that $k > n$ and $d(x, x_k) < \frac{1}{2}r_n$. So

$$d(x, x_n) \leq d(x, x_k) + d(x_k, x_n) < \tfrac{1}{2}r_n + \tfrac{1}{2}r_n = r_n,$$

that is, $x \in B_n$. Therefore

$$x \in \bigcap_{n=1}^{\infty} B_n \subset \bigcap_{n=1}^{\infty} A_n^c = \left(\bigcup_{n=1}^{\infty} A_n \right)^c = S^c = \emptyset,$$

which is a contradiction.   □

The theorem is named after René-Louis Baire who studied sets of real numbers in the paper [2] published in 1899. The theorem is often called the Baire category theorem. The metric space formulation of the theorem appeared in 1914 in Hausdorff [22].

Theorem 11.2 (Uniform boundedness principle). *Let X be a Banach space and $\{T_\alpha : \alpha \in D\}$ a family of bounded linear operators on X to a normed space Y. Suppose that the family is pointwise bounded, that is, for each fixed $x \in X$,*

$$\sup_{\alpha \in D} \|T_\alpha x\| = M(x) < \infty.$$

Then the family is uniformly bounded, that is,

$$\sup_{\alpha \in D} \|T_\alpha\| = M < \infty.$$

Proof. For each integer k define

$$A_k = \{x \in X : \|T_\alpha x\| \le k \text{ for all } \alpha \in D\}.$$

Each set A_k is closed (exercise). Given $x \in X$, by hypothesis there is k such that $\|T_\alpha x\| \le M(x) \le k$ for all $\alpha \in D$. So $X = \bigcup_{k=1}^{\infty} A_k$. By Baire's theorem, one of these sets, say A_m, contains an open ball $B_0 = B(x_0; r)$.

Let x be an arbitrary nonzero vector in X. For a sufficiently small $\lambda > 0$, say $\lambda = r/(2\|x\|)$, $z = x_0 + \lambda x \in B_0$. Write $\mu = \lambda^{-1}$. Then

$$\|T_\alpha x\| = \|T_\alpha(\mu(z - x_0))\| = \mu\|T_\alpha z - T_\alpha x_0\|$$

$$\le \mu(\|T_\alpha z\| + \|T_\alpha x_0\|) \le 2\mu m = \frac{4m}{r}\|x\|$$

for all $\alpha \in D$. Hence $\|T_\alpha\| \le M$, where $M = 4m/r$. $\qquad\square$

Theorem 11.3 (Banach–Steinhaus theorem). *Let X be a Banach space and let (T_n) be a sequence of bounded linear operators on X to a normed space Y which is pointwise convergent on X. Then*

$$Tx = \lim_{n \to \infty} T_n x, \quad x \in X,$$

defines a bounded linear operator on X.

Proof. The linearity of T follows from the linearity of the T_n and the linearity of the limit operation. The family $\{T_n : n \in \mathbb{N}\}$ is pointwise bounded (being pointwise convergent), and so it is uniformly bounded by the preceding theorem, say $\|T_n\| \le M$ for all $n \in \mathbb{N}$. Then

$$\|Tx\| \le \|T_n x\| + \|Tx - T_n x\| \le M\|x\| + \|Tx - T_n x\|$$

for all n; taking the limit as $n \to \infty$, we get $\|Tx\| \le M\|x\|$. $\qquad\square$

11.2   The open mapping theorem

Before stating our first result, it will be convenient to introduce the following notation. Let $A, B \subset X$, and let α be a scalar. We define

$$A + B = \{x + y : x \in A, \ y \in B\}, \quad \alpha A = \{\alpha x : x \in A\};$$

$A + (-B)$ will be written as $A - B$. If $B = \{w\}$, we write $A + w$ in place
of $A + \{w\}$. It can be proved that $\overline{A + w} = \overline{A} + w$, $\overline{\alpha A} = \alpha \overline{A}$. This
notation enables us to express any ball in X in terms of the open unit ball
$U = \{x \in X : \|x\| < 1\}$:

$$B(x; r) = x + rU.$$

Lemma 11.4. *Let T be a bounded linear surjection between Banach spaces
X and Y; let U, V be the open unit balls in X, Y, respectively. Then the
set TU contains an open ball αV for some $\alpha > 0$.*

Proof.   Since $X = \bigcup_{n=1}^{\infty} nU$ and $Y = TX$, we have $Y = \bigcup_{n=1}^{\infty} nTU = \bigcup_{n=1}^{\infty} n\overline{TU}$. By Baire's theorem, one of the sets $n\overline{TU}$ contains a ball, say
$y_0 + \rho V \subset p\overline{TU}$. Then

$$\rho V \subset p\overline{TU} - y_0 \subset p\overline{TU} - p\overline{TU} \subset 2p\overline{TU}.$$

If $r = \rho/2p$, then

$$rV \subset \overline{TU}.$$

The proof will be completed when we show that

$$\tfrac{1}{2}rV \subset TU.$$

Let $y \in \tfrac{1}{2}rV$. Since $y \in \tfrac{1}{2}\overline{TU}$, there is $x_1 \in \tfrac{1}{2}U$ such that $\|y - Tx_1\| < (\tfrac{1}{2})^2 r$. Similarly, as

$$y - Tx_1 \in (\tfrac{1}{2})^2 rV \subset (\tfrac{1}{2})^2 \overline{TU} = \overline{T((\tfrac{1}{2})^2 U)},$$

we can choose $x_2 \in (\tfrac{1}{2})^2 U$ such that $\|(y - Tx_1) - Tx_2\| < (\tfrac{1}{2})^3 r$. Pro-
ceeding by induction, we obtain a sequence (x_n) such that $x_n \in (\tfrac{1}{2})^n U$
and

$$\left\| y - \sum_{i=1}^{n} Tx_i \right\| < (\tfrac{1}{2})^{n+1} r.$$

Then $y = \sum_{i=1}^{\infty} Tx_i$. The series $\sum_{n=1}^{\infty} x_n$ is norm-absolutely convergent
and, since X is complete, it is convergent in X, say $\sum_{n=1}^{\infty} x_n = x$. By the
continuity of T,

$$y = \sum_{i=1}^{\infty} Tx_i = T\left(\sum_{i=1}^{\infty} x_i \right) = Tx.$$

Further, $\|x\| \leq \sum_{n=1}^{\infty} \|x_n\| < \sum_{n=1}^{\infty} (\tfrac{1}{2})^n = 1$ and $y = Tx \in TU$. $\qquad\square$

Theorem 11.5 (Open mapping theorem). *A bounded linear surjection $T\colon X \to Y$ between Banach spaces is an open mapping, that is, TG is open in Y whenever G is open in X.*

Proof. Let U, V be the open unit balls in X, Y, respectively. If G is an open set in X and $y \in TG$, then $y = Tx$ for some $x \in G$ and $x + \rho U \subset G$ for some ρ. By the preceding lemma, $TU \supset \alpha V$ for some $\alpha > 0$. Then

$$y + \rho\alpha V \subset y + \rho TU = Tx + \rho TU = T(x + \rho U) \subset TG,$$

which shows that TG is open. □

We now turn to two important consequences of the open mapping theorem.

Theorem 11.6 (Banach's inverse mapping theorem).
A bounded linear bijection $T\colon X \to Y$ between Banach spaces is a homeomorphism.

Proof. By the previous theorem, T is an open mapping. Let $S\colon Y \to X$ be the inverse of T. Then S is linear. If G is open in X, then $S^{-1}(G) = TG$ is open in Y, and so S is continuous. □

For any mapping $f\colon X \to Y$ between two sets, the graph $\Gamma(f)$ is the subset of the cartesian product $X \times Y$ defined by

$$\Gamma(f) = \{(x, f(x)) : x \in X\}.$$

If the spaces X, Y are each equipped with a topology, then we consider $\Gamma(f)$ as a subset of the product space $X \times Y$. The product topology for normed spaces X, Y can be generated by the norm $\|(x, y)\| = \|x\| + \|y\|$. Suppose that $T\colon X \to Y$ is a linear operator between normed spaces whose graph is closed. This can be expressed in terms of sequential convergence as follows:

$$(x_n, Tx_n) \to (x, y) \implies y = Tx,$$

or,

$$x_n \to x \text{ and } Tx_n \to y \implies y = Tx.$$

We observe that $\Gamma(T)$ is a normed subspace of the product space $X \times Y$. A linear operator whose graph is closed is usually called a *closed linear operator*.

Theorem 11.7 (The closed graph theorem). *If $T\colon X \to Y$ is a closed linear operator between Banach spaces, then T is bounded.*

Proof.   Define a mapping $A\colon \Gamma(T) \to X$ by $A(x, Tx) = x$ for all $(x, Tx) \in \Gamma(T)$. Then A is linear (exercise) and bounded:

$$\|A(x, Tx)\| = \|x\| \le \|x\| + \|Tx\| = \|(x, Tx)\|.$$

In fact, A is a bijection (exercise). Since the normed spaces X, Y are complete, their product space $X \times Y$ is also complete, and $\Gamma(T)$ is complete being a closed subspace of a complete space. Thus we can apply the preceding inverse mapping theorem to A to conclude that A^{-1} is continuous and hence bounded. So, for every $x \in X$,

$$\|Tx\| \le \|x\| + \|Tx\| = \|(x, Tx)\| = \|A^{-1}x\| \le \|A^{-1}\| \, \|x\|. \qquad \square$$

It is interesting to note that the last three theorems—the deepest of all the results so far encountered in functional analysis—are in fact all equivalent (see Problem 11.16).

11.3   Problems for Chapter 11

1.   Let (α_n) be a complex sequence such that $\sum_{n=1}^{\infty} \alpha_n \xi_n$ converges for every complex sequence (ξ_n) with $\lim_{n \to \infty} \xi_n = 0$. Show that $\sum_{n=1}^{\infty} |\alpha_n|$ converges.

Suggestion. Uniform boundedness principle.

2.   Let $1 \le p < \infty$ and let (α_n) be a complex sequence such that $\sum_{n=1}^{\infty} \alpha_n \xi_n$ converges for every complex sequence (ξ_n) with $\sum_{n=1}^{\infty} |\xi_n|^p$ convergent.

   (i)   If $p = 1$, show that (α_n) is bounded.

   (ii)   If $p > 1$, show that $\sum_{n=1}^{\infty} |\alpha_n|^q$ converges, where $q = p/(p-1)$.

Suggestion. Uniform boundedness principle.

3.   Let A be a subset of a complex normed space X such that, for each $x' \in X'$, $x'(A)$ is a bounded subset of the complex plane. Show that A is a bounded subset of X.

4.   Show that the hypothesis that X is complete cannot be dropped from the uniform boundedness principle. For a counterexample consider the subspace X of ℓ^∞ consisting of all $x = (\xi_1, \xi_2, \dots)$ such that only finite many of the ξ_i are nonzero. Define (bounded linear) operators $T_n \colon X \to X$ by

$$T_n(\xi_1, \xi_2, \xi_3, \dots) = (\xi_1, 2\xi_2, 3\xi_3, \dots, n\xi_n, 0, 0, \dots).$$

Show that the family $\{T_n\}$ is pointwise, but not uniformly, bounded.

5.   *Toeplitz theorem.* Let $A = [\alpha_{ij}]$ be an infinite matrix $(i, j = 1, 2, \dots)$. A sequence (ξ_j) of scalars is *A-summable* to ξ if the series $\sum_{j=1}^{\infty} \alpha_{ij} \xi_j$ converges to η_i

for $i = 1, 2 \ldots$ and the sequence (η_i) converges to ξ. Prove that every convergent sequence with the limit ξ is A-summable to ξ if and only if the following are true:

(i) $\sum_{j=1}^{\infty} |\alpha_{ij}| \leq M$ for $i \in \mathbb{N}$;

(ii) $\lim_{i \to \infty} \alpha_{ij} = 0$ for $j \in \mathbb{N}$;

(iii) $\lim_{i \to \infty} \sum_{j=1}^{\infty} \alpha_{ij} = 1$.

Suggestion. Uniform boundedness principle.

6. If $x_n \rightharpoonup x$, then $\|x\| \leq \liminf_{n \to \infty} \|x_n\|$.

7. Let (x_n), (y_n) be sequences in a Hilbert space H. Prove that

$$x_n \rightharpoonup x \text{ and } y_n \to y \implies \langle x_n, y_n \rangle \to \langle x, y \rangle.$$

(Here (x_n) converges weakly and (y_n) strongly.) Generalize to normed spaces.

8. Let X be a normed space and N a closed subspace of X. Let the quotient space X/N be equipped with the quotient norm $\|x + N\| = \text{dist}(x, N)$. If $\pi \colon X \to X/N$ is the quotient map defined by $\pi(x) = x + N$, prove that π is linear and continuous. Show from first principles that π is an open mapping.

Suggestion. For the open mapping property observe that $\pi(B_X(0; r)) = B_{X/N}(0; r)$ and show that a set U is open in X/N if and only if $\pi^{-1}(U)$ is open in X.

9. Let $T \colon X \to Y$ be a bounded linear operator between Banach spaces with a closed range $R(T)$. Prove that there is a constant $c > 0$ such that for each $y \in R(T)$ there is $x \in X$ with $y = Tx$ and $\|x\| \leq c\|y\|$.

10. Let $T \colon X \to Y$ be a bounded linear injection between Banach spaces, and let $S \colon X \to R(T)$ be the codomain restriction of T. (Note that S is surjective while T may not be.) Show that S has a continuous inverse if and only if $R(T)$ is closed in Y.

11. Let X be a vector space complete with respect to norms $\| \ \|$ and $\| \ \|_1$. Suppose $\|x_n\| \to 0$ implies $\|x_n\|_1 \to 0$ for all sequences (x_n) in X. Prove that the two norms are equivalent, that is, there are constants α, β such that

$$\alpha\|x\| \leq \|x\|_1 \leq \beta\|x\| \text{ for all } x \in X.$$

Suggestion. Inverse mapping theorem.

12. Let $\sum_{i=1}^{\infty} |\alpha_i| < \infty$ with $\alpha_i \neq 0$ for all i, and let T be the (bounded linear) operator on ℓ^{∞} given by $Tx = T(\xi_1, \xi_2, \ldots) = (\alpha_1 \xi_1, \alpha_2 \xi_2, \ldots)$. Prove that T cannot be surjective.

Suggestion. Inverse mapping theorem.

13. (i) Use Problem 9.7 to show that the hypothesis that the initial space X is Banach cannot be dropped in the closed graph theorem.

(ii) Show that the hypothesis that the target space Y is Banach cannot be dropped in the closed graph theorem.

Suggestion. (ii) Let $(X, \|\cdot\|)$ be a separable Banach space and $\{x_\alpha : \alpha \in D\}$ a Hamel basis for X, normalized so that $\|x_\alpha\| = 1$.

14. Let H be a Hilbert space and T, S two linear (not necessarily bounded) operators on H to H. Suppose that, for any $x, y \in H$,

$$\langle Tx, y \rangle = \langle x, Sy \rangle.$$

Prove that T (and S) is a bounded linear operator on H.

Suggestion. Closed graph theorem.

15. Let X be a Banach space and let $X = M \oplus N$ with the subspaces M and N closed. Let P be the projection operator onto M along N. Prove that P is a bounded linear operator.

16. Prove this interesting result: The open mapping theorem (OMT), the Banach inverse mapping theorem (IMT) and the closed graph theorem (CGT) are all equivalent. We have proved that OMT implies IMT, which in turn implies CGT. Assuming that X, Y are Banach spaces, prove the following:

CGT $\implies$ IMT: Let $T \in \mathcal{B}(X, Y)$ be bijective, and let $S \colon Y \to X$ be the algebraic inverse of T. Show that S is linear and closed (by considering the graph of S).

IMT $\implies$ OMT: Let $T \in \mathcal{B}(X, Y)$ be surjective and let $N = N(T)$ (a closed subspace of X). Define $A \colon X/N \to Y$ by $A(x + N) = Tx$, $x \in X$. Show that A is a bijective linear operator between Banach spaces X/N and Y. Let $\pi \colon X \to X/N$ be the quotient map (see Problem 11.8); then $T = A\pi$. If $G \subset X$ is open, consider

$$T(G) = A\pi(G) = (A^{-1})^{-1}(\pi(G))$$

taking into account that π is an open mapping.

PART 3
Integration

Chapter 12

Measure Spaces

As we know from elementary calculus, the definite integral can be used to calculate the area of a plane figure or a volume of a solid in the three dimensional space. The integral is then a real valued function whose argument is a set. Such functions are called *set functions*. Even before we can evaluate integrals, we may be able to measure the length or the area or the volume of certain sets. A set function μ that behaves in a sensible way on a family of 'measurable' sets will be called a *measure*.

But what do we mean by a 'sensible' behaviour? For example, we want the measure to be nonnegative, and to obey the additivity requirement

$$\mu(A \cup B) = \mu(A) + \mu(B)$$

if the sets A, B are measurable and disjoint. This condition already imposes some restrictions on the family of measurable sets. We need $A \cup B$ measurable when A, B are measurable, otherwise we could not calculate the measure of $A \cup B$.

In this chapter we study families of measurable sets, the so called σ-algebras, and set functions, or measures, on these σ-algebras that obey certain basic rules including the countable additivity. We need to study measurable sets and measures in sufficient abstraction that we can define the abstract Lebesgue integral, which is required in various applications, notably in probability theory. Restricting our study only to the real line or finite dimensional Euclidean spaces would deprive us from the use of the integral in these powerful applications.

12.1   Measures and measure spaces

Let S be a nonempty set. As we will be mostly dealing with subsets of S, we will write A^c for the complement of a set $A \subset S$ in S, that is, $A^c = \{s \in S : s \notin A\}$.

Definition 12.1. A *σ-algebra* in S is a family Σ of subsets of S which has the following properties:

   (i) $\emptyset \in \Sigma$.

   (ii) If A is an element of Σ, then A^c is also an element of Σ.

   (iii) If (A_n) is a sequence of elements of Σ, then their union $\bigcup_{n=1}^{\infty} A_n$ is also an element of Σ.

Let us add that an *algebra* satisfies conditions (i) and (ii) of Definition 12.1, but (iii) is satisfied only for finite unions. The 'σ' in 'σ-algebra' refers to countably infinite unions. In measure theory the prefix 'σ' is often used to describe countable infinity.

Definition 12.2. A *measurable space* is a pair (S, Σ), where S is a nonempty set and Σ is a σ-algebra in S. A set $A \subset S$ is called *measurable* (or more precisely Σ-measurable) if $A \in \Sigma$.

From properties (i) and (ii) of a σ-algebra we deduce that the set S itself is measurable. It is also true that the intersection of a sequence of measurable sets is a measurable set (Problem 12.1), and that a finite union and a finite intersection of measurable sets are measurable sets (Problem 12.2).

Example 12.3. Let S be an arbitrary nonempty set. The family $\{\emptyset, S\}$ is the smallest possible σ-algebra in S.

Example 12.4. The family $\mathcal{P}(S)$ of all subsets of S is the largest possible σ-algebra in S. ($\mathcal{P}(S)$ is the so-called *power set* of S.)

Definition 12.5. Let $\mathcal{F}$ be a nonempty family of subsets of S. Then there is a smallest σ-algebra $\mathcal{A}$ in S which contains $\mathcal{F}$, called the σ-algebra *generated by* $\mathcal{F}$. ($\mathcal{A}$ can be taken to be the intersection of all σ-algebras in S containing $\mathcal{F}$. This intersection is nonempty as $\mathcal{P}(S)$ is one such σ-algebra.)

Definition 12.6. Let $(S, \mathcal{T})$ be a topological space. The σ-algebra generated by the family of all open sets in S is called the *Borel σ-algebra* of S, written $\mathcal{B}(S)$. Any set belonging to this σ-algebra is called a *Borel set* in S.

Borel sets are very important in measure theory, so we should be able to recognize them in some standard spaces. Remember that a complement of a Borel set is a Borel set, and that the union (and intersection) of a sequence of Borel sets is again a Borel set. It is not hard to show that any interval is a Borel set in the Euclidean space $\mathbb{R}$. On the real line, the σ-algebra of Borel sets is generated by the family of all semiclosed intervals $(a, b]$ (see Problem 12.5).

Example 12.7. The set $A = [0, 1) \times [0, 1)$ is a Borel set in the Euclidean space $\mathbb{R}^2$. We observe that A is neither closed nor open. Define

$$B = \{(\xi_1, \xi_2) : \xi_1 \geq 0 \text{ and } \xi_2 \geq 0\}, \qquad C = \{(\xi_1, \xi_2) : \xi_1 < 1 \text{ and } \xi_2 < 1\}.$$

Then B is closed and C is open in $\mathbb{R}^2$ (sketch); hence B and C are Borel sets in $\mathbb{R}^2$, and so is A as $A = B \cap C$.

Example 12.8. The set $A = \{1/n : n \in \mathbb{N}\}$ is a Borel set in $\mathbb{R}$.

Recall that A is not a closed set, and that its closure is the set $\overline{A} = A \cup \{0\}$. However, $A = \bigcup_{n=1}^{\infty} \{1/n\}$, where each set $\{1/n\}$ is closed, and hence Borel, in $\mathbb{R}$. So A is Borel.

Before we introduce positive measures, we adjoin an additional element to the set $\mathbb{R}_+$ of all nonnegative real numbers. It is customary to denote this element by ∞. We extend the ordering on reals by stipulating that $a < \infty$ for each $a \in \mathbb{R}_+$. We define the interval $[0, \infty]$ by

$$[0, \infty] = \mathbb{R}_+ \cup \{\infty\} = \{x \in \mathbb{R} : x \geq 0\} \cup \{\infty\}.$$

We extend addition and multiplication to $[0, \infty]$ by stipulating

$$\infty \cdot 0 = 0 \cdot \infty = 0,$$

$$\infty \cdot a = a \cdot \infty = \infty \quad \text{if} \quad a \neq 0,$$

$$\infty + a = a + \infty = \infty \quad \text{if} \quad 0 \leq a \leq \infty.$$

Metrically, the set $[0, \infty]$ is the completion of the set $\{x \in \mathbb{R} : x \geq 0\}$ in the metric space $(\mathbb{R}, \rho)$, where $\rho(x, y) = |f(x) - f(y)|$ and $f(x) = x/(1 + |x|)$. Recall that the metric ρ is equivalent to the Euclidean metric $d(x, y) = |x - y|$ on $\mathbb{R}$; this means that the convergence with respect to ρ and d is the same.

Definition 12.9. Let (S, Σ) be a measurable space with Σ the family of measurable sets. A *positive measure* (or just *measure*) on (S, Σ) is a set function $\mu : \Sigma \to [0, \infty]$ such that

(i) $\mu(\emptyset) = 0$,

(ii)   for any sequence (A_n) of measurable mutually disjoint sets $(A_i \cap A_j = \emptyset$ when $i \neq j)$ we have

$$\mu\left(\bigcup_{n=1}^{\infty} A_n\right) = \sum_{n=1}^{\infty} \mu(A_n).$$

Property (ii) is usually described by saying that μ is a *countably additive set function* on Σ. Note that ∞ is allowed as a summand; the sum is then ∞.

We have to exercise caution when dealing with equations which may contain infinity. There are two important rules to remember for such equations:

(i)   *No subtractions in the possible presence of infinity.*

(ii)   *No cancellations in the possible presence of infinity.*

Example 12.10. *Dirac measure.* Let S be a nonempty set and let a be a selected point of S. For any subset A of S we define $\mu(A) = 1$ if $a \in A$, and $\mu(A) = 0$ otherwise. Then μ is a positive measure on the measurable space $(S, \mathcal{P}(S))$; this measure is usually known as the *Dirac measure* concentrated at the point a.

Example 12.11. *Counting measure.* On any nonempty set S we can define a *counting measure* ν by setting $\nu(A)$ to be equal to the number of elements of A if $A \subset S$ is finite, and to ∞ if A is infinite. Then ν is a positive measure on the measurable space $(S, \mathcal{P}(S))$.

In the following three examples we give descriptive definitions of the Lebesgue measure on $\mathbb{R}$ and $\mathbb{R}^n$, and the Lebesgue–Stieltjes measures on $\mathbb{R}$. Most of the properties of these measures can be deduced from their respective definitions. For each of these examples we need a theorem (printed in italics) which will be proved in Chapter 15.

Example 12.12. *Lebesgue measure in* $\mathbb{R}$. Consider the measurable space $(\mathbb{R}, \mathcal{B})$, where $\mathcal{B}$ is the σ-algebra of all Borel sets on $\mathbb{R}$.
There exists a unique positive measure m on $\mathcal{B}$ which coincides with Euclidean length on semiclosed intervals, that is

$$m((a, b]) = b - a$$

for each interval $(a, b]$ in $\mathbb{R}$.

This unique positive measure on $(\mathbb{R}, \mathcal{B})$ is known as the *Lebesgue measure* on the real line; its construction is discussed in Chapter 15.

The uniqueness means that if μ is another measure on $(\mathbb{R}, \mathcal{B})$ such that $\mu((a, b]) = b - a$ for each interval $(a, b]$, then $\mu = m$.

Example 12.13. *Lebesgue measure in $\mathbb{R}^k$. There exists a unique positive measure m on the σ-algebra $\mathcal{B}_k = \mathcal{B}(\mathbb{R}^k)$ of all Borel sets in $\mathbb{R}^k$ which coincides with the Euclidean volume on semiclosed k-cells, that is,*

$$m(A) = m((a_1, b_1] \times (a_2, b_2] \times \cdots \times (a_k, b_k]) = (b_1 - a_1)(b_2 - a_2) \cdots (b_k - a_k)$$

for each semiclosed cell A.

This unique positive measure is called the *Lebesgue measure on $\mathbb{R}^k$*. (Semiclosed cells are chosen here because they form a semiring of sets. See Chapter 15.)

Example 12.14. *Lebesgue–Stieltjes measures. Let $g : \mathbb{R} \to \mathbb{R}$ be an increasing function which is right continuous at each point $x \in \mathbb{R}$ (that is, $g(x+) = g(x)$ at each point $x \in \mathbb{R}$). There exists a unique positive measure m_g on the σ-algebra $\mathcal{B}(\mathbb{R})$ such that*

$$m_g((a, b]) = g(b) - g(a)$$

for each interval $(a, b]$ in $\mathbb{R}$.

This unique positive measure is known as the *Lebesgue–Stieltjes measure* generated by g; its construction will be discussed in Chapter 15.

Definition 12.15. A *measure space* is a triple (S, Σ, μ), where S is a nonempty set, Σ a σ-algebra of subsets of S, and μ a positive measure on Σ. The sets in Σ are called *measurable* (Σ-*measurable*) *sets*.

We now define some set theoretical concepts that are often used in measure theory. We say that a sequence (A_n) of subsets of S is *expanding* if $A_n \subset A_{n+1}$ for all $n \in \mathbb{N}$. If $A = \bigcup_{n \in \mathbb{N}} A_n$, we write

$$A = \lim_{n \to \infty} A_n, \quad \text{or} \quad A_n \nearrow A.$$

Similarly, we say that (A_n) is *contracting* if $A_n \supset A_{n+1}$ for all $n \in \mathbb{N}$. If $A = \bigcap_{n \in \mathbb{N}} A_n$, we write

$$A = \lim_{n \to \infty} A_n, \quad \text{or} \quad A_n \searrow A.$$

In fact, the limit of a sequence of sets can be defined in full generality as follows: First we define $\limsup_{n \to \infty} A_n$ and $\liminf_{n \to \infty} A_n$, and then say that the sequence (A_n) converges to the limit A if

$$\limsup_{n \to \infty} A_n = A = \liminf_{n \to \infty} A_n.$$

The upper and lower limits of (A_n) are defined by

$$\limsup_{n \to \infty} A_n = \bigcap_{n=1}^{\infty} \bigcup_{k=n}^{\infty} A_k, \qquad \liminf_{n \to \infty} A_n = \bigcup_{n=1}^{\infty} \bigcap_{k=n}^{\infty} A_k.$$

Theorem 12.16. *Let (S, Σ, μ) be a measure space. Then:*

(i)   *If A, B are measurable and $A \subset B$, then $\mu(A) \leq \mu(B)$.*

(ii)   *If (A_n) is an expanding sequence of measurable sets with $A_n \nearrow A$, then A is measurable and $\mu(A_n) \to \mu(A)$.*

(iii)   *If (A_n) is a contracting sequence of measurable sets satisfying $A_n \searrow A$ and if $\mu(A_k) < \infty$ for some $k \in \mathbb{N}$, then A is measurable and $\mu(A_n) \to \mu(A)$.*

Proof.   See Problems 12.11 and 12.14.   □

Example 12.17. Let $(\mathbb{R}, \mathcal{B}, m)$ be the measure space from Example 12.12; $\mathcal{B}$ is the σ-algebra of Borel sets, and m is the Lebesgue measure. A point set $\{a\}$ is measurable, that is, a Borel set. We show that

$$m(\{a\}) = 0.$$

Indeed, by Theorem 12.16,

$$m(\{a\}) = \lim_{n \to \infty} m((a - 1/n, a]) = \lim_{n \to \infty} (a - (a - 1/n)) = \lim_{n \to \infty} (1/n) = 0.$$

Example 12.18. Let $g : \mathbb{R} \to \mathbb{R}$ be an increasing right continuous function and let m_g be the Lebesgue–Stieltjes measure from Example 12.14. Then

$$m_g(\{a\}) = g(a) - g(a-).$$

Indeed,

$$m_g(\{a\}) = \lim_{n \to \infty} m_g((a - 1/n, a]) = \lim_{n \to \infty} (g(a) - g(a - 1/n)) = g(a) - g(a-).$$

This shows that a singleton $\{a\}$ may have nonzero measure provided the function g is not continuous at a. As an example consider the function g defined on $\mathbb{R}$ by $g(x) = 0$ if $x < 0$ and $g(x) = 1$ if $x \geq 0$. Then g is increasing and right continuous with $m_g(\{0\}) = g(0) - g(0-) = 1 - 0 = 1$.

Definition 12.19. Let μ be a positive measure on a σ-algebra Σ.

(i) μ is *finite* if it does not take value ∞. (Since $\mu(A) \leq \mu(S)$ for every measurable set A, μ is finite if and only if $\mu(S) < \infty$; a finite measure μ is in fact bounded.)

(ii) μ is *σ-finite* if there is a sequence (A_n) of measurable sets with $S = \bigcup_{n=1}^{\infty} A_n$ and $\mu(A_n) < \infty$ for all $n \in \mathbb{N}$.

(iii) μ is *complete* if every subset of a set of measure zero is measurable.

Example 12.20. Lebesgue measure m on $\mathcal{B}(\mathbb{R})$ is σ-finite: $\mathbb{R}$ can be expressed as the countable union of the intervals $(n, n+1]$, where n is an arbitrary integer, and $m((n, n+1]) = 1$.

Example 12.21. Lebesgue measure on $\mathcal{B}(\mathbb{R}^k)$ is σ-finite: $\mathbb{R}^k$ can be expressed as the countable union of the cells
$$S_n = (-n, n]^k = (-n, n] \times \cdots \times (-n, n]$$
each of which has a finite Lebesgue measure $m(S_n) = (2n)^k$.

Theorem 12.22. *For every measure space* (S, Σ, μ) *there is a least complete measure space* $(S, \overline{\Sigma}, \overline{\mu})$ *such that* $\overline{\Sigma} \supset \Sigma$ *and that* $\overline{\mu}$ *coincides with* μ *on* Σ. *This measure space is called a* measure completion *of the original measure space.*

Proof. Problem 12.29. By definition, $E \in \overline{\Sigma}$ if and only if there are sets $A, B \in \Sigma$ such that $A \subset E \subset B$ with $\mu(B \backslash A) = 0$. The measure $\overline{\mu}(E)$ is then defined by $\overline{\mu}(E) = \mu(A) = \mu(B)$. $\qquad\qquad\square$

We can ask whether the Lebesgue measure m on the σ-algebra of all Borel sets in $\mathbb{R}$ (or $\mathbb{R}^k$) is complete. This can be settled by a cardinality argument. From Problem 12.5 we can infer that $\mathcal{B}(\mathbb{R})$ is generated by the *countable* family of semiclosed intervals with rational end points. According to Example E.12 in Appendix E, $\mathcal{B}(\mathbb{R})$ has cardinality $\aleph_0^{\aleph_0} = \mathfrak{c}$. Let C be the Cantor singular set introduced in Problem 12.27; then C is an uncountable Borel set of Lebesgue measure zero. It has $2^{\mathfrak{c}}$ subsets; since $2^{\mathfrak{c}} > \mathfrak{c}$, most of these sets are not Borel.

12.1.1 *Lebesgue measurable sets in* $\mathbb{R}^k$

When we construct the measure-completion of the σ-algebra of Borel sets in $\mathbb{R}^k$ with respect to the Lebesgue measure m in $\mathbb{R}^k$, we obtain the σ-algebra $\Lambda(\mathbb{R}^k)$ of *Lebesgue measurable* sets. Explicitly, a set $E \subset \mathbb{R}^k$ is Lebesgue measurable if there are Borel sets A, B such that $A \subset E \subset B$ and $m(B \backslash A) = 0$. (The measure of E can be defined as the infimum of

$$\sum_{n=1}^{\infty} m(K_n),$$

where $\{K_n\}$ is a countable family of semiclosed cells of the form

$$K = (a_1, b_1] \times (a_2, b_2] \times \cdots \times (a_k, b_k]$$

in $\mathbb{R}^k$ which covers E. See Chapter 15.)

12.1.2   Sets in $\mathbb{R}^k$ which are not Lebesgue measurable

The following construction in $\mathbb{R}$ was given in 1905 by the Italian mathematician Giuseppe Vitali (1875–1932). For real numbers x, y define $x \sim y$ if and only if their difference $x - y$ is rational. Then $\sim$ is an equivalence relation (see Definition A.3). Let T be the subset of $(0, 1)$ which contains exactly one point in each equivalence class. (The existence of such a set depends on the Axiom of Choice—see E.13.) T has the following properties the proof of which is left as an exercise:

(i)   For each $x \in (0, 1)$ there is a rational number $r \in (-1, 1)$ such that $x \in T + r$. ($T + r$ is the translation of T by r.)

(ii)   If r, s are distinct rationals, then $T + r$ and $T + s$ are disjoint.

Suppose that T is Lebesgue measurable. All rationals in $(-1, 1)$ can be ordered in a sequence (r_i). The set $A = \bigcup_{i=1}^{\infty} (T + r_i)$ is then Lebesgue measurable. Since Lebesgue measure is translation invariant (Problem 12.25), $m(T + r_i) = m(T)$ for all i, and

$$m(A) = \sum_{i=1}^{\infty} m(T + r_i) = \sum_{i=1}^{\infty} m(T)$$

by the countable additivity of m. If $m(T) = 0$, we have $m(A) = 0$ which is impossible as $(0, 1) \subset A$ and $m(A) \geq 1$. If $m(T) > 0$, we have $m(A) = \infty$, which is again impossible as $A \subset (-1, 2)$ and $m(A) \leq 3$. This contradiction means that T cannot be Lebesgue measurable.

In $\mathbb{R}^k$ we can show that the set $T \times (0, 1)^{k-1} = T \times (0, 1) \times \cdots \times (0, 1)$ cannot be Lebesgue measurable (exercise).

12.1.3   Lebesgue measurable sets in $\mathbb{R}^k$ which are not Borel

We give an argument similar to the one we used previously to show that Lebesgue measure is not complete on Borel sets. Let $E = C \times [0, 1]^{k-1} = C \times [0, 1] \times \cdots \times [0, 1]$ in $\mathbb{R}^k$, where C is the Cantor set (Problem 12.27). Then E is uncountable with cardinality $\mathfrak{c} = \text{card}\,(\mathbb{R})$ and with Lebesgue measure 0. So there are $2^{\mathfrak{c}}$ subsets of E, each Lebesgue measurable. On the other hand, as we saw earlier, the σ-algebra of Borel sets in $\mathbb{R}^k$ has cardinality $\mathfrak{c} = \aleph_0^{\aleph_0}$. Since $\mathfrak{c} < 2^{\mathfrak{c}}$, there exist Lebesgue measurable sets which are not Borel.

12.1.4   Hausdorff measure in $\mathbb{R}^k$

For any nonnegative integer $m \leq k$ we define the so called m-dimensional *Hausdorff measure* $\mathcal{H}^m$ on the σ-algebra $\mathcal{B}$ of all Borel sets in $\mathbb{R}^k$; $\mathcal{H}^m$ gives the area of m-dimensional manifolds in $\mathbb{R}^k$ (Example 15.12). The measure $\mathcal{H}^0$ is the counting measure, and $\mathcal{H}^k$ coincides with Lebesgue measure on Borel sets.

12.2 Problems for Chapter 12

In what follows, S denotes a nonempty set, Σ a σ-algebra of subsets of S, and μ a positive measure on Σ. A set $A \subset S$ is said to be *measurable* if $A \in \Sigma$.

1. Let (A_n) be a sequence of measurable sets. Show that the intersection $A = \bigcap_{n=1}^{\infty} A_n$ is measurable.

Suggestion. De Morgan's law.

2. Let $A_1, A_2, \ldots, A_k$ be a finite family of measurable sets. Show that their union (and intersection) is measurable.

3. If S is an uncountable set, let Σ be the family of subsets of S which are either countable or have countable complements in S. Show that Σ is a σ-algebra in S.

4. Let $(S, \mathcal{T})$ be a topological space. Show that:

(i) Every closed set is a Borel set in S.

(ii) Every set which is an intersection of a closed and an open set is a Borel set in S.

(iii) Every set which is a countable union of closed sets (an F_σ set) is a Borel set in S.

(iv) Every set which is a countable intersection of open sets (a G_δ set) is a Borel set in S.

5. Prove the following:

(i) The family of all semiclosed intervals of the form $(a, b]$ generates the σ-algebra of Borel sets in $\mathbb{R}$.

(ii) The family of all semiclosed k-cells of the form $(a_1, b_1] \times \cdots \times (a_k, b_k]$ generates the σ-algebra of Borel sets in $\mathbb{R}^k$.

6. Prove the following.

(i) The set $\mathbb{Q}$ of all rational numbers is a Borel set in $\mathbb{R}$.

(ii) The set of all irrational numbers is a Borel set in $\mathbb{R}$.

7. A family $\mathcal{D}$ of subsets of S is called a *Dynkin system* if it has the following properties: (i) $\emptyset \in \mathcal{D}$, (ii) $A \in \mathcal{D} \implies A^c \in \mathcal{D}$, (iii) if A_n are disjoint sets in $\mathcal{D}$, then $\bigcup_{n=1}^{\infty} A_n \in \mathcal{D}$. Prove the following facts about a Dynkin system $\mathcal{D}$:

(i) If $A, B \in \mathcal{D}$ and $A \subset B$, then $B \setminus A \in \mathcal{D}$.

(ii) If $\mathcal{D}$ has an additional property that $B \setminus A \in \mathcal{D}$ whenever $A, B \in \mathcal{D}$, then $\mathcal{D}$ is a σ-algebra.

8. Let $S = \{1, 2, \ldots, 2n - 1, 2n\}$ for some $n \in \mathbb{N}$, and let $\mathcal{D}$ consist of all subsets of S with an even number of elements. Show that $\mathcal{D}$ is a Dynkin system which is not a σ-algebra if $n > 1$.

9. If Σ consists of the sets $\emptyset$, $\{1, 3, 5, \ldots\}$, $\{2, 4, 6, \ldots\}$, $\mathbb{N}$, show that Σ is a σ-algebra in $\mathbb{N}$.

10. If $A, B \in \Sigma$, $\mu(A) + \mu(B) = \mu(A \cup B) + \mu(A \cap B)$. (Avoid subtractions!)

11. *Monotonicity of positive measure.* Let $A, B \in \Sigma$ and let $A \subset B$. Prove that $\mu(A) \leq \mu(B)$. Can we conclude that $\mu(B \setminus A) = \mu(B) - \mu(A)$?

12. *Subadditivity of positive measure.* Let (S, Σ, μ) be a measure space and let $A_n \in \Sigma$. Prove that $\mu\left(\bigcup_{n=1}^{\infty} A_n\right) \leq \sum_{n=1}^{\infty} \mu(A_n)$. ($A_n$ need not be disjoint.)

13. Let $\mathcal{A}$ be an algebra of subsets of S such that $A_n \in \mathcal{A}$ and $A_n \nearrow A \implies A \in \mathcal{A}$. Prove that $\mathcal{A}$ is a σ-algebra.

14. Prove the following properties of μ on a measure space (S, Σ, μ):

(i) $(A_n \in \Sigma$ and $A_n \nearrow A) \implies (\mu(A_n) \to \mu(A))$.

(ii) $(A_n \in \Sigma$, $A_n \searrow A$ and $\mu(A_1) < \infty) \implies (\mu(A_n) \to \mu(A))$. Show by example that this may fail if the condition $\mu(A_1) < \infty$ is not fulfilled.

15. Let (S, Σ) be a measurable space and $\mu : \Sigma \to [0, \infty]$ a finitely additive function with $\mu(\emptyset) = 0$ and $(A_n \in \Sigma$ and $A_n \nearrow A) \implies (\mu(A_n) \to \mu(A))$. Prove that μ is a measure on Σ.

16. Let m_1 be the Lebesgue measure on $(\mathbb{R}, \mathcal{B}(\mathbb{R}))$. Prove that $m_1(\mathbb{R}) = \infty$. More generally, if m_k is the Lebesgue measure on $\mathbb{R}^k$, prove that $m_k(\mathbb{R}^k) = \infty$. (Problem 12.14.)

17. Let m_2 be the Lebesgue measure on $(\mathbb{R}^2, \mathcal{B}(\mathbb{R}^2))$. Prove that $m_2(\mathbb{R}) = 0$ considering $\mathbb{R}$ as the set of all $x \in \mathbb{R}^2$ with $\xi_2 = 0$. More generally, if m_k is the Lebesgue measure on $\mathbb{R}^k$ prove that $m_k(\mathbb{R}^{k-1}) = 0$ considering $\mathbb{R}^{k-1}$ as the set of all $x \in \mathbb{R}^k$ with $\xi_k = 0$. (Problem 12.14.)

18. Let (S, Σ, μ) be a measure space, and let A be a set in Σ. Define $\nu : \Sigma \to [0, \infty]$ by $\nu(E) = \mu(E \cap A)$ for all $E \in \Sigma$. Show that ν is a positive measure on Σ.

19. If μ_1 and μ_2 are positive measures on Σ, show that $\mu_1 + \mu_2$ is a positive measure on Σ.

20. Let m be the Lebesgue measure on $\mathbb{R}$. Show that:

(i) $m((a, b)) = m([a, b)) = m([a, b]) = b - a$.

(ii) $m((a, \infty)) = \infty$.

(iii) If A is a countable subset of $\mathbb{R}$, then $m(A) = 0$.

21. Let $A = \{(x,y) \in \mathbb{R}^2 : x > 1 \text{ and } 0 \le y < 1/x\}$. Sketch A, and show that it is a Borel subset of $\mathbb{R}^2$ with infinite Lebesgue measure $m(A)$.

22. Let $B = \{(x,y) \in \mathbb{R}^2 : 0 < x \le 1 \text{ and } 1 \le y \le 1/\sqrt{x}\}$. Sketch B and show that it is a Borel set with finite Lebesgue measure $m(B)$.

Suggestion. For each $n \ge 1$ let V_n be the rectangle $[0, 1/n^2] \times [n, n+1]$. Consider $V = \bigcup_{n=1}^{\infty} V_n$.

23. Let $g : \mathbb{R} \to \mathbb{R}$ be an increasing right continuous function, and let μ_g be Lebesgue–Stieltjes measure generated by g. Prove that

$$\mu_g((a,b)) = g(b-) - g(a), \qquad \mu_g([a,b)) = g(b-) - g(a-),$$
$$\mu_g([a,b]) = g(b) - g(a-), \qquad \mu_g((a,\infty)) = \lim_{t \to \infty} g(t) - g(a).$$

24. Let S be a normed space, and let u be a nonzero vector in S. For each set $A \subset S$ define its *translation* $A + u$ as the set $A + u = \{x + u : x \in A\}$. Prove that if E is a Borel set in S, then $E + u$ is also a Borel set.

Suggestion. Keep u fixed and show that $\mathcal{A} = \{E : E + u \in \mathcal{B}\}$ is a σ-algebra, and that $\mathcal{A}$ contains open sets.

25. Let m be the Lebesgue measure on the Borel σ-algebra $\mathcal{B}$ in $\mathbb{R}^k$. Show that m is *translation invariant* on $\mathcal{B}$, that is, $m(E + u) = m(E)$ for all $E \in \mathcal{B}$ and for all $u \in \mathbb{R}^k$.

26. Let (a_n) be a sequence of nonnegative real numbers. For each set $A \subset \mathbb{N}$ define $\mu(A) = \sum_{n \in A} a_n$. Show that μ is a positive measure on $\mathcal{P}(\mathbb{N})$. Conversely, show that every positive measure on $\mathcal{P}(\mathbb{N})$ is obtained in this way for some sequence (a_n) in $[0, \infty]$.

27. *The Cantor set.* We construct a sequence (C_k) of closed subsets of the interval $[0, 1]$ as follows. Let $C_0 = [0, 1]$. C_1 is obtained from C_0 by deleting the open middle third of the interval $[0, 1]$; C_2 from C_1 by deleting the open middle thirds of the intervals $[0, \frac{1}{3}]$ and $[\frac{2}{3}, 1]$, and so on. If C_k denotes the union of the 2^n closed intervals of length $(\frac{1}{3})^n$ which remain at the nth stage, set $C = \bigcap_{k=1}^{\infty} C_k$. Prove that:

(i) C is uncountable,

(ii) C is a compact Borel set,

(iii) C has Lebesgue measure 0.

28. Let (S, Σ, μ) be a measure space, and Σ_0 the family of all measurable sets A with $\mu(A) = 0$. Show that Σ_0 is a σ-*ring*, that is, (i) $\emptyset \in \Sigma_0$, (ii) $A, B \in \Sigma_0 \implies A \backslash B \in \Sigma_0$, (iii) $A_n \in \Sigma_0 \implies \bigcup_{n=1}^{\infty} A_n \in \Sigma_0$. Show that Σ_0 need not be a σ-algebra.

29. *Measure completion.* Let (S, Σ, μ) be a measure space and let Σ_0 be the σ-ring of all sets of measure zero. Let $\overline{\Sigma}$ be the family of all sets $E \subset S$ for which there are sets $A, B \in \Sigma$ such that $A \subset E \subset B$ and $B \backslash A \in \Sigma_0$. Define $\overline{\mu}(E) = \mu(A) = \mu(B)$. Show that $\overline{\Sigma}$ is a σ-algebra containing Σ, and $\overline{\mu}$ a complete

measure on $\overline{\Sigma}$ which coincides with μ on Σ. If (S, Σ_1, μ_1) is another complete measure space such that $\Sigma \subset \Sigma_1$ and μ_1 coincides with μ on Σ, show that $\Sigma_1 \supset \overline{\Sigma}$ and μ_1 is an extension of $\overline{\mu}$.

Suggestion. First verify that if $A \subset B$ and $B \setminus A \in \Sigma_0$, then $\mu(A) = \mu(B)$, regardless whether the measures are finite or infinite. To prove that $\overline{\mu}$ is well defined, assume that A_i, B_i are Σ-measurable sets with $A_i \subset E \subset B_i$ and $B_i \setminus A_i \in \Sigma_0$ for $i = 1, 2$. Show that $A_1 \cup A_2 \subset E \subset B_1 \cap B_2$, and conclude that $\mu(B_2) = \mu(A_2) = \mu(A_1) = \mu(B_1)$.

30.   This problem is more involved, and can be treated as a miniproject: Give a proof that the Lebesgue measure in $\mathbb{R}^k$ is *rotation invariant* based on the following points in which ρ is a rotation about the origin and A a semiclosed k-cell in $\mathbb{R}^k$.

(i)   Show that an open set $G \subset \mathbb{R}^k$ can be expressed as a countable union of disjoint semiclosed k-cells, each similar to A. (B is *similar* to A if $B = u + \lambda A$, where $u \in \mathbb{R}^k$ and $\lambda \in \mathbb{R}$.)

(ii)   Show that the σ-algebra $\Lambda(\mathbb{R}^k)$ of all Lebesgue measurable sets in $\mathbb{R}^k$ is invariant under ρ.

(iii)   Show that for every k-cell B similar to A, $m(\rho B) = \alpha m(B)$, where $\alpha = m(\rho A)/m(A)$.

(iv)   Let $S = \{x \in \mathbb{R}^k : \|x\|_{\ell^2} < 1\}$. Express S as a countable union of disjoint semiclosed k-cells A_k, each similar to A, and use $m(\rho S) = m(S)$ to conclude that the constant $\alpha = \alpha(\rho, A)$ from point (iii) satisfies $\alpha = 1$.

(v)   The measure $\mu(E) := m(\rho E)$ coincides with the Euclidean volume on k-cells.

Chapter 13

The Abstract Lebesgue Integral

A definite integral is a very important tool in pure and applied mathematics as well as in physics, where many quantities, such as volumes, surface areas, lengths, moments and centers of mass, are expressed as definite integrals. In most elementary expositions, notably in calculus texts, the integral used owes its origin to the French mathematician Augustin Cauchy and to the German mathematician Bernhard Riemann (1826–1866). However, this integral, usually referred to as the Riemann integral, has several deficiencies. In particular, it does not handle efficiently taking limits 'under the integral sign'; for instance the formula

$$\int \lim_{n \to \infty} f_n(x)\, dx = \lim_{n \to \infty} \int f_n(x)\, dx$$

is valid for the Riemann integral only under very restrictive conditions, such as uniform convergence.

Toward the end of the nineteenth century many mathematicians tried to replace the Riemann integral by a more general integral which would be better suited for dealing with limit processes. The French mathematician Henri Lebesgue (1875–1941) gave a successful construction of such an integral that now bears his name. The first exposition was given in Lebesgue's doctoral dissertation in 1902. The same year his 130 page thesis appeared as a journal article [32], and then Lebesgue published his treatise [33] *Leçons sur l'integration et la recherche des fonctions primitives* in 1904. There is also an English version [34].

While the Riemann integral relies on a partition of the *domain* of a given function, Lebesgue's crucial idea was to partition the *range* of the real valued function f. For each subinterval $(y_{k-1}, y_k]$ of this partition a representative value $w_k \in (y_{k-1}, y_k]$ is chosen, which is then multiplied by the 'length' of the set $E_k = \{x : f(x) \in (y_{k-1}, y_k]\}$. But E_k need not be an interval, or even a union of intervals, but must be e in the sense of Lebesgue, with measure $m(E_k)$. The integral of f is then approximated by the sum

$$\sum_{k=1}^{n} w_k m(E_k).$$

The construction given below differs considerably from the one given by Lebesgue himself, as (following Johann Radon) we define integration on abstract measure spaces, rather than just on the real line, and rely on the concept of metric space completion for the construction.

We want the integral sufficiently general to handle the following situations:

(i)   Integration of unbounded as well as bounded functions.

(ii)   Integration on a variety of spaces; where applicable, integration over regions which may be bounded or unbounded.

(iii)   Integration with respect to a variety of measures.

(iv)   Taking limits under the integral sign under minimal restrictions.

For most of the theory we assume that a measure is already present, and accept the existence of Lebesgue measure in $\mathbb{R}^n$ as the unique measure on Borel sets which coincides with the Euclidean volume on semiclosed n-cells.

13.1   The integral of simple functions

The definition of integral adopted in this book is applicable equally well to scalar valued functions as it is to functions with values in Banach spaces. For most applications it is enough to study complex valued functions. In this section we assume that (S, Σ, μ) is a measure space.

For any set $A \subset S$ we define its *characteristic function* χ_A by

$$\chi_A(t) = \begin{cases} 1 & \text{if } t \in A, \\ 0 & \text{if } t \notin A. \end{cases}$$

It is easily verified that

$$\chi_{A \cup B} = \chi_A + \chi_B \ \text{ if } \ A \cap B = \emptyset,$$

and

$$\chi_{A \cap B} = \chi_A \cdot \chi_B, \qquad \chi_{A^c} = 1 - \chi_A.$$

If A, B are not disjoint, we have

$$\chi_{A \cup B} = \chi_A + \chi_B - \chi_A \cdot \chi_B.$$

Definition 13.1. A complex valued function f on S is a *simple function* if it takes only a finite number of values, each on a measurable set. This means that a simple function can be written as a linear combination of characteristic functions of measurable sets:

$$f = \sum_{k=1}^{l} a_k \chi_{A_k}, \qquad a_k \in \mathbb{C}. \qquad (13.1)$$

Here $\{A_1, \ldots, A_l\}$ is a so called *measurable f-partition* of S; that is, each A_k is measurable,

$$A_1 \cup \ldots \cup A_l = S, \qquad A_i \cap A_j = \emptyset \text{ if } i \neq j,$$

and f is constant on each A_k.

Definition 13.2. A simple function $f \colon S \to \mathbb{C}$ is said to be *integrable* (or more precisely μ-integrable) if the set $N_f = \{t \in S : f(t) \neq 0\}$ has a finite measure $\mu(N_f)$; N_f is measurable being a finite union of measurable sets. Observe that this condition implies that $a_k = 0$ whenever $\mu(A_k) = \infty$ in (13.1). The set of all integrable simple functions will be denoted by $E(S, \Sigma, \mu)$ or by $E(\mu)$ if S and Σ are understood.

Theorem 13.3. *The set $E(\mu)$ has the following properties:*

(i) *$f, g \in E(\mu) \implies f + g \in E(\mu)$ and $f \cdot g \in E(\mu)$.*

(ii) *$f \in E(\mu) \implies cf \in E(\mu)$ for any complex number c.*

(iii) *$f \in E(\mu) \implies |f| \in E(\mu)$ and $\overline{f} \in E(\mu)$.*

Proof.   Exercise.   □

Definition 13.4. Let $f = \sum_{k=1}^{l} a_k \chi_{A_k}$ be an integrable simple function with respect to a measurable f-partition $\{A_1, \ldots A_l\}$. We define the *integral* $\int_S f \, d\mu$ of f to be

$$\int_S f \, d\mu = \sum_{k=1}^{l} a_k \mu(A_k). \qquad (13.2)$$

First we check that the sum on the right is defined: Since f is integrable, $\mu(A_k) = \infty$ only if $a_k = 0$. In this case $a_k \mu(A_k) = 0 \cdot \infty = 0$ by definition. So the sum on the right in (13.2) is a complex number. Secondly, we have to verify that the sum is independent of the choice of an f-partition (Problem 13.2).

In the next theorem we summarize basic properties of the integral of integrable simple functions. Note that the first two properties say that the mapping $f \mapsto \int_S f \, d\mu$ is a complex linear functional on the vector space $E(\mu)$.

Theorem 13.5. *Let $f, g \in E(\mu)$ and let $c \in \mathbb{C}$. Then:*

(i) $\displaystyle \int_S (f + g) \, d\mu = \int_S f \, d\mu + \int_S g \, d\mu$

(ii) $\displaystyle \int_S (cf) \, d\mu = c \int_S f \, d\mu$

(iii) $\displaystyle \overline{\int_S f \, d\mu} = \int_S \overline{f} \, d\mu$

(iv) $\displaystyle \int_S f \, d\mu \leq \int_S g \, d\mu$ *if f, g are real valued and $f \leq g$*

(v) $\displaystyle \left| \int_S f \, d\mu \right| \leq \int_S |f| \, d\mu$

Proof. As a sample we prove (i). We choose a measurable partition $\{E_1, \ldots, E_p\}$ which is simultaneously an f- and a g-partition; e.g. $\{A_k \cap B_j\}$ where $\{A_k\}$ is an f-partition and $\{B_j\}$ a g-partition. Then

$$\int_S (f + g) \, d\mu = \sum_{i=1}^p (a_i + b_i) \, \mu(E_i) = \int_S f \, d\mu + \int_S g \, d\mu.$$

□

We define a seminorm $\| \cdot \|_{L^1}$ on the space $E(\mu)$ of integrable simple functions by letting

$$\|f\|_{L^1} = \int_S |f| \, d\mu.$$

We refer to this seminorm as the L^1-*seminorm*. A proof of the seminorm properties is left as an exercise. (A seminorm differs from a norm in only one respect: It may admit $\|f\|_{L^1} = 0$ for some nonzero f.)

To find all integrable simple functions f for which $\|f\|_{L^1} = 0$ we express f in the form (13.1), and consider the equation

$$\|f\|_{L^1} = \sum_{k=1}^l |a_k| \mu(A_k) = 0.$$

For this to be true, we must have $\mu(A_k) = 0$ whenever $a_k \neq 0$. Then

$$\mu(N_f) = \mu\left(\bigcup_{a_k \neq 0} A_k\right) = \sum_{a_k \neq 0} \mu(A_k) = 0.$$

This means that $\|f\|_{L^1} = 0$ if and only if $f = 0$ outside a set of measure zero.

Sets of measure zero play a very important role in measure theory. The family of all these sets will be denoted by Σ_0. Then Σ_0 is a σ-ring (Problem 12.28); a σ-ring differs from a σ-algebra by admitting sets whose complements are not in the σ-ring. We say that a property applicable to points of S holds μ-*almost everywhere* if there is a set $A \in \Sigma_0$ such that the property holds for all $s \notin A$. We can then say that an integrable simple function f is of L^1-seminorm zero if and only if $f = 0$ μ-almost everywhere; we often write $f = 0$ μ-a.e. Symbolically,

$$\|f\|_{L^1} = 0 \iff f = 0 \ \mu\text{-a.e.} \tag{13.3}$$

Similarly we say that a sequence (f_n) of functions on S *converges μ-almost everywhere* to f if there is a set $A \in \Sigma_0$ such that $f_n(s) \to f(s)$ for all $s \notin A$. We write

$$f_n \to f \ \mu\text{-a.e.}$$

If the measure μ is understood, we often say 'almost everywhere' instead of 'μ-almost everywhere'.

We observe that the equality μ-a.e. is an equivalence relation, that is, it is reflexive, symmetric and transitive (see A.3). If we adopt the equality μ-a.e. as the new equality relation on $E(\mu)$, the L^1-seminorm becomes a norm. Then we have the following result:

Theorem 13.6. *If the space $E(\mu)$ is equipped with the equality μ-a.e. as the new equality, then it becomes a normed space. The functional $\varphi\colon E(\mu) \to \mathbb{C}$ defined by $\varphi(f) = \int_S f \, d\mu$ is linear and bounded with functional norm 1.*

Proof. To show that the functional φ is defined unambiguously under the new equality, we have to prove that

$$f, g \in E(\mu) \text{ and } f = g \ \mu\text{-a.e.} \implies \int_S f \, d\mu = \int_S g \, d\mu. \tag{13.4}$$

Indeed, if $f = g$ μ-a.e., then $f - g = 0$ μ-a.e. Then $\|f - g\|_{L^1} = 0$ by (13.3), and

$$\left|\int_S f \, d\mu - \int_S g \, d\mu\right| = \left|\int_S (f - g) \, d\mu\right| \leq \int_S |f - g| \, d\mu = \|f - g\|_{L^1} = 0.$$

It is important to note that the converse implication in (13.4) is false.

For φ, only a proof of boundedness is required. Let $f \in E(\mu)$. By Theorem 13.5 (v),

$$|\varphi(f)| = \left| \int_S f \, d\mu \right| \le \int_S |f| \, d\mu = \|f\|_{L^1}.$$

The equality is attained for any simple function f which takes only real positive values.   □

Suppose A is a measurable set and χ_A its characteristic function. For any integrable simple function f on S we define the integral $\int_A f \, d\mu$ by

$$\int_A f \, d\mu = \int_S f \cdot \chi_A \, d\mu.$$

Observe that the product $f \cdot \chi_A$ is also an integrable simple function.

Theorem 13.7. *For any measurable set A and any integrable simple function f we have*

$$\left| \int_A f \, d\mu \right| \le \int_A |f| \, d\mu \le \|f\|_\infty \cdot \mu(A).$$

If f is positive, then

$$\int_A f \, d\mu \le \int_S f \, d\mu.$$

Proof.   Left to the reader as an exercise.   □

13.2   The completion of $E(\mu)$

We are going to define integrable functions as the elements of the metric completion of the space of integrable simple functions (with respect to the L^1 norm). We know that a normed space is complete if and only if each norm-absolutely convergent series converges in the space (Theorem 2.33).

The space $E(\mu)$ of all simple integrable functions (with equality μ-a.e. as the new equality) is a normed space with the L^1 norm

$$\|f\|_{L^1} = \int_S |f| \, d\mu.$$

To obtain the completion of $(E(\mu), \|\cdot\|_{L^1})$, we consider all norm-absolutely convergent series $\sum_{n=1}^\infty f_n$ of integrable simple functions, and the equivalence relation

$$\sum_{n=1}^\infty f_n \sim \sum_{n=1}^\infty g_n \iff \lim_{n\to\infty} \left\| \sum_{k=1}^n (f_k - g_k) \right\|_{L^1} = 0.$$

An ideal point of the completion is an equivalence class $\boldsymbol{f}$ of norm-absolutely convergent series $\sum_{n=1}^{\infty} f_n$ which do not converge in $(E(\mu), \| \cdot \|_{L^1})$. It is important that the ideal points of the completion can be identified with certain complex valued functions on S. The following theorem paves the path to this result.

Theorem 13.8. *Let $\sum_{n=1}^{\infty} f_n$ be a norm-absolutely convergent series of integrable simple functions. Given $\varepsilon > 0$, there exists a set $A \in \Sigma$ such that $\mu(A) < \varepsilon$ and the series converges absolutely and uniformly on A^c. Consequently the series converges absolutely μ-a.e. in S.*

Proof. Since $\sum_{n=1}^{\infty} \|f_n\|_{L^1} < \infty$, there exists a strictly increasing sequence $(k(n))$ of positive integers such that $\sum_{i=k(n)+1}^{\infty} \|f_i\|_{L^1} < (\frac{1}{2})^{2n}$ for all $n \in \mathbb{N}$. Set $g_n = \sum_{i=k(n)+1}^{k(n+1)} |f_i|$; then $\|g_n\|_{L^1} \leq (\frac{1}{2})^{2n}$. For each n, the set $E_n := \{t \in S : |g_n(t)| \geq (\frac{1}{2})^n\}$ is measurable, and

$$\mu(E_n) \leq (\tfrac{1}{2})^{-n} \int_{E_n} g_n \, d\mu \leq (\tfrac{1}{2})^{-n} \|g_n\|_{L^1} \leq (\tfrac{1}{2})^n.$$

For each $n \in \mathbb{N}$ set further $A_n = \bigcup_{j=n+1}^{\infty} E_j$; then $A_n \in \Sigma$, and $\mu(A_n) \leq \sum_{j=n+1}^{\infty} (\frac{1}{2})^j = (\frac{1}{2})^n$. If $t \in A_n^c$, then $0 \leq g_j(t) < (\frac{1}{2})^j$ for all $j > n$, which proves that

$$\sum_{i=1}^{\infty} |f_i| = \sum_{i=1}^{k(1)} |f_i| + \sum_{j=1}^{\infty} g_j$$

converges uniformly on the set A_n^c whose complement A_n has measure less than $(\frac{1}{2})^n$. If $\varepsilon > 0$, we can choose n so that $(\frac{1}{2})^n < \varepsilon$, and set $A = A_n$. This proves the first part of the theorem.

Let $D := \bigcap_{j=1}^{\infty} A_j$. Then D is a measurable set with $\mu(D) = 0$ as $\mu(D) \leq \mu(A_j) \leq (\frac{1}{2})^j$ for all j. Thus the series $\sum_{n=1}^{\infty} |f_n(t)|$ converges for all t in D^c, where $\mu(D) = 0$. $\square$

Definition 13.9. If f is a complex valued function on S such that $f = \sum_{n=1}^{\infty} f_n$ μ-a.e., where $\sum_{n=1}^{\infty} f_n$ is a norm-absolutely convergent series of simple functions, we call $\sum_{n=1}^{\infty} f_n$ an *approximating series* for f.

Motivated by the preceding theorem, we consider the pairs $(\sum_n g_n, g)$, where $\sum_n g_n$ is a norm-absolutely convergent representative of an ideal point $\boldsymbol{g}$ of the completion of $E(\mu)$, and an approximating series for g. To show that there is a one-to-one correspondence between $\boldsymbol{g}$ and g (modulo equality μ-a.e. for g), we need to prove that if $(\sum_n g_n, g)$ and $(\sum_n h_n, h)$

are two such pairs, then $\sum_n g_n$ and $\sum_n h_n$ determine the same ideal point g if and only if $g = h$ μ-a.e. This can be deduced from the following result.

Theorem 13.10. *Let $\sum_n f_n$ be a norm-absolutely convergent series in $E(\mu)$. Then*

$$\lim_{n \to \infty} \left\| \sum_{k=1}^n f_k \right\|_{L^1} = 0 \iff \sum_{n=1}^\infty f_n = 0 \ \mu\text{-a.e.} \tag{13.5}$$

Proof.   Throughout this proof we write $s_n = \sum_{i=1}^n f_i$.

Assume that $\|s_n\|_{L^1} \to 0$. By Theorem 13.8, there exists a function $f \colon S \to \mathbb{C}$ and a set D of measure 0 such that $s_n(t) \to f(t)$ for all $t \in D^c$. For any positive integers n, k let $A_{nk} = \{t \in S : |s_n(t)| > 1/k\}$. Then each A_{nk} is measurable with

$$\mu(A_{nk}) \leq k \int_{A_{nk}} |s_n| \, d\mu \leq k \|s_n\|_{L^1},$$

where $\|s_n\|_{L^1} \to 0$ by hypothesis. If $f(t) \neq 0$ for some $t \in D^c$, then there exists k and m such that $|s_n(t)| > 1/k$ for all $n \geq m$. Hence $t \in B_k = \bigcup_{m=1}^\infty \bigcap_{n=m}^\infty A_{kn}$, and $\mu(B_k) = 0$. This shows that for any $t \in D^c$, $f(t) \neq 0$ if and only if $t \in B = \bigcup_{k=1}^\infty B_k$, where $\mu(B) = 0$.

Conversely assume that $s_n \to 0$ μ-a.e. Let $\varepsilon > 0$ be given. There exists $p \in \mathbb{N}$ such that $\sum_{i=n}^\infty \|f_i\|_{L^1} < \varepsilon$ for all $n > p$. There is a set $A \in \Sigma$ of finite measure such that $s_p = 0$ on A^c. If $n > p$, then

$$\int_{A^c} |s_n| \, d\mu = \int_{A^c} |s_n - s_p| \, d\mu \leq \|s_n - s_p\|_{L^1} \leq \sum_{i=p+1}^\infty \|f_i\|_{L^1} < \varepsilon.$$

Let $\eta > 0$. By Theorem 13.8 there exists a set $B \subset A$ with $\mu(B) < \eta$ such that $s_n \to 0$ uniformly on $A \setminus B$. Hence there exists k such that $|s_n(t)| < \eta$ for all $t \in A \setminus B$ and all $n > k$. If $n > \max\{p, k\}$, then

$$\int_{A \setminus B} |s_n| \, d\mu \leq \eta \, \mu(A),$$

and

$$\int_B |s_n| \, d\mu \leq \int_B |s_n - s_p| \, d\mu + \int_B |s_p| \, d\mu \leq \varepsilon + \mu(B) \|s_p\|_\infty < \varepsilon + \eta \|s_p\|_\infty.$$

Choose η so that $\|s_p\|_\infty \, \eta < \varepsilon$ and $\mu(A) \, \eta < \varepsilon$. Then

$$\|s_n\|_{L^1} = \int_{A^c} |s_n| \, d\mu + \int_{A \setminus B} |s_n| \, d\mu + \int_B |s_n| \, d\mu < 4\varepsilon.$$

This proves the theorem. $\square$

We now return to the situation discussed prior to Theorem 13.10. Let $\sum_n g_n$ and $\sum_n h_n$ be approximating series for functions $g, h\colon S \to \mathbb{C}$, respectively. According to the preceding theorem,

$$\sum_{n=1}^{\infty} g_n \sim \sum_{n=1}^{\infty} h_n \iff g = h \ \mu\text{-a.e.}$$

This proves that each point in the completion of $E(\mu)$ is associated with one and only one complex valued function on S (modulo equality μ-a.e.) possessing an approximating series.

13.3 Integrable functions and their integral

Definition 13.11. Let (S, Σ, μ) be a measure space. A function $f\colon S \to \mathbb{C}$ is *integrable*, or more precisely μ-*integrable*, if there is an approximating series $\sum_{n=1}^{\infty} f_n$ for f, that is, a norm-absolutely convergent series $\sum_{n=1}^{\infty} f_n$ of integrable simple functions such that

$$f = \sum_{n=1}^{\infty} f_n \quad \mu\text{-a.e.} \tag{13.6}$$

The set of all integrable functions will be denoted by $L^1(S, \Sigma, \mu)$. Notation $L^1(\mu)$ or L^1 is also used when the measure space is understood.

Any integrable simple function f is integrable in the sense of Definition 13.11 since $f + 0 + 0 + \cdots$ is norm-absolutely convergent.

According to Theorem 13.10 there is a one-to-one correspondence between μ-integrable functions on S and the elements of the completion of $(E(\mu), \| \cdot \|_{L^1})$; in other words, $L^1(\mu)$ is the completion of $E(\mu)$. It will be convenient to use both the series and sequence representation for the elements of $L^1(\mu)$. Series were essential for the construction of the space $L^1(\mu)$, but sequences are more convenient in proofs. We rely on the following alternative characterization of integrable functions which follows from the construction of $L^1(\mu)$ by completion.

Theorem 13.12. *A function $f\colon S \to \mathbb{C}$ is integrable if and only if there is a sequence (f_n) of integrable simple functions such that*

(i) *(f_n) is a Cauchy sequence in $(E(\mu), \| \cdot \|_{L^1})$,*

(ii) *$f_n \to f$ μ-a.e.*

Such a sequence is called an approximating sequence *for f.*

Definition 13.13. For any integrable function f we define its integral by

$$\int_S f\, d\mu = \lim_{n \to \infty} \int_S f_n\, d\mu, \tag{13.7}$$

where (f_n) is any approximating sequence for f.

We have to show that the limit exists and is independent of the choice of the approximating sequence. The existence follows from the fact that the integrals of f_n form a Cauchy sequence in $\mathbb{C}$:

$$\left| \int_S f_m\, d\mu - \int_S f_n\, d\mu \right| \le \|f_m - f_n\|_{L^1}.$$

For the independence of the integral of the choice of an approximating series we consider two approximating sequences (f_n) and (g_n) for f. Since the sequences define the same ideal point, they are equivalent in the sense that $\|f_n - g_n\|_{L^1} \to 0$. Then

$$\left| \int_S f_n\, d\mu - \int_S g_n\, d\mu \right| \le \|f_n - g_n\|_{L^1} \to 0.$$

The integral of $f \in L^1(\mu)$ can be also described with the help of an approximating series $\sum_n h_n$:

$$\int_S f\, d\mu = \sum_{n=1}^{\infty} \int_S h_n\, d\mu.$$

The integrable functions defined above are usually referred to as *Lebesgue integrable functions* on a measure space (S, Σ, μ), and the integral defined by (13.7) is called the *Lebesgue integral* on (S, Σ, μ).

We start exploring properties of Lebesgue integrable functions. The first property of great importance in the theory of Lebesgue integral is that the integrability of f always implies the integrability of $|f|$; any integral with this property is called an *absolute integral*.

Theorem 13.14. *Let $\|f\|$ be the norm of $f \in L^1(\mu)$ defined from the completion of $(E(\mu), \|\cdot\|_{L^1})$ as $\|f\| = \lim_{n \to \infty} \|f_n\|_{L^1}$, where (f_n) is an approximating sequence for f. Then $(|f_n|)$ is an approximating sequence for $|f|$, and*

$$\|f\| = \int_S |f|\, d\mu.$$

Proof. The formula holds in the case that f is a simple function. Let f be an arbitrary element of $L^1(\mu)$ with an approximating sequence (f_n). Then $(|f_n|)$ is Cauchy in $(E(\mu), \|\cdot\|_{L^1})$ and $|f_n| \to |f|$ μ-a.e. (exercise). Hence

$$\|f\| = \lim_{n\to\infty} \|f_n\|_{L^1} = \lim_{n\to\infty} \int_S |f_n| \, d\mu = \int_S |f| \, d\mu.$$

□

In view of the preceding theorem we will use the notation $\|f\|_{L^1}$ for the norm (arising from the completion) of any function $f \in L^1(\mu)$. Combining Theorem 13.14 with Theorem 13.10, we obtain the following result.

Theorem 13.15. *Let $f \in L^1(\mu)$. Then*

$$\int_S |f| \, d\mu = 0 \iff f = 0 \quad \mu\text{-a.e.} \tag{13.8}$$

The following theorem summarizes basic facts about $L^1(\mu)$.

Theorem 13.16. *The set $L^1(\mu)$ has the following properties:*

(i) *$(L^1(\mu), \|\cdot\|_{L^1})$ is a complete normed space (with equality μ-a.e.).*

(ii) *If $f \in L^1(\mu)$, then $|f| \in L^1(\mu)$.*

(iii) *If $f, g \in L^1(\mu)$ are real valued, then $\max(f, g) \in L^1(\mu)$ and $\min(f, g) \in L^1(\mu)$.*

The only property needing a proof, (iii), follows from

$$f \vee g := \max(f, g) = \tfrac{1}{2}(f + g + |f - g|), \quad f \wedge g := \min(f, g) = \tfrac{1}{2}(f + g - |f - g|).$$

As for simple functions, for any integrable function f and any measurable set A we define the integral $\int_A f \, d\mu$ by

$$\int_A f \, d\mu = \int_S f \cdot \chi_A \, d\mu,$$

where χ_A is the characteristic function of A; the function $f \cdot \chi_A$ is integrable in view of Problem 13.12.

The next theorem summarizes basic properties of the integral.

Theorem 13.17. *Let $f, g \in L^1(\mu)$ and let $c \in \mathbb{C}$. Then:*

(i) *$f + g \in L^1(\mu)$ and $\int_S (f + g) \, d\mu = \int_S f \, d\mu + \int_S g \, d\mu$.*

(ii) *$cf \in L^1(\mu)$ and $\int_S (cf) \, d\mu = c \int_S f \, d\mu$.*

(iii) *If f is real valued, then $\int_S f \, d\mu$ is a real number.*

(iv) $\mathsf{Re} \int_S f \, d\mu = \int (\mathsf{Re} \, f) \, d\mu$ *and* $\mathsf{Im} \int_S f \, d\mu = \int (\mathsf{Im} \, f) \, d\mu.$

(v) $\int_S f \, d\mu \leq \int_S g \, d\mu$ *if f, g are real valued and $f \leq g$ μ-a.e.*

(vi) $|f| \in L^1(\mu)$ *and* $\left| \int_S f \, d\mu \right| \leq \int_S |f| \, d\mu.$

(vii) $\left| \int_A f \, d\mu \right| \leq \|f\|_\infty \cdot \mu(A)$ *if f is bounded, $A \in \Sigma$ and $\mu(A) < \infty.$*

(viii) *The integral is a continuous linear functional on $L^1(\mu)$, that is, if $f_n \to f$ in the norm of $L^1(\mu)$, then* $\int_S f \, d\mu = \lim_{n \to \infty} \int_S f_n \, d\mu.$

Proof. (i) and (ii) follow from the corresponding theorems for simple functions by taking limits.

(iii) If (f_n) is an approximating sequence for f, then $(\mathsf{Re} \, f_n)$ is also an approximating sequence for f. Thus real valued integrable functions have approximating sequences of real valued simple functions. Property (iv) follows from (iii).

(v) It is enough to prove that if $h \in L^1(\mu)$ and $h \geq 0$, then $\int_S h \, d\mu \geq 0$. Let (h_n) be an approximating sequence of real valued functions for $h \geq 0$. Then also the sequence of $h_n^+ = \max(h_n, 0)$ is an approximating sequence for h. Hence $\int_S h \, d\mu = \lim_{n \to \infty} \int_S h_n^+ \, d\mu \geq 0$.

(vi) The inequality is known to hold for simple functions. If (f_n) is an approximating sequence for an integrable function f, then

$$\left| \int_S f \, d\mu \right| = \lim_{n \to \infty} \left| \int_S f_n \, d\mu \right| \leq \lim_{n \to \infty} \int_S |f_n| \, d\mu = \int_S |f| \, d\mu.$$

Properties (vii) and (viii) follow from (vi). □

In the next example we consider an unbounded Lebesgue integrable function.

Example 13.18. Apply the definition to show that the function $f(t) = 1/\sqrt{t}$ is Lebesgue integrable on $(0, 1]$ and calculate the integral $\int_{(0,1]} f \, dm$ relative to Lebesgue measure m.

The measure space here is $((0, 1], \Lambda(0, 1], m)$, where $\Lambda(0, 1]$ is the σ-algebra of all Lebesgue measurable subsets of $(0, 1]$. We construct an approximating sequence for f. Given $n \in \mathbb{N}$, define the intervals

$$A_{nk} = \left(\left(\frac{k-1}{n}\right)^2, \left(\frac{k}{n}\right)^2\right], \quad k = 1, \ldots, n,$$

and let

$$f_n = \sum_{k=1}^{n} f\left(\left(\frac{k}{n}\right)^2\right) \chi_{A_{nk}}.$$

For each fixed n, the intervals A_{nk} $(k = 1, \ldots, n)$ form a measurable partition of $(0, 1]$. Hence each f_n is a simple integrable function with

$$\int_{(0,1]} f_n \, dm = \sum_{k=1}^{n} f\left(\left(\frac{k}{n}\right)^2\right) m(A_{nk}) = \sum_{k=1}^{n} \frac{n}{k} \frac{2k-1}{n^2} = 2 - \frac{1}{n} \sum_{k=1}^{n} \frac{1}{k} = 2 - \alpha_n,$$

where $\alpha_n = n^{-1}(\sum_{k=1}^{n} k^{-1})$. We leave it to the reader to prove that the sequence $g_n = f_{2^n}$ is increasing and pointwise convergent to f on the interval $(0, 1]$. Using the inequality

$$\frac{1}{2} + \frac{1}{3} + \cdots + \frac{1}{n} \leq \log n \quad (n = 1, 2, \ldots)$$

(or otherwise) we deduce that α_n converges to zero as $n \to \infty$; for $k > l$,

$$\|g_k - g_l\|_{L^1} = \int_{(0,1]} g_k \, dm - \int_{(0,1]} g_l \, dm = \alpha_{2^l} - \alpha_{2^k} \to 0 \text{ as } \min\{k, l\} \to \infty.$$

Hence (g_n) is an approximating sequence for f, and

$$\int_{(0,1]} f \, dm = \lim_{n \to \infty} \int_{(0,1]} g_n \, dm = \lim_{n \to \infty} (2 - \alpha_{2^n}) = 2.$$

Since singletons have Lebesgue measure zero, the interval $(0, 1]$ in this example can be replaced by any of the intervals $(0, 1)$, $[0, 1)$, $[0, 1]$.

It is interesting to observe that the product of two integrable functions need not be integrable:

Example 13.19. In the preceding example we showed that the function $f(x) = x^{-1/2}$ is integrable with respect to the Lebesgue measure on the interval $(0, 1]$. We show that the product function $g = f \cdot f$, that is, $g(x) = x^{-1}$, is not integrable on $(0, 1]$:

For each $n \in \mathbb{N}$ construct the intervals $B_{nk} = ((k-1)/n, k/n]$, and the simple integrable functions $g_n = \sum_{k=1}^{n} g(k/n) \chi_{B_{nk}}$ $(k = 1, \ldots, n)$. Then

$$\int_{(0,1]} g_n \, dm = \sum_{k=1}^{n} g\left(\frac{k}{n}\right) \frac{1}{n} = \sum_{k=1}^{n} \frac{1}{k}.$$

Assume that g is integrable. Since $g \geq g_n$ for all $n \in \mathbb{N}$, we have

$$\int_{(0,1]} g \, dm \geq \int_{(0,1]} g_n \, dm = \sum_{k=1}^{n} \frac{1}{k} \quad \text{for all } n;$$

this is impossible as the harmonic series diverges.

While approximating series and sequences are needed to define the Lebesgue integral and derive its properties, calculating concrete integrals from the definition is technically quite difficult and not very efficient, as seen in Example 13.18 above. For this reason we need to develop other techniques, such as convergence theorems for the integral, a relation between the integral and derivative, the reduction of multiple integrals to repeated single integrals, etc. In order to have available a supply of concrete examples, we briefly consider integration of continuous functions on $\mathbb{R}$ before we return to the topic in full generality in the next chapter.

13.3.1   *Integration of continuous functions on $[a, b]$*

We consider the measure space $([a, b], \Lambda[a, b], m)$, where $\Lambda[a, b]$ is the σ-algebra of all Lebesgue measurable subsets of $[a, b]$, and m is the Lebesgue measure.

Theorem 13.20. *Any function $f \colon [a, b] \to \mathbb{C}$ continuous (or quasicontinuous) on $[a, b]$ is Lebesgue integrable on $[a, b]$.*

Proof.   We say that $f \colon [a, b] \to \mathbb{C}$ is *quasicontinuous* (see Problem 2.27) if the one sided limits $f(x+)$ and $f(x-)$ exist at each $x \in [a, b]$, postulating $f(a-) = f(a)$ and $f(b+) = f(b)$. This happens if and only if there exists a sequence (f_n) of complex valued step functions on $[a, b]$ uniformly convergent to f on $[a, b]$. Then (f_n) is an approximating sequence for the quasicontinuous function f (check). Every continuous function is quasicontinuous, and the result follows.   □

Theorem 13.21. *Let $f \colon [a, b] \to \mathbb{C}$ be continuous on $[a, b]$. Then the function*

$$F(x) = \int_{[a,x]} f \, dm, \quad a \le x \le b,$$

is differentiable on $[a, b]$ with $F'(x) = f(x)$ for all $x \in [a, b]$ (the derivatives at the end points are one sided).

Proof.   The argument is essentially the same as the first year calculus proof for the Riemann integral. Let $x \in (a, b]$. Given $\varepsilon > 0$ there exists $\delta > 0$ such that $x - \delta < s \le x$ implies $|f(s) - f(x)| < \varepsilon$. Let $x - \delta < t \le x$. Then

$$\left| \frac{F(t) - F(x)}{t - x} - f(x) \right| = \frac{1}{x - t} \left| \int_{[t,x]} (f(s) - f(x)) \, dm(s) \right|$$

$$\le \frac{1}{x - t} \int_{[t,x]} |f(s) - f(x)| \, dm(s)$$

$$\le \frac{1}{x - t} \int_{[t,x]} \varepsilon \, dm(s) = \frac{1}{x - t} \varepsilon (x - t) = \varepsilon,$$

which proves that $F'_-(x) = f(x)$ (derivative from the left). Similarly we prove that $F'_+(x) = f(x)$ for any $x \in [a, b)$ (derivative from the right). □

A *primitive* to $f: [a, b] \to \mathbb{C}$ is a function $F: [a, b] \to \mathbb{C}$ satisfying $F'(x) = f(x)$ for all $x \in [a, b]$ (with one sided derivatives at the end points). Theorem 13.21 shows that a continuous function on $[a, b]$ always has a primitive. The following theorem is a special case of the fundamental theorem of calculus for the Lebesgue integral.

Theorem 13.22. *Let $f: [a, b] \to \mathbb{C}$ be continuous on $[a, b]$. Then for any primitive F of f,*

$$\int_{[a,b]} f \, dm = F(b) - F(a).$$

Proof.   Let F be a primitive of f on $[a, b]$ and let $G(x) = \int_{[a,x]} f \, dm$ for $x \in [a, b]$. By the preceding theorem G is a primitive of f on $[a, b]$. The mean value theorem implies that $F(x) = G(x) + C$, where C is a complex constant. Then

$$F(b) - F(a) = (G(b) + C) - (G(a) + C) = G(b) = \int_{[a,b]} f \, dm.$$

□

Theorem 13.22 can be extended to quasicontinuous functions (see Problem 13.6).

It is customary to write

$$\int_a^b f(t) \, dt = \int_{[a,b]} f \, dm, \qquad \int_b^a f(t) \, dt = -\int_a^b f(t) \, dt$$

if f is Lebesgue integrable on $[a, b]$. If $f, g: [a, b] \to \mathbb{C}$ are continuous functions with primitives F, G, then we have the following useful formula for integration by parts:

$$\int_a^b F(t)g(t) \, dt = (FG)(b) - (FG)(a) - \int_a^b f(t)G(t) \, dt.$$

Example 13.23. The function $F(x) = \log x$ is a primitive to $f(x) = x^{-1}$ on $(0, \infty)$. For $x > 1$, $\int_1^x t^{-1} \, dt = F(x) - F(1) = \log x$; for $0 < x < 1$, $\int_1^x t^{-1} \, dt = -\int_x^1 t^{-1} \, dt = -(F(1) - F(x)) = \log x$.

Example 13.24. Let $z \neq 0$ be a complex number. Then $f(t) = \exp(zt)$ has a primitive $F(t) = z^{-1} \exp(zt)$ for all $t \in \mathbb{R}$, and $\int_a^b \exp(zt) \, dt = z^{-1}(\exp(zb) - \exp(za))$. In particular, $\int_0^{2\pi} \exp(imt) \, dt = (im)^{-1}(\exp(2m\pi i) - 1) = 0$ for any nonzero integer m.

13.3.2 *Integration of continuous functions on rectangles*

Let $A = [a_1, b_1]$ and $B = [a_2, b_2]$ be compact intervals. A complex valued function g on $A \times B$ is a *step function* if there exist partitions of A and B into subintervals $A_1, \ldots, A_n$ and $B_1, \ldots, B_s$, respectively, such that g is constant on the interior of each rectangle $A_j \times B_k$. Any continuous function $f \colon A \times B \to \mathbb{C}$ is uniformly continuous on the compact set $A \times B$, and therefore is the limit of a uniformly convergent sequence (f_n) of step functions on $A \times B$. We can show that (f_n) is an approximating sequence for f, and so f is integrable on the measure space $(A \times B, \Lambda(A \times B), m_2)$, where m_2 is the Lebesgue measure on $\mathbb{R}^2$.

If $f = \chi_{A_j \times B_k}$, then $\int_{A \times B} f\, dm_2 = \int_{A_j \times B_k} dm_2 = m(A_j)m(B_k)$. In this case

$$\int_{A \times B} f(x,y) dm_2(x,y) = \int_A \left(\int_B f(x,y)\, dm(y) \right) dm(x)$$

$$= \int_B \left(\int_A f(x,y)\, dm(x) \right) dm(y). \qquad (13.9)$$

The equation (13.9) remains true if f is any step function on $A \times B$, and consequently if f is any function continuous on $A \times B$ (the uniform limit of a sequence of step functions on $A \times B$).

Example 13.25. Calculate

$$\iint_{[0,1] \times [0,1]} \exp(x^2 + y^2) xy\, dm_2(x,y).$$

The integrand $f(x,y) = \exp(x^2 + y^2) xy$ is continuous on the compact square $[0,1] \times [0,1]$. Observe that, for a fixed y, $F(x) = \frac{1}{2} \exp(x^2 + y^2)$ is a primitive to $x \mapsto \exp(x^2 + y^2)x$, and that $G(y) = \frac{1}{2}(\exp(1 + y^2) - \exp(y^2))$ is a primitive to $(\exp(1 + y^2) - \exp(y^2))y$. Apply (13.9):

$$\iint_{[0,1] \times [0,1]} \exp(x^2 + y^2) xy\, dm_2(x,y) = \int_0^1 \left(\int_0^1 \exp(x^2 + y^2) xy\, dx \right) dy$$

$$= \int_0^1 (F(1) - F(0)) y\, dy = \tfrac{1}{2} \int_0^1 (\exp(1 + y^2) - \exp(y^2)) y\, dy$$

$$= \tfrac{1}{2}(G(1) - G(0)) = \tfrac{1}{4} e^2 - \tfrac{1}{2} e + \tfrac{1}{4}.$$

13.4 Measurable functions

The concept of a measurable function is important in integration as in many cases it enables us to decide whether a function is integrable. For instance, continuous functions are Borel measurable over Hausdorff topological spaces, and Lebesgue measurable over Euclidean spaces. We will show that every integrable function is equal μ-a.e. to a measurable function. One of the most useful results is the theorem stating that a measurable function dominated by an integrable function is itself integrable.

In several proofs involving a simple function f we used the fact that sets such as

$$\{t \in S : \alpha < f(t) \leq \beta\} = f^{-1}((\alpha, \beta])$$

are measurable. This observation motivates the following definition.

Definition 13.26. A function $f : S \to \mathbb{C}$ on a measurable space (S, Σ) is said to be *measurable* if

$$f^{-1}(B) \in \Sigma \qquad \text{for every Borel set} \quad B \subset \mathbb{C}.$$

This means that the inverse image of every Borel subset of the complex plane is a measurable set in S (Figure 13.1).

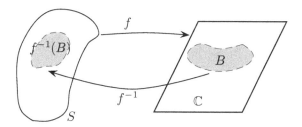

Figure 13.1

Clearly, every simple function is measurable. It is important to know that for measurability of f it is enough to check that the inverse images of open sets are in the given σ-algebra (Problem 13.18).

Example 13.27. Let S be a topological space, and $\mathcal{B}(S)$ the σ-algebra of all Borel subsets of S. Then every continuous function $f : S \to \mathbb{C}$ is measurable.

Indeed, by the topological criterion of continuity, $f^{-1}(G)$ is open in S for every open subset G of $\mathbb{C}$. So $f^{-1}(G) \in \mathcal{B}(S)$ for every G open, and that is enough to ensure that f is measurable.

Theorem 13.28. *Let (f_n) be a sequence of measurable functions convergent pointwise to a function f on S. Then f is measurable.*

Proof. We need the definition of $\limsup_{n\to\infty} A_n$ for a sequence of sets—see page 181 and Appendix A, Section A.3.

If G is an open set in $\mathbb{C}$, it can be verified that

$$f^{-1}(G) \subset \limsup_{n\to\infty} f_n^{-1}(G). \qquad (13.10)$$

Similarly, if $F \subset \mathbb{C}$ is closed, then

$$f^{-1}(F) \supset \limsup_{n\to\infty} f_n^{-1}(F). \qquad (13.11)$$

Let G be a fixed open set in $\mathbb{C}$. For each $p \in \mathbb{N}$ define an open set G_p by

$$G_p = \{z \in \mathbb{C} : d(z, G^c) > 1/p\}.$$

By (13.11) we have

$$f^{-1}(G) = \bigcup_{p=1}^{\infty} f^{-1}(\overline{G_p}) \supset \bigcup_{p=1}^{\infty} \limsup_{n\to\infty} f_n^{-1}(\overline{G_p}) \supset \bigcup_{p=1}^{\infty} \limsup_{n\to\infty} f_n^{-1}(G_p),$$

and by (13.10),

$$f^{-1}(G) = \bigcup_{p=1}^{\infty} f^{-1}(G_p) \subset \bigcup_{p=1}^{\infty} \limsup_{n\to\infty} f_n^{-1}(G_p).$$

Thus

$$f^{-1}(G) = \bigcup_{p=1}^{\infty} \limsup_{n\to\infty} f_n^{-1}(G_p) = \bigcup_{p=1}^{\infty} \bigcap_{n=1}^{\infty} \bigcup_{k=n}^{\infty} f_k^{-1}(G_p) \in \Sigma,$$

which proves that f is measurable. $\qquad\square$

Theorem 13.29. *For each measurable function $f: S \to \mathbb{C}$ there is a sequence (f_n) of simple functions on S such that $f_n \to f$ pointwise. The functions f_n can be chosen so that $|f_n| \leq |f|$ for all n.*

Proof. First we assume that f is bounded. The set $f(S)$ is then bounded and hence totally bounded in $\mathbb{C}$. Given $\varepsilon > 0$, there is a finite family of open balls $B_1, \ldots, B_s$ of radius $\frac{1}{2}\varepsilon$ which covers $f(S)$. Set

$$A_1 = B_1, \quad A_k = B_k \setminus (B_1 \cup \cdots \cup B_{k-1}), \quad k = 2, \ldots, s.$$

Then $A_1, \ldots, A_s$ are disjoint Borel sets in $\mathbb{C}$ covering $f(S)$. If $A_k \neq \emptyset$, the set $\overline{A_k}$ is compact (closed and bounded in $\mathbb{C}$), and so there is a point $z_k \in \overline{A_k}$ such that $|z_k| = \inf\{|z| : z \in A_k\}$; if $A_k = \emptyset$, choose $z_k = 0$. Set $E_k = f^{-1}(A_k)$ and define $g = \sum_{k=1}^{s} z_k \chi_{E_k}$. Since $E_1, \ldots, E_s$ are disjoint measurable, the function g is simple, and $|f(t) - g(t)| \leq \varepsilon$ for all $t \in S$ as each A_k has a diameter not exceeding ε. In addition, $|g| \leq |f|$. Thus g approximates f uniformly on S satisfying $\|f - g\|_\infty \leq \varepsilon$.

Let f be measurable and unbounded. For each n, the set $B_n = \{t \in S : |f(t)| \leq n\}$ is the inverse image under f of the set $\{z : |z| \leq n\}$, and so it is measurable. Define g_n by $g_n(t) = f(t)$ if $t \in B_n$, and $g_n(t) = 0$ otherwise. Each g_n is bounded, and $g_n \to f$ pointwise. It can be verified that each g_n is also measurable. By the first part of the proof, there is f_n simple with $\|f_n - g_n\|_\infty < 1/n$. So $f_n \to f$ pointwise in S. $\qquad\square$

It is useful to observe the following specialization of the preceding theorem.

Corollary 13.30. *If f is a bounded measurable function, then there is a sequence of simple functions which converges to f uniformly on S.*

Theorem 13.31. *The set $\mathcal{M} = \mathcal{M}(S, \Sigma)$ of all measurable functions on S is a complex vector space. If $f, g \in \mathcal{M}$, then $|f| \in \mathcal{M}$, $\overline{f} \in \mathcal{M}$ and $f \cdot g \in \mathcal{M}$.*

Proof. Follows from Theorems 13.28 and 13.29. $\qquad\square$

Note that the concept of measurability is defined in any measurable space (S, Σ), and that it does not require measure. However, for most applications in integration we need the concept of a μ-measurable function.

Definition 13.32. A function $f : S \to \mathbb{C}$ is said to be *μ-measurable* (or *measurable μ-a.e.*) if there exists a measurable function g such that $f = g$ μ-a.e.

Theorem 13.33. *A function $f : S \to \mathbb{C}$ is μ-measurable if and only if there exists a sequence (f_n) of simple functions such that $f_n \to f$ μ-a.e.*

Proof. The preceding definition and Theorem 13.29. $\qquad\square$

The preceding theorem and the definition of integrability show that every μ-integrable function is μ-measurable. In measure spaces in which the measure μ is complete, there is no difference between measurable and μ-measurable functions (Problem 13.26). In particular, if μ is the Lebesgue measure in $\mathbb{R}^n$ defined on the σ-algebra of Lebesgue measurable sets, every μ-measurable function is measurable.

Remark 13.34. We observe that Theorems 13.8 and 13.10, originally proved for series of simple functions, are valid for integrable functions. For this we examine the proofs of these two theorems. We assume first that the functions are measurable. Then all the sets defined in the proofs of those theorems are measurable, and the proofs go through. For the general case the functions are μ-measurable being μ-integrable. The results then follow since for every μ-measurable function there is a measurable function equal to it μ-a.e.

13.5   Convergence theorems for the integral

In this section we study conditions on the sequence (f_n) of integrable functions that will guarantee the convergence of integrals

$$\int_S f_n \, d\mu \to \int_S f \, d\mu. \tag{13.12}$$

The problem is often referred to as 'taking the limit under the integral sign' since it can be written in the form

$$\lim_{n\to\infty} \int_S f_n \, d\mu = \int_S \left(\lim_{n\to\infty} f_n \right) d\mu.$$

Remember that $\|f_n - f\|_{L^1} \to 0$ implies the convergence of the integrals.

The main result of this section is Lebesgue's dominated convergence theorem, the single most important theorem in the theory of Lebesgue integration. The condition that guarantees the limit passage (13.12) is that the sequence (f_n) converges μ-a.e. and is *dominated* by a positive integrable function. Not only is the theorem one of the most often used in theoretical arguments, but it is also a rich source of many applications. It gives a useful criterion of integrability of measurable functions and provides a technique for taking limits and for differentiating under the integral sign when the integrands depend on a real parameter.

The other two useful convergence theorems are the monotone convergence theorem and Fatou's lemma. The latter can often be used when Lebesgue's dominated convergence theorem is not immediately applicable. The following theorem on term by term integration is an important result in its own right, and also serves as a preparatory result for other convergence theorems.

Theorem 13.35 (Term-by-term integration of series). *Let $\sum_n f_n$ be a norm-absolutely convergent series of integrable functions, that is, let*

$\sum_n \int_S |f_n| \, d\mu < \infty$. *Then the series converges absolutely μ-a.e. to an integrable function f, and*

$$\int_S f \, d\mu = \int_S \left(\sum_{n=1}^{\infty} f_n \right) d\mu = \sum_{n=1}^{\infty} \int_S f_n \, d\mu. \tag{13.13}$$

Proof. Since $(L^1(\mu), \| \cdot \|_{L^1})$ is a complete space, there exists a function $f \in L^1(\mu)$ such that $\lim_{n \to \infty} \| f - \sum_{k=1}^{n} f_k \| = 0$. According to Remark 13.34 we can apply Theorems 13.8 and 13.10 to μ-integrable f_n; hence $\sum_n f_n = f$ μ-a.e. Equation (13.13) is proved as follows:

$$\left| \int_S f \, d\mu - \sum_{k=1}^{n} \int_S f_k \, d\mu \right| = \left| \int_S \left(f - \sum_{k=1}^{n} f_k \right) d\mu \right| \leq \left\| f - \sum_{k=1}^{n} f_k \right\|_{L^1} \to 0. \qquad \square$$

Theorem 13.36. *Let (f_n) be a Cauchy sequence of integrable functions. Then there exists a subsequence (g_n) of (f_n) and a function $f \in L^1(\mu)$ such that $g_n \to f$ μ-a.e. In addition, $\|f_n - f\|_{L^1} \to 0$.*

Proof. Extracting a suitable subsequence convert to a norm-absolutely convergent series. Then apply the preceding theorem. $\qquad \square$

It follows from the preceding theorem that if (f_n) is a Cauchy sequence of integrable functions such that $f_n \to f$ μ-a.e., then f is integrable and $\|f_n - f\|_{L^1} \to 0$.

Theorem 13.37 (Monotone convergence theorem). *Let (f_n) be a monotone sequence of real valued functions in $L^1(\mu)$ such that the integrals*

$$\int_S f_n \, d\mu$$

form a bounded real sequence. Then there exists $f \in L^1(\mu)$ such that $f_n \to f$ μ-a.e. and $\|f_n - f\|_{L^1} \to 0$.

Proof. We prove the theorem under the assumption that the sequence is increasing, that is, that $f_n \leq f_{n+1}$ for all n. By the monotonicity of the integral (Theorem 13.17 (v)), the real sequence $\int_S f_n \, d\mu$ is increasing. If $m < n$,

$$\|f_n - f_m\|_{L^1} = \int_S f_n \, d\mu - \int_S f_m \, d\mu,$$

which shows that (f_n) is Cauchy in $L^1(\mu)$. By Theorem 13.36, a subsequence converges almost everywhere; since (f_n) is monotone, it itself converges almost everywhere to an integrable function f for which $\|f_n - f\|_{L^1} \to 0$. The final result follows from Theorem 13.17 (vii). $\qquad \square$

Corollary 13.38. *Let (f_n) be a sequence of real valued functions in $L^1(\mu)$. If there is a real valued function $g \in L^1(\mu)$ such that $g \geq 0$ μ-a.e. and $|f_n| \leq g$ μ-a.e., then the functions*

$$u = \sup_{n \in \mathbb{N}} f_n \quad and \quad v = \inf_{n \in \mathbb{N}} f_n$$

are integrable.

Proof.   The functions $g_n = \max(f_1, f_2, \ldots, f_n)$ are integrable by Theorem 13.16 (iii), and form an increasing sequence. We have $g_n \to u$ μ-a.e. Also,

$$\left| \int_S g_n \, d\mu \right| \leq \int_S |g_n| \, d\mu \leq \int_S g \, d\mu;$$

consequently $u \in L^1(\mu)$ by the monotone convergence theorem. The infimum is treated similarly. $\qquad \square$

Theorem 13.39 (Lebesgue's dominated convergence theorem).
Let (f_n) be a sequence of integrable functions for which there exists an integrable dominant, that is, a function $g \in L^1(\mu)$ such that $g \geq 0$ and $|f_n| \leq g$ μ-a.e. for all n. If (f_n) converges μ-a.e. to a function f, then f is integrable and $\|f_n - f\|_{L^1} \to 0$.

Proof.   For a fixed integer n define

$$g_n = \sup_{i,k \geq n} |f_i - f_k|.$$

Since $|f_i - f_k| \leq 2g$ μ-a.e., each function g_n is integrable by Corollary 13.38. Also, (g_n) is a decreasing sequence of real valued integrable functions with $0 \leq \int_S g_n \, d\mu \leq 2\|g\|_{L^1}$. By the monotone convergence theorem, (g_n) is convergent both in L^1 and μ-a.e. to a function h in $L^1(\mu)$. But $h = 0$ μ-a.e. as

$$g_n = \sup_{i,k \geq n} |f_i - f_k| \leq \sup_{i \geq n} |f_i - f| + \sup_{k \geq n} |f - f_k|.$$

Therefore $\|g_n\|_{L^1} \to 0$. Since

$$\|f_i - f_k\|_{L^1} \leq \|g_n\|_{L^1} \text{ for all } i, k \geq n,$$

(f_n) is an L^1-Cauchy sequence. By the remark following Theorem 13.36, the μ-a.e. limit f of (f_n) is in $L^1(\mu)$, and $\|f_n - f\|_{L^1} \to 0$. The result then follows. $\qquad \square$

Recall that for a sequence of real numbers (a_n) we define

$$\liminf_{n\to\infty} a_n = \lim_{n\to\infty} \inf_{k\geq n} a_k$$

provided the limit exists.

Theorem 13.40 (Fatou's lemma). *Let (f_n) be a sequence of real valued nonnegative functions in $L^1(\mu)$ such that*

$$\liminf_{n\to\infty} \int_S f_n \, d\mu = c < \infty. \tag{13.14}$$

Then $f(t) = \liminf_{n\to\infty} f_n(t)$ exists for almost all $t \in S$, the function f is integrable, and

$$\int_S \liminf_{n\to\infty} f_n \, d\mu \leq \liminf_{n\to\infty} \int_S f_n \, d\mu = \liminf_{n\to\infty} \|f_n\|_{L^1}.$$

Proof. Fix an integer n, and put

$$f_{nm} = \min(f_n, f_{n+1}, \cdots f_{n+m}).$$

We recall that the functions f_{nm} are integrable in view of Theorem 13.16 (iii). The sequence (f_{nm}) is decreasing with m, and

$$0 \leq \int_S f_{nm} \, d\mu \leq \int_S f_n \, d\mu, \quad m = 1, 2, \ldots$$

By the monotone convergence theorem (Theorem 13.37), there is $g_n \in L^1(\mu)$ such that

$$g_n = \lim_{m\to\infty} f_{nm} = \inf_{k\geq n} f_k \quad \mu\text{-a.e.}$$

Since $g_n \leq f_k$ μ-a.e. for all $k \geq n$, we have

$$0 \leq \int_S g_n \, d\mu \leq \inf_{k\geq n} \int_S f_k \, d\mu \leq c. \tag{13.15}$$

We observe that $g_n \leq g_{n+1}$ μ-a.e. for all n. Hence we may apply the monotone convergence theorem to the μ-a.e. monotone sequence (g_n) to conclude that there is $f \in L^1(\mu)$ such that

$$f = \lim_{n\to\infty} g_n = \lim_{n\to\infty} \inf_{k\geq n} f_k = \liminf_{n\to\infty} f_n \quad \mu\text{-a.e.}$$

and that $\int_S g_n \, d\mu \to \int_S f \, d\mu$. In view of (13.14),

$$\int_S f \, d\mu = \lim_{n\to\infty} \int_S g_n \, d\mu \leq \lim_{n\to\infty} \inf_{k\geq n} \int_S f_k \, d\mu = \liminf_{n\to\infty} \int_S f_n \, d\mu. \qquad \square$$

It is interesting to observe that Fatou's lemma may fail if the assumption $f_n \geq 0$ (or $f_n \geq 0$ μ-a.e.) is not satisfied (see Problem 13.17 (v)). A special case of Fatou's lemma which is often useful arises when the sequence of the integrals in the lemma is bounded above. We then have a guarantee that $\liminf_n \int_S f_n \, d\mu$ is finite.

Seemingly, Fatou's lemma is very restrictive, as it applies only to non-negative functions. However, in many situations we cannot use Lebesgue's dominated convergence directly as we do not have an integrable dominant, but may apply Fatou's lemma to the sequence $(|f_n|)$.

The following theorem is one of the main reasons for our interest in measurable functions. It gives an important criterion of integrability, as in many cases μ-measurability is given or is easy to verify. We already know that the integrability of f implies the integrability of $|f|$. The following theorem can be viewed as a partial converse of this fact.

Theorem 13.41. *A function f is μ-integrable if and only if f is μ-measurable and there exists an integrable function $g \geq 0$ such that $|f| \leq g$ μ-a.e. (Such a function g is called an integrable dominant for f.)*

Proof. Suppose that f is integrable. Then f is μ-measurable by the definition of integrability. In addition, $|f|$ is also integrable, and $|f| \leq g$ is satisfied with $g = |f|$.

Suppose f is μ-measurable, and $|f| \leq g$ μ-a.e. for some integrable function g.

Case 1: f is simple (not necessarily integrable). If A is the set on which f is nonzero, then A is measurable and $0 < c = \min_{t \in A} |f(t)| \leq g(t)$ for almost all $t \in A$. Then $c\mu(A) \leq \int_A g \, d\mu$, which shows that A has finite measure. So f is integrable.

Case 2: f is measurable. Then there exists a sequence of simple functions f_n such that $f_n \to f$ and $|f_n| \leq |f| \leq g$ μ-a.e. Each f_n is integrable by Case 1 of this proof, and f is integrable by Lebesgue's dominated convergence theorem.

Case 3: f is μ-measurable. Then there exists a measurable function h such that $f = h$ μ-a.e. Then $|h| \leq g$ μ-a.e., and h is integrable by Case 2 of this proof. Consequently f is integrable. $\qquad\square$

This theorem has an important corollary.

Corollary 13.42. *A function f is μ-integrable if and only if f is μ-measurable and $|f|$ is μ-integrable.*

In general, from the integrability of $|f|$ we cannot deduce the integrability of f. This is demonstrated in the following example.

Example 13.43. Let T be the nonmeasurable Vitali set in $\mathbb{R}$ constructed in Section 12.1.2. Define $f\colon [0,1] \to \mathbb{C}$ by $f(t) = 1$ if $t \in [0,1] \cap T$ and $f(t) = -1$ if $t \in [0,1] \cap T^c$. Then $|f| = 1$ is a Lebesgue integrable function on $[0,1]$. However, f is not Lebesgue measurable since $f^{-1}(\{1\}) = [0,1] \cap T$ is not Lebesgue measurable. Since the Lebesgue measure is complete, m-measurability coincides with measurability. Consequently f is not integrable in view of Theorem 13.41.

From the definition we know that a simple integrable function vanishes outside a set of finite measure. This is reflected in an important property of μ-integrable functions given in the following theorem whose proof depends on μ-measurability of integrable functions.

Theorem 13.44. *Any μ-integrable function f vanishes outside a σ-finite set.*

Proof. (Problem 13.28). $\qquad\qquad\qquad\qquad\qquad\qquad\qquad\qquad\square$

Theorem 13.41 enables us to weaken the hypotheses of several theorems on integration by replacing integrable functions by functions measurable μ-a.e. The following is a sample.

Theorem 13.45 (Lebesgue's dominated convergence theorem II).
Let (f_n) be a sequence of μ-measurable functions for which there exists an integrable dominant $g \in L^1(\mu)$. If (f_n) converges μ-a.e. to a function f, then f is integrable and

$$\int_S f \, d\mu = \lim_{n\to\infty} \int_S f_n \, d\mu.$$

13.6 Integrals dependent on a parameter

In this section we look at functions $f\colon S \times M \to \mathbb{C}$, where (S, Σ, μ) is a measure space and M a metric space, and study integrals of the form $F(t) = \int_S f(s,t) \, d\mu(s)$, where the integrand depends on a parameter $t \in M$. (Here the notation $d\mu(s)$ is used to indicate that s is the variable with respect to which we integrate.)

Theorem 13.46 (Limit under the integral sign). *Let M be a metric space and let $f\colon S \times M \to \mathbb{C}$ be such that:*

(i) $s \mapsto f(s,t)$ *is μ-integrable for each $t \in M$.*

(ii) $\lim_{t \to t_0} f(s,t) = f(s,t_0)$ *for some $t_0 \in M$ and all $s \in S$.*

(iii) *There exists $g \in L^1(\mu)$ such that $|f(s,t)| \le g(s)$ for all $s \in S$ and all $t \in M$.*

Then

$$\lim_{t \to t_0} \int_S f(s,t)\,d\mu(s) = \int_S \lim_{t \to t_0} f(s,t)\,d\mu(s) = \int_S f(s,t_0)\,d\mu(s).$$

Proof.   Select a sequence (t_n) in M convergent to t_0, define $f_n(s) = f(s,t_n)$, and apply Lebesgue's dominated convergence theorem.   □

Theorem 13.47 (Differentiation under the integral sign). *Let I be an interval in $\mathbb{R}$ and let $f\colon S \times I \to \mathbb{C}$ be such that:*

(i) $s \mapsto f(s,t)$ *is μ-integrable for each $t \in I$.*

(ii) *The partial derivative $(\partial/\partial t)f(s,t)$ exists for each $t \in I$ and each $s \in S$.*

(iii) *There exists $g \in L^1(\mu)$ such that $|(\partial/\partial t)f(s,t)| \le g(s)$ for all $s \in S$ and all $t \in I$.*

Then $F(t) := \int_S f(s,t)\,d\mu(s)$ is differentiable on I with

$$F'(t) = \int_S \frac{\partial f}{\partial t}(s,t)\,d\mu(s) \quad \text{for all } s \in S \text{ and all } t \in I.$$

Proof.   Let $t \in I$ be fixed, let (t_n) be a sequence in I such that $t_n \to t$, $t_n \ne t$, and define $f_n(s) = (f(s,t_n) - f(s,t))/(t_n - t)$. Then each f_n is μ-integrable, while

$$\frac{\partial f}{\partial t}(s,t) = \lim_{n \to \infty} f_n(s), \quad s \in S.$$

Hence $\partial f/\partial t$ is μ-measurable. By the mean value theorem, for each n and each $s \in S$ there exists a point $w_n(s)$ between t_n and t such that $f_n(s) = (\partial f/\partial t)(w_n(s))$; hence $|f_n(s)| \le g(s)$ for all $s \in S$, and Lebesgue's dominated convergence theorem can be applied to conclude that

$$F'(t) = \lim_{n \to \infty} \frac{F(t_n) - F(t)}{t_n - t} = \lim_{n \to \infty} \int_S f_n(s)\,d\mu(s) = \int_S \frac{\partial f}{\partial t}(s,t)\,d\mu(s).$$

□

In the preceding theorem it is not always possible to find an integrable dominant $g(s)$ for $(\partial/\partial t)f(s,t)$ for all $t \in I$ if I is unbounded; however, since differentiation is a 'local' operation, for a given $t_0 \in I$ we can restrict t to some finite subinterval (t_1, t_2) of I containing t_0, and find an integrable dominant on (t_1, t_2). This is illustrated in the following example. We shall anticipate the fundamental theorem of calculus for Lebesgue integration (Theorem 14.11). $\qquad\bullet$

Example 13.48. Using repeated differentiation under the integral sign, calculate the Lebesgue integrals

$$\int_0^1 s^t (\log s)^n \, ds, \quad t > -1, \quad n = 0, 1, 2 \ldots$$

We start with $F(t) = \int_0^1 s^t \, ds$ and apply Theorem 13.47 to the function $f(s,t) = s^t$ on $(0,1) \times (t_1, t_2)$ with a suitable selection of $-1 < t_1 < t_2 < \infty$. First verify the conditions of the theorem.

(i) For any fixed $t > -1$, $f(s,t) = s^t$ is integrable on $(0,1)$ with

$$F(t) = \int_0^1 s^t \, ds = \frac{1}{t+1}$$

(by the fundamental theorem of calculus).

(ii) The partial derivative of f exists for each $s \in (0,1)$ and each $t > -1$:

$$\frac{\partial}{\partial t} f(s,t) = s^t \log s.$$

(iii) Let us restrict the parameter t to an interval (t_1, t_2), where $-1 < t_1 < t_2 < \infty$. Since $s \in (0,1)$, the function $t \mapsto s^t$ is decreasing on (t_1, t_2), so $s^t |\log s| \le s^{t_1} |\log s|$. Then $g(s) := s^{t_1} |\log s|$ is an integrable dominant for $\partial f/\partial t$ on $(0,1)$ independent of $t \in (t_1, t_2)$ (observe that $s^{t_1} |\log s| = O(s^\alpha)$ as $s \to 0+$ for any α satisfying $-1 < \alpha < \min\{0, t_1\}$).

Having satisfied the three conditions we can apply Theorem 13.47, and for any t satisfying $t_1 < t < t_2$ obtain

$$\int_0^1 s^t \log s \, ds = \int_0^1 \frac{\partial}{\partial t} s^t \, ds = \frac{d}{dt} \int_0^1 s^t \, ds = F'(t) = -\frac{1}{(t+1)^2}.$$

But this is in fact true for an *arbitrary* $t > -1$ as for any such t we can select t_1, t_2 so that $-1 < t_1 < t < t_2 < \infty$.

In the general case we get for each $t > -1$,

$$\int_0^1 s^t (\log s)^n \, ds = \int_0^1 \left(\frac{\partial}{\partial t}\right)^n s^t \, ds = \left(\frac{d}{dt}\right)^n \int_0^1 s^t \, ds$$

$$= \left(\frac{d}{dt}\right)^n \frac{1}{t+1} = \frac{(-1)^n n!}{(t+1)^{n+1}}$$

(exercise).

Theorem 13.47 can be used to calculate integrals in several ways. Let $F(t) := \int_S f(s,t) \, d\mu(s)$.

(a) If we can calculate $\int_S f(s,t) \, d\mu(s)$, then by differentiating with respect to t we can find

$$\int_S \left(\frac{\partial}{\partial t} \right)^n f(s,t) \, d\mu(s) = F^{(n)}(t), \quad n \in \mathbb{N}$$

(as in Example 13.48).

(b) If we can calculate $F'(t)$, then $\int_S f(s,t) \, d\mu(s)$ can be found as a suitable primitive to $F'(t)$. (Problem 13.35.)

(c) If $F(t)$ satisfies a differential equation, we can calculate $\int_S f(s,t) \, d\mu(s)$ by solving that equation. (Problem 13.36.)

13.7   Problems for Chapter 13

1. Show that equality μ-a.e. is an equivalence relation (see A.3).

2. Prove that the definition of the integral of an integrable simple function f is independent of an f-partition.

Suggestion. Let $\{A_k\}_{k=1}^n$ and $\{B_j\}_{j=1}^m$ be two f-partitions of S with f taking the value c_k on A_k and d_j on B_j. (Note that $c_k = d_j$ on $A_k \cap B_j$.) Express the integral in two ways using the expansion $A_k = \bigcup_{j=1}^m (A_k \cap B_j)$.

3. Let $f_n \to f$ μ-a.e. and let $g_n = f_n$ μ-a.e. for all n. Show that $g_n \to f$ μ-a.e.

4. If $f \in L^1(\mu)$ and $g = f$ μ-a.e. show that $g \in L^1(\mu)$.

5. Consider the measure space $(\mathbb{N}, \mathcal{P}(\mathbb{N}), \nu)$ with ν the counting measure on $\mathbb{N}$.

(i) Show that f is an integrable simple function if and only if there exists $n \in \mathbb{N}$ such that $f(k) = 0$ for all $k > n$.

(ii) Show that f is integrable if and only if the series $\sum_{n=1}^\infty |f(n)|$ is convergent. In this case show also that

$$\int_\mathbb{N} f \, d\nu = \sum_{n=1}^\infty f(n).$$

6. Extend Theorem 13.22 to quasicontinuous functions: Let $f : [a,b] \to \mathbb{C}$ be quasicontinuous on $[a,b]$. Then there exists a function $F : [a,b] \to \mathbb{C}$ continuous on $[a,b]$ and such that $F'(x) = f(x)$ for all $x \in [a,b]$ at which f is continuous.

For any such function F,

$$\int_{[a,b]} f \, dm = F(b) - F(a).$$

Suggestion. Look up Corollary B.26 in Appendix B.

7. If $f \in L^1(\mu)$ is a bounded function and μ a finite measure, show that

$$\left| \int_S f \, d\mu \right| \leq \int_S |f| \, d\mu \leq \|f\|_\infty \cdot \mu(S).$$

8. Let (S, Σ, μ) be a finite measure space, and let (f_n) be a sequence in $L^1(\mu)$ which converges uniformly to f. Show that f is in $L^1(\mu)$ and that

$$\int_S f \, d\mu = \lim_{n \to \infty} \int_S f_n \, d\mu.$$

Suggestion. Preceding problem.

9. Show that f is integrable if and only if $\mathrm{Re}\, f$ and $\mathrm{Im}\, f$ are integrable.

10. If f is an integrable function, show that $\overline{f}$ is also integrable and that $\int_S \overline{f} \, d\mu = \overline{\int_S f \, d\mu}$.

11. Let u, v be real valued integrable functions. Show that the function $w = \sqrt{u^2 + v^2}$ is integrable, even though u^2, v^2 themselves need not be integrable.

12. Let f be an integrable function and A a measurable set. Prove that the function $f \cdot \chi_A$ is integrable.

13. If f is an integrable function and A a measurable set of measure zero, show that $\int_A f \, d\mu = 0$.

14. Let $E_n \nearrow E$ in a σ-algebra Σ, and let f be integrable on E. Show that

$$\int_E f \, d\mu = \lim_{n \to \infty} \int_{E_n} f \, d\mu.$$

15. Let $A = \bigcup_{n=1}^\infty A_n$ with A_n disjoint sets in a σ-algebra Σ, and let f be integrable on A. Show that

$$\int_A f \, d\mu = \sum_{n=1}^\infty \int_{A_n} f \, d\mu.$$

16. Prove the monotone convergence theorem replacing the assumption that the sequence (f_n) is monotone by the assumption that it is monotone μ-a.e.

Suggestion. For each n there is a set A_n of measure zero such that $f_n(t) \leq f_{n+1}(t)$ for all $t \in A_n^c$. Consider $A = \bigcup_{n=1}^\infty A_n$.

17.   Consider the measure space $(\mathbb{R}, \mathcal{B}, m)$, where $\mathcal{B}$ is the σ-algebra of all Borel sets in $\mathbb{R}$, and m is the Lebesgue measure.

(i)   Show that the functions $f_n = \chi_{[0,n]}$ are integrable, and the sequence (f_n) is monotonically increasing to $f = \chi_{[0,\infty)}$. Show that f is not integrable. Why does the monotone convergence theorem not apply?

(ii)   Let $f_n = n\chi_{[0,1/n]}$. Show that the condition $|f_n| \leq g$ cannot be dropped in Lebesgue's dominated convergence theorem.

(iii)   Let $f_n = (1/n)\chi_{[0,n]}$ and $f = 0$. Show that the sequence (f_n) converges uniformly to f, but

$$\int_{\mathbb{R}} f \, dm \neq \lim_{n \to \infty} \int_{\mathbb{R}} f_n \, dm.$$

Why does this not contradict the monotone convergence theorem? Does Fatou's lemma apply?

(iv)   Let $g_n = n\chi_{[1/n,2/n]}$ and $g = 0$. Show that

$$\int_{\mathbb{R}} g \, dm \neq \lim_{n \to \infty} \int_{\mathbb{R}} g_n \, dm.$$

Does the monotone convergence theorem or Fatou's lemma apply?

(v)   If $f_n = (-1/n)\chi_{[0,n]}$, show that $f_n \to 0$ uniformly on the interval $[0, \infty)$. Use this sequence to show that Fatou's lemma may fail if we drop the requirement that $f_n \geq 0$ for all n.

18.   Show that in the definition of a measurable function it is enough to check that $f^{-1}(G) \in \Sigma$ for every *open* set $G \subset \mathbb{C}$.
Suggestion. Show that the family of all subsets E of $\mathbb{C}$ for which $f^{-1}(E) \in \Sigma$ is a σ-algebra.

19.   Let (S, Σ) be a measurable space, and let $S = A \cup B$ for some $A, B \in \Sigma$. The families $\Sigma_A = \{A \cap E : E \in \Sigma\}$ and $\Sigma_B = \{B \cap E : E \in \Sigma\}$ are σ-algebras contained in Σ. Let $f: S \to \mathbb{C}$ be such that the restrictions f_A and f_B are Σ_A-measurable and Σ_B-measurable, respectively. Prove that f is Σ-measurable.

20.   Show that a bounded μ-measurable function on a finite measure space is integrable.

21.   Let $g: [a, b] \to \mathbb{R}$ be monotonic. Prove that g is Lebesgue integrable on $[a, b]$.

22.   Let g be a nonnegative measurable function on S. For each $A \in \Sigma$ define $\nu(A) = \int_A g \, d\mu$ if the integral exists, and $\nu(A) = \infty$ otherwise.

(i)   Show that ν is a positive measure on Σ.

(ii)   If f is a simple function, show that $\int_S f \, d\nu = \int_S fg \, d\mu$ if one of the integrals exists.

(iii) Repeat the preceding exercise for a measurable function f.

23. Show that a function f continuous on a compact set $K \subset \mathbb{R}^k$ is Lebesgue integrable over K.

24. Let f be a function on the interval $[a, b]$ defined by $f(t) = 1$ if t is irrational, and $f(t) = 0$ otherwise. Show that f is Lebesgue integrable over $[a, b]$ and find $\int_{[a,b]} f \, dm$.

25. Let $E \subset \mathbb{R}^k$ be a Lebesgue measurable set of finite measure, and let the function $f \colon E \to \mathbb{C}$ be bounded and continuous almost everywhere in E. Show that f is Lebesgue integrable over E.

26. If μ is a complete measure, show that every μ-measurable function is measurable.

27. (Important.) *Absolute continuity of the integral.* Let $f \in L^1(\mu)$. Prove that to each $\varepsilon > 0$ there exists $\delta > 0$ such that

$$A \in \Sigma, \ \mu(A) < \delta \implies \int_A |f| \, d\mu < \varepsilon.$$

28. (Important.) Let f be a μ-integrable function. Show that there is a σ-finite set A such that $f = 0$ outside A.

Suggestion. If (f_n) is an approximating sequence for f, then for each n, f_n vanishes outside a set $B_n \in \Sigma$ of finite measure. Consider $B = \bigcup_{n=1}^{\infty} B_n$.

29. *Extension lemma.* Let $E_n \nearrow E$ in a σ-algebra Σ, let a function $f \colon S \to \mathbb{C}$ be integrable on E_n for all n, and let $\sup_n \int_{E_n} |f| \, d\mu$ be finite. Then f is integrable over E, and

$$\int_E f \, d\mu = \lim_{n \to \infty} \int_{E_n} f \, d\mu.$$

30. *Extension lemma for unions.* Let $\bigcup_{n=1}^{\infty} A_n = A$ in a σ-algebra Σ with disjoint A_n, let f be integrable on A_n for all n, and let $\sum_{n=1}^{\infty} \int_{A_n} |f| \, d\mu$ be convergent. Then f is integrable over A and

$$\int_A f \, d\mu = \sum_{n=1}^{\infty} \int_{A_n} f \, d\mu.$$

$\star$ In the remaining problems you may anticipate the fundamental theorem of calculus for Lebesgue integration (Theorem 14.11) stated and proved in the next section and/or the relations between the Lebesgue and Newton integrals (Theorems 14.12 and 14.13). $\star$

31. Differentiating

$$F(t) = \int_0^1 \frac{dx}{1 + tx}, \quad t \in (-1, \infty),$$

under the integral sign, calculate the integrals

$$\int_0^1 \frac{x^n \, dx}{(1+x)^{n+1}}, \quad n = 1, 2, \dots$$

32. Prove that the following functions are well defined and continuous:

(i) $J(t) = \displaystyle\int_0^\infty \frac{s^{t-1}}{1+s} \, ds, \quad 0 < t < 1,$

(ii) $\Gamma(t) = \displaystyle\int_0^\infty e^{-s} s^{t-1} \, ds, \quad t > 0 \quad$ (the Gamma function).

33. Use differentiation under the integral sign to evaluate

$$\int_0^\infty s e^{-ts} \sin s \, ds, \quad t > 0.$$

Suggestion. First use the complex exponential and the fundamental theorem of calculus to show that

$$\int_0^\infty e^{-ts} \sin s \, ds = (1 + t^2)^{-1}, \quad t > 0.$$

34. Show that, for any $t > 0$,

$$\int_0^\infty e^{-tx} \, dx = \frac{1}{t}.$$

Using differentiation under the integral sign, prove that

$$\int_0^\infty x^n e^{-tx} \, dx = \frac{n!}{t^{n+1}}, \quad t > 0, \quad n = 1, 2, \dots$$

35. Using differentiation under the integral sign, show that

$$F(t) := \int_0^\infty e^{-ts} \frac{\sin s}{s} \, ds = \frac{\pi}{2} - \arctan t, \quad t > 0.$$

Suggestion: Start with $F'(t)$.

36. Let

$$F(b) := \int_0^\infty e^{-as^2} \cos bs \, ds$$

for a fixed $a > 0$ and a variable $b > 0$. Differentiating under the integral sign show that $F'(b) = -(b/2a)F(b)$. Find $F(b)$ by solving the differential equation.

Chapter 14

Integral on the Real Line

14.1 Differentiation of integrals

In this section we study the Lebesgue integral on the real line. We will use the abbreviation 'a.e.' to mean 'almost everywhere with repect to the Lebesgue measure'. The σ-algebra of Lebesgue measurable sets on $\mathbb{R}$ will be denoted by Λ. Lebesgue measurable sets can be constructed from a generating pair $(\mathcal{N}, m)$, described in the section on construction of measures. Alternatively, they are obtained from Borel sets by a measure completion: A set E is Lebesgue measurable if and only if there are Borel sets A, B such that $A \subset E \subset B$ and $m(B \backslash A) = 0$. The measure $m(E)$ is then defined as the infimum of

$$\sum_{n=1}^{\infty} (b_n - a_n)$$

over all countable covers of E by intervals $(a_n, b_n]$.

The Lebesgue measure is *translation invariant*, that is

$$m(E + u) = m(E),$$

where $E \in \Lambda$ and $E + u = \{x + u : x \in E\}$ is the translation of E by u (Problems 12.24 and 12.25).

The Lebesgue integral of f over an interval I with end points $a < b$ will be often written as

$$\int_I f \, dm = \int_a^b f(t) \, dt;$$

since singletons have Lebesgue measure zero, it is irrelevant whether the integral is over $[a, b]$, $[a, b)$, $(a, b]$ or (a, b). This peculiarity of the Lebesgue integral is not shared by integrals over other measures. We also make the following convention: If $b < a$, we define

$$\int_a^b f(t)\, dt = -\int_b^a f(t)\, dt.$$

In view of the translation invariance of m, we can show that

$$\int_a^b f(t + u)\, dt = \int_{a+u}^{b+u} f(t)\, dt;$$

this can be first verified for simple functions and then made general by a limit passage.

Since the Lebesgue measure m is complete, every m-measurable function is in fact Lebesgue measurable (Problem 13.26). In particular, every Lebesgue integrable function is Lebesgue measurable.

We discuss the important concept of absolute continuity which is at the heart of the relation between the Lebesgue integral and the inversion of derivatives.

Definition 14.1. A function $F\colon I \to \mathbb{C}$ is *absolutely continuous* on an interval I if, for each $\varepsilon > 0$, there is $\delta > 0$ such that for any finite pairwise disjoint family of subintervals (a_k, b_k) of I for which

$$\sum_{k=1}^n (b_k - a_k) < \delta$$

we have

$$\sum_{k=1}^n |F(b_k) - F(a_k)| < \varepsilon.$$

We observe that an absolutely continuous function is also uniformly continuous (exercise); the real and imaginary parts of an absolutely continuous function are absolutely continuous (exercise).

Theorem 14.2. *An absolutely continuous real valued function F can be written as the difference $F = F_1 - F_2$ of two absolutely continuous increasing functions F_1, F_2. So every complex valued absolutely continuous function F can be written as*

$$F = F_1 - F_2 + i(F_3 - F_4), \tag{14.1}$$

where F_1, F_2, F_3, F_4 are absolutely continuous real valued increasing functions.

Proof. Let F be an absolutely continuous real valued function. For each $x \in [a, b]$ define

$$V_a^x(F) = \sup \sum_{k=1}^{n} |F(b_k) - F(a_k)|, \tag{14.2}$$

where the supremum is taken over all finite pairwise disjoint families of subintervals (a_k, b_k) of $[a, x]$. (The number $V_a^x(F)$ is called the *total variation of F on $[a, x]$*; any function F for which the supremum (14.2) exists is called a *function of bounded variation on $[a, x]$*.) The absolute continuity of F ensures that the supremum always exists. Prove as an exercise that the function $V(x) = V_a^x(F)$ is absolutely continuous, and then check that the functions V and $V - F$ are increasing. Finally,

$$F = V - (V - F).$$

The statement about a complex valued F then easily follows. $\qquad \square$

Theorem 14.3. *Let $f : [a, b] \to \mathbb{C}$ be Lebesgue integrable on $[a, b]$ and let F be defined by*

$$F(x) = \int_a^x f(t)\, dt \ \ \text{for all } x \in [a, b]. \tag{14.3}$$

Then F is absolutely continuous on $[a, b]$.

Proof. Problem 13.27 implies that to each $\varepsilon > 0$ there exists $\delta > 0$ such that, for each Lebesgue measurable set $A \subset [a, b]$,

$$m(A) < \delta \implies \int_A |f(t)|\, dt < \varepsilon.$$

The results follows on considering sets of the form $A = \bigcup_{k=1}^{n} (a_k, b_k)$. $\quad \square$

The following theorem, one of the most important results in theory of functions of a real variable, was proved by Lebesgue in 1904 in the last chapter of his book [33] on integration.

Lebesgue's differentiation theorem *A real valued monotonic function g on $[a, b]$ possesses a derivative g' a.e. in $[a, b]$.*

Section B.5 of Appendix B is devoted to the proof of this remarkable result (Theorem B.29). We also need the fact that *every real valued monotonic function on $[a, b]$ is Lebesgue integrable* (see Problem 13.21).

Theorem 14.4. *If g is an increasing real valued function on $[a, b]$, then g' is integrable with*

$$\int_a^b g'(t)\, dt \le g(b) - g(a).$$

Proof.   Define $g(t) = g(b)$ if $t > b$ and for each $n \in \mathbb{N}$ set

$$g_n(t) = \frac{g(t + \frac{1}{n}) - g(t)}{\frac{1}{n}} \quad \text{for } t \in [a, b].$$

In view of Lebesgue's differentiation theorem, $g_n \to g'$ a.e. on $[a, b]$. Each g_n is Lebesgue integrable over $[a, b]$ being the difference of real valued monotonic functions. We have

$$\int_a^b g_n(t) \, dt = n \int_a^b \left(g(t + \tfrac{1}{n}) - g(t) \right) dt$$

$$= n \int_{a+1/n}^{b+1/n} g(t) \, dt - n \int_a^b g(t) \, dt$$

$$= n \int_b^{b+1/n} g(b) \, dt - n \int_a^{a+1/n} g(t) \, dt$$

$$\leq n \int_b^{b+1/n} g(b) \, dt - n \int_a^{a+1/n} g(a) \, dt \quad (g \text{ is increasing})$$

$$= g(b) - g(a).$$

We can apply Fatou's lemma as the functions g_n are nonnegative and the real sequence $\left(\int_a^b g_n(t) \, dt \right)$ is bounded above:

$$\int_a^b g'(t) \, dt = \int_a^b \lim_{n \to \infty} g_n(t) \, dt \leq \liminf_{n \to \infty} \int_a^b g_n(t) \, dt \leq g(b) - g(a).$$

This completes the proof. $\square$

Combining Theorems 14.2 and 14.3 with the decomposition (14.1) of absolutely continuous functions, we see that the function F defined by (14.3) has a derivative a.e. The following result will help to show that $F' = f$ a.e.

Lemma 14.5. *Let $g \colon [a, b] \to \mathbb{C}$ be a Lebesgue integrable function such that $\int_a^x g(t) \, dt = 0$ for all $x \in [a, b]$. Then $g = 0$ a.e.*

Proof.   We show that $\int_E g = 0$ for every Lebesgue measurable set $E \subset [a, b]$. Let $\mathcal{D}$ be the family of all Lebesgue measurable subsets A of $[a, b]$ for which $\int_A g = 0$. It can be checked that $\mathcal{D}$ is a σ-algebra and that $\mathcal{D}$ contains all open subsets of $[a, b]$ (this follows from the fact that $\int_u^v g(t) \, dt = 0$ for every interval (u, v) and that open sets in $[a, b]$ are countable unions of open intervals). Consequently, $\mathcal{D}$ contains all Borel subsets of $[a, b]$. Let $E \subset [a, b]$

be Lebesgue measurable. There is a Borel set A such that $E \subset A \subset [a, b]$ and $m(A \backslash E) = 0$. Therefore

$$\int_E g = \int_A g - \int_{A \backslash E} g = 0.$$

The function g is measurable since Lebesgue measure is complete. Let a be a nonzero complex number and let $A = g^{-1}(G)$, where G is the disc $|z - a| < \frac{1}{2}|a|$. Then A is measurable, $A \in \mathcal{D}$, and

$$|a|\, m(A) = \left| \int_A (g(t) - a)\, dt \right| \leq \int_A |g(t) - a|\, dt \leq \tfrac{1}{2}|a|\, m(A);$$

this shows that $m(A) = 0$. The set $\mathbb{C} \backslash \{0\}$ can be covered by a countable union of such discs; from this we deduce that $g = 0$ a.e. $\qquad \square$

The following two results give the relation between Lebesgue integration and differentiation. They appear as the culmination of the measure and integration theory in Lebesgue's doctoral dissertation [32].

Theorem 14.6 (Differentiation of integrals). *If a function $f \colon [a, b] \to \mathbb{C}$ is Lebesgue integrable on $[a, b]$ and*

$$F(x) = \int_a^x f(t)\, dt,$$

then F is absolutely continuous on $[a, b]$ and $F'(x) = f(x)$ for almost all x in $[a, b]$.

Proof. (i) Assume first that f is bounded, say $|f| \leq c$. Extend F to $[a, b + 1]$ by setting $F(x) = F(b)$ if $b \leq x \leq b + 1$. For each $n \in \mathbb{N}$ define

$$f_n(x) = \frac{F(x + \frac{1}{n}) - F(x)}{\frac{1}{n}} = n \int_x^{x + 1/n} f(t)\, dt.$$

The derivative F' exists a.e. by Theorems 14.2, 14.3 and Lebesgue's differentiation theorem. So $f_n \to F'$ a.e. in $[a, b]$, while $|f_n| \leq n \int_x^{x+1/n} |f(t)|\, dt \leq c \chi_{[a,b]}$. Writing $G(s) = \int_a^s F(t)\, dt$ and applying Lebesgue's dominated convergence theorem, we get

$$\int_a^x F'(t)\, dt = \lim_{n \to \infty} \int_a^x f_n(t)\, dt$$

$$= \lim_{n \to \infty} n \int_a^x \left(F(t + \tfrac{1}{n}) - F(t) \right) dt$$

$$= \lim_{n \to \infty} \left(n \int_x^{x+1/n} F(t)\, dt - n \int_a^{a+1/n} F(t)\, dt \right)$$

$$= G'(x) - G'(a) = F(x) - F(a) \qquad \text{(Theorem 13.21)}$$

$$= \int_a^x f(t)\, dt.$$

Applying Lemma 14.5, we get $F' = f$ a.e.

(ii) Assume that f is nonnegative but possibly unbounded. For each $n \in \mathbb{N}$ define

$$f_n(x) = \min\left(f(x), n\right), \quad F_n(x) = \int_a^x f_n(t)\, dt, \quad G_n(x) = \int_a^x (f(t) - f_n(t))\, dt.$$

By part (i) of the proof, $F_n' = f_n$ a.e. So, for each n,

$$F' = G_n' + F_n' = G_n' + f_n \geq f_n \quad \text{a.e.}$$

as $G_n' \geq 0$ a.e., and

$$F' \geq \lim_{n \to \infty} f_n = f \quad \text{a.e.}$$

So

$$\int_a^x F'(t)\, dt \geq \int_a^x f(t)\, dt = F(x) - F(a).$$

F is increasing as f is nonnegative. In view of Theorem 14.4, $\int_a^x F'(t)\, dt \leq F(x) - F(a)$. Therefore

$$\int_a^x F'(t)\, dt = F(x) - F(a) = \int_a^x f(t)\, dt \text{ for all } x \in [a, b];$$

by Lemma 14.5, $F' = f$ a.e.

(iii) If f is arbitrary complex valued, we can write $f = u + iv$ with u, v real valued, and further write $u = u^+ - u^-$, $v = v^+ - v^-$, where $g^+ = \max(g, 0) = \frac{1}{2}(|g| + g)$, $g^- = \max(-g, 0) = \frac{1}{2}(|g| - g)$. Then part (ii) of the proof applies to nonnegative functions u^+, u^-, v^+ and v^-. □

Theorem 14.7 (Integration of derivatives). *Let $F \colon [a, b] \to \mathbb{C}$ be absolutely continuous on $[a, b]$. Then the derivative $F'(x)$ exists for almost all x in $[a, b]$, F' is Lebesgue integrable, and*

$$\int_a^b F'(t)\, dt = F(b) - F(a).$$

Proof. F is differentiable a.e. in view of Theorems 14.2 and B.29. Define G by $G(x) = \int_a^x F'(t)\, dt$ for all $x \in [a, b]$. By the preceding theorem, $G' = F'$ a.e. The function $H = G - F$ is absolutely continuous and $H' = 0$ a.e. Applying Theorem B.27, we conclude that H is constant. Then

$$\int_a^b F'(t)\, dt = G(b) - G(a) = (F(b) + H) - (F(a) + H) = F(b) - F(a).$$

This completes the proof. □

Note. If H is merely continuous on $[a, b]$ and $H' = 0$ a.e., H need not be constant. See Problem 14.2 which exhibits a nonconstant increasing function, (uniformly) continuous on the interval $[0, 1]$, whose derivative is equal to 0 almost everywhere.

14.1.1 *The Newton integral*

We saw earlier (Theorem 13.21) that for a function f continuous on $[a, b]$, the integral can be calculated from the fundamental theorem of calculus

$$\int_a^b f(t)\, dt = F(b) - F(a), \tag{14.4}$$

where F is a primitive of f. In the elementary calculus, the fundamental theorem of calculus is extended to the case when f is Riemann integrable on $[a, b]$. In this section we extend this theorem to the case of a Lebesgue integrable function f.

For all their power, Lebesgue's theorems on differentiation of integrals and integration of derivatives leave us stranded in the case when f is Lebesgue integrable and we want to calculate $\int_a^b f(t)\, dt$ using (14.4). Indeed, from these theorems we cannot directly conclude that an arbitrary primitive F to a Lebesgue integrable function f on $[a, b]$ is absolutely continuous. As we shall see from Lemma 14.10, it is true, but far from obvious. The approach adopted here follows the author's exposition given in [27] and [28].

The concept of a primitive can be relaxed a little by assuming that the derivative of F exists outside a countable set, as long as F remains continuous. We say that a statement is true *nearly everywhere* in a set S if it is true in S except for a countable subset of S. (Compare with the concept 'almost everywhere'.) We can generalize the concept of a primitive function as follows:

Definition 14.8. Let $-\infty \le a < b \le \infty$ and let $f\colon (a, b) \to \mathbb{C}$ be given. Then $F\colon (a, b) \to \mathbb{C}$ is a *generalized primitive of f* if it is continuous on (a, b) and $F' = f$ nearly everywhere in (a, b).

For convenience we introduce the following terminology.

Definition 14.9. Let $-\infty \le a < b \le \infty$. A function $f\colon (a, b) \to \mathbb{C}$ is said to be *Newton integrable* if f has a generalized primitive F in (a, b) and the one sided limits $F(a+)$ and $F(b-)$ exist; it is *absolutely Newton integrable* if both f and $|f|$ are Newton integrable. The complex number

$$(\mathcal{N}) \int_a^b f = F(b-) - F(a+)$$

is the *Newton integral* of f over (a, b).

By Corollary B.26, if H is continuous in (a, b) and $H'(x) = 0$ nearly everywhere in (a, b), then H is constant in (a, b). This ensures that the preceding definition is independent of the choice of a generalized primitive.

The Newton integral has several advantages over the other elementary integral of calculus, the Riemann integral. Whereas the Riemann integral is defined only for bounded functions on bounded intervals, the Newton integral applies also to unbounded functions and to unbounded intervals. If a primitive or a generalized primitive is known, the Newton integral is given by a very simple formula. The Newton integral is discussed in detail in the textbook by Černý and Rokyta [14], in which the authors assume that a primitive F of f is continuous and satisfies $F' = f$ except for a finite number of points.

In many situations we are able to find a (generalized) primitive to a given function, so it is of great practical value to know the relation between the Lebesgue and Newton integrals. In this section, we aim to prove the following two properties of the Newton integral.

(a) The Newton integral is compatible with the Lebesgue integral: If $f: [a, b] \to \mathbb{C}$ is both Lebesgue and Newton integrable, the two integrals are equal.

(b) If a function $f: [a, b] \to \mathbb{C}$ is absolutely Newton integrable, it is also Lebesgue integrable.

First a preparatory result (see [28]).

Lemma 14.10. *Let $f: [a, b] \to \mathbb{C}$ be Lebesgue integrable on $[a, b]$, let $F: [a, b] \to \mathbb{C}$ be continuous on $[a, b]$, and let $F'(t) = f(t)$ nearly everywhere in $[a, b]$. Then*

$$|F(b) - F(a)| \le \int_a^b |f(t)|\, dt. \tag{14.5}$$

Proof.   Let $F'(t) = f(t)$ for all $t \in A = [a, b] \setminus D$, where D is countable. We may assume that the (one sided) derivatives exist at the end points of $[a, b]$, otherwise we consider intervals $[a_n, b_n]$ which have this property and satisfy $[a_n, b_n] \nearrow [a, b]$.

Let $\varepsilon > 0$ be given. Set $c_i = i\varepsilon/(b - a)$, $i = 0, 1, 2, \ldots$, and define

$$E_i = \{t \in A : c_{i-1} \le |f(t)| < c_i\}, \ i = 1, 2, \ldots$$

Since f and $|f|$ are Lebesgue integrable on $[a, b]$, the sets E_i are Lebesgue measurable, and A is the disjoint union of the E_i. Hence the Lebesgue measure of A is $m(A) = b - a = \sum_{i=1}^{\infty} m(E_i)$, and

$$c_{i-1} m(E_i) \leq \int_{E_i} |f(t)| \, dt \leq c_i m(E_i), \quad i \in \mathbb{N},$$

which gives

$$0 \leq c_i m(E_i) - \int_{E_i} |f(t)| \, dt \leq \frac{\varepsilon}{b - a} m(E_i).$$

From the countable additivity of the Lebesgue integral we conclude that

$$\sum_{i=1}^{\infty} c_i m(E_i) \leq \int_a^b |f(t)| \, dt + \varepsilon. \tag{14.6}$$

For each $i \in \mathbb{N}$ there exists a bounded open set $G_i \subset \mathbb{R}$ such that

$$G_i \supset E_i \text{ and } m(G_i) \leq m(E_i) + c_i^{-1} (\tfrac{1}{2})^i \varepsilon, \quad i \in \mathbb{N}. \tag{14.7}$$

Define functions $H, M : [a, b] \to \mathbb{R}$ by $H(a) = M(a) = 0$ and

$$H(t) = \sum_{i=1}^{\infty} c_i m(G_i \cap [a, t]), \quad M(t) = \sum_{u_j \in [a, t)} (\tfrac{1}{2})^j \varepsilon, \quad a < t \leq b, \tag{14.8}$$

where $\{u_j : j \in \mathbb{N}\}$ is an enumeration of D. Both H and M are increasing.

Let x be the supremum of all $t \in [a, b]$ such that $|F(t) - F(a)| \leq H(t) + M(t)$. For a proof by contradiction assume that $x < b$. Suppose first that $x \in A$. Then $x \in E_k$ for some $k \in \mathbb{N}$, and from $|f(x)| < c_k$ it follows that there exists $x_1 \in (x, b)$ such that

$$[x, x_1] \subset G_k \text{ and } |F(x_1) - F(x)| < c_k(x_1 - x).$$

Since F is continuous, $|F(x) - F(a)| \leq H(x) + M(x)$. Then

$$|F(x_1) - F(a)| \leq |F(x_1) - F(x)| + |F(x) - F(a)| \leq c_k(x_1 - x) + H(x) + M(x),$$

while $H(x) + c_k(x_1 - x) \leq H(x_1)$. Then $|F(x_1) - F(a)| \leq H(x_1) + M(x_1)$, which contradicts the definition of x.

Suppose that $x \in D$. Then $x = u_m$ for some $m \in \mathbb{N}$. Since F is continuous, there exits $x_2 \in (x, b)$ such that $|F(x_2) - F(x)| < (\tfrac{1}{2})^m \varepsilon$, and

$$|F(x_2) - F(a)| \leq |F(x_2) - F(x)| + |F(x) - F(a)| \leq (\tfrac{1}{2})^m \varepsilon + H(x) + M(x);$$

since $M(x) + (\tfrac{1}{2})^m \varepsilon \leq M(x_2)$, we have $|F(x_2) - F(a)| \leq H(x_2) + M(x_2)$, which again contradicts the definition of x. This proves that $x = b$. Hence, by (14.6),

(14.7) and (14.8),

$$|F(b) - F(a)| \leq H(b) + M(b) \leq \int_a^b |f(t)| \, dt + 3\varepsilon.$$

Since ε was arbitrary, (14.5) holds. $\hfill\square$

Theorem 14.11 (Fundamental theorem of calculus). *Let $f \colon [a, b] \to \mathbb{C}$ be a Lebesgue integrable function on $[a, b]$ which has a primitive (or a generalized primitive) F in $[a, b]$. Then F is absolutely continuous on $[a, b]$, and*

$$\int_a^b f(t) \, dt = F(b) - F(a). \tag{14.9}$$

Proof. By Lemma 14.10, $|F(v) - F(u)| \leq \int_u^v |f(t)| \, dt$ for any subinterval $[u, v]$ of $[a, b]$. Since the Lebesgue integral is absolutely continuous (Problem 13.27), so is F, and (14.9) holds by Lebesgue's theorem on integration of derivatives of absolutely continuous functions. $\hfill\square$

In a recent paper [44], Volintiru proved a stronger version of the Fundamental theorem of calculus for the Lebesgue integral, in which he assumed that F is continuous on $[a, b]$ and $F'(t) = f(t)$ outside a set $B \subset [a, b]$ of Lebesgue measure zero, and that $m^*(F(B)) = 0$, where m^* is the Lebesgue outer measure.

From Theorem 14.11 we deduce the following consistency result.

Theorem 14.12 (Consistency of the two integrals). *Let $-\infty \leq a < b \leq \infty$. If $f \colon (a, b) \to \mathbb{C}$ is both Newton and Lebesgue integrable on (a, b), then*

$$\int_a^b f(t) \, dt = (\mathcal{N}) \int_a^b f. \tag{14.10}$$

Proof. Choose a sequence $[a_n, b_n]$ of subintervals of (a, b) such that $[a_n, b_n] \nearrow (a, b)$, and set $f_n = f\chi_{[a_n, b_n]}$, where $\chi_{[a_n, b_n]}$ is the characteristic function of $[a_n, b_n]$. Then $f_n \to f$ pointwise on (a, b), and $|f_n| \leq |f|$ for all $n \in \mathbb{N}$. By Lebesgue's dominated convergence theorem and by Theorem 14.11,

$$\int_a^b f(t) \, dt = \lim_{n \to \infty} \int_{a_n}^{b_n} f(t) \, dt = \lim_{n \to \infty} (F(b_n) - F(a_n)) = F(b-) - F(a+).$$
$\hfill\square$

We show that an absolutely Newton integrable complex valued function is also Lebesgue integrable, and the two integrals are consistent.

Theorem 14.13. *Let* $-\infty \leq a < b \leq \infty$ *and let* $f: (a, b) \to \mathbb{C}$ *be absolutely Newton integrable on* (a, b). *Then* f *is Lebesgue integrable, and*

$$\int_a^b f(t)\, dt = (\mathcal{N}) \int_a^b f. \tag{14.11}$$

Proof. Assume first that f is Newton integrable and nonnegative. By Theorem B.25, a generalized primitive F to f is an increasing function on (a, b) and so, by Lebesgue's theorem on differentiation of monotonic functions, f is Lebesgue integrable on any compact subinterval of (a, b). Choose a sequence $[a_n, b_n]$ of subintervals of (a, b) such that $[a_n, b_n] \nearrow (a, b)$. By Theorem 14.11,

$$0 \leq \int_{a_n}^{b_n} f(t)\, dt = F(b_n) - F(a_n) \leq F(b-) - F(a+).$$

Writing $f_n = f\chi_{[a_n, b_n]}$, we have $f_n \nearrow f$, and the monotonic convergence theorem ensures that f is Lebesgue integrable on (a, b).

Let f be complex valued and absolutely Newton integrable. We observe that f is Lebesgue measurable as it is the limit of continuous functions

$$F_n(t) = n(F(t + \tfrac{1}{n}) - F(t))$$

convergent nearly (and therefore almost) everywhere in (a, b). By the first part of the proof, $|f|$ is Lebesgue integrable on (a, b). According to Theorem 13.41, so is f, and Theorem 14.12 applies to complete the proof. $\square$

Theorem 14.12 enables us to obtain the following version of integration by parts for the Lebesgue integral.

Theorem 14.14 (Integration by parts). *Let* $-\infty \leq a < b \leq \infty$, *and let the functions* $f, g: (a, b) \to \mathbb{C}$ *be continuous on* (a, b), *differentiable nearly everywhere in* (a, b), *and such that* $f'g$ *and* fg' *are Lebesgue integrable on* (a, b) *and the limits* $(fg)(a+)$ *and* $(fg)(b-)$ *exist (in* $\mathbb{C}$). *Then*

$$\int_a^b f'g\, dm = \left[(fg)(b-) - (fg)(a+)\right] - \int_a^b fg'\, dm. \tag{14.12}$$

Proof. The function $f'g + fg'$ has a generalized primitive $F = fg$ in (a, b); hence $f'g + fg'$ is Newton integrable. By assumption, $f'g + fg'$ is also Lebesgue integrable. By Theorem 14.12 the two integrals are equal, that is,

$$\int_a^b (f'g + fg') \, dm = F(b-) - F(a+).$$

The result then follows.   □

The condition that both f and $|f|$ have generalized primitives in Theorem 14.13 cannot be dropped. The Lebesgue integral is often described as an absolutely convergent integral in view of the fact that the integrability of f implies the integrability of $|f|$. If f has a generalized primitive F and f changes sign very often, then the limit

$$\lim_{(u,v) \nearrow (a,b)} \int_{(u,v)} f(t) \, dt = F(b-) - F(a+)$$

may exist but f need not be Lebesgue integrable. A classical example follows.

Example 14.15. Let

$$f(t) = (t^2 \cos^2(1/t^2))', \quad 0 < t < 1.$$

Then f has a primitive $F(t) = t^2 \cos^2(1/t^2)$ on $(0,1)$, and the one-sided limits $F(0+) = 0$ and $F(1-) = \cos^2 1$ exist. Hence

$$(\mathcal{N}) \int_0^1 f(t) \, dt = F(1) - F(0+) = \cos^2 1.$$

However, we show that f is not Lebesgue integrable on $(0,1)$. For $n \geq 1$ choose

$$a_n = \frac{1}{\sqrt{(n + \frac{1}{2})\pi}}, \qquad b_n = \frac{1}{\sqrt{n\pi}}.$$

Then $F(b_n) = 1/(n\pi)$ and $F(a_n) = 0$, so that

$$\int_{[a_n,1]} |f| \, dm \geq \int_{[a_n,b_n]} |f| \, dm \geq \int_{[a_n,b_n]} f \, dm = F(b_n) - F(a_n) = \frac{1}{n\pi},$$

which leads to $\lim_{\varepsilon \to 0+} \int_{[\varepsilon,1]} |f| \, dm = \infty$. Observe that the primitive F of f is absolutely continuous on each interval $(\varepsilon, 1)$, where $\varepsilon > 0$, but not on $(0,1)$.

Example 14.16. Confirm the existence of the Lebesgue integral $\int_0^1 (1/\sqrt{t})\,dt$ and evaluate it.

The function $f(t) = 1/\sqrt{t}$ has a primitive $F(t) = 2\sqrt{t}$ on $(0,1)$, and the one sided limits $F(1-) = 2$ and $F(0+) = 0$ exist. Since $f \geq 0$, f is absolutely Newton integrable on $(0,1)$. By Theorem 14.13, f is Lebesgue integrable on $(0,1)$, and

$$\int_{(0,1)} \frac{dt}{\sqrt{t}} = (\mathcal{N})\int_0^1 \frac{dt}{\sqrt{t}} = F(1-) - F(0+) = 2.$$

Compare the ease of the preceding calculation with Example 13.18.

Example 14.17.

Investigate the Lebesgue integral $\int_0^1 (1+t)^{-1}\log t\,dt$.

The function f has a pointwise series expansion

$$f(t) = \frac{\log t}{1+t} = \sum_{n=0}^{\infty}(-1)^n t^n \log t, \quad 0 < t < 1.$$

Put $f_n(t) = (-1)^n t^n \log t$. We note that

$$\int_0^1 |f_n(t)|\,dt = -\int_0^1 t^n \log t\,dt = \frac{1}{(n+1)^2},$$

where the series $\sum_{n=0}^{\infty}(n+1)^{-2}$ converges. Hence we can apply the term by term integration: By Theorem 13.35, f is Lebesgue integrable over $(0,1)$, and

$$\int_0^1 \frac{\log t\,dt}{1+t} = \sum_{n=0}^{\infty}(-1)^n \int_0^1 (-t^n \log t)\,dt = -\sum_{n=0}^{\infty}\frac{(-1)^n}{(n+1)^2} = -\frac{\pi^2}{12}.$$

Example 14.18. The limit

$$\lim_{t\to\infty}\int_0^t \frac{\sin x}{x}\,dx$$

of Lebesgue integrals exists (and equals $\frac{1}{2}\pi$), but the Lebesgue integral

$$\int_0^\infty \frac{\sin x}{x}\,dx$$

does not.

First we establish the existence of the limit:

$$\int_0^t \frac{\sin x}{x}\,dx = \int_0^1 \frac{\sin x}{x}\,dx + \int_1^t \frac{\sin x}{x}\,dx;$$

so, by 'integration by parts' for primitives, we get

$$\int_1^t \frac{\sin x}{x}\, dx = -\frac{\cos x}{x}\Big|_1^t - \int_1^t \frac{\cos x}{x^2}\, dx = \cos 1 - \frac{\cos t}{t} - \int_1^t \frac{\cos x}{x^2}\, dx.$$

Note that $x^{-2}\cos x$ is Lebesgue measurable on $(1,\infty)$; for any sequence (b_n) with $b_n \nearrow \infty$ we see that the limit

$$\lim_{n\to\infty} \int_1^{b_n} \frac{\cos x}{x^2}\, dx$$

exists (as a real number). So

$$\lim_{t\to\infty} \int_1^t \frac{\cos x}{x^2}\, dx$$

exists. We then have

$$\lim_{t\to\infty} \int_0^t \frac{\sin x}{x}\, dx = \int_0^1 \frac{\sin x}{x}\, dx + \cos 1 - \lim_{t\to\infty} \int_0^t \frac{\cos x}{x^2}\, dx.$$

We prove that the Lebesgue integral $\int_0^\infty x^{-1}\sin x\, dx$ does not exist by showing that $\lim_{t\to\infty} \int_0^t |x^{-1}\sin x|\, dx = \infty$: For any $n \in \mathbb{N}$,

$$\int_\pi^{n\pi} \left|\frac{\sin x}{x}\right|\, dx = \sum_{k=2}^n \int_{(k-1)\pi}^{k\pi} \frac{|\sin x|}{x}\, dx \geq \sum_{k=2}^n \int_{(k-1)\pi}^{k\pi} \frac{|\sin x|}{k\pi}\, dx$$

$$\geq \sum_{k=2}^n \frac{1}{k\pi} \int_0^\pi \sin x\, dx \geq \frac{2}{\pi} \sum_{k=2}^n \frac{1}{k} \to \infty \text{ as } n \to \infty.$$

We observe, however, that $x^{-1}\sin x$ is Newton integrable on $(0,\infty)$ with

$$(\mathcal{N})\int_0^\infty \frac{\sin x}{x}\, dx = \lim_{t\to\infty} \int_0^t \frac{\sin x}{x}\, dx.$$

The value of the limit is $\frac{1}{2}\pi$—see Problem 14.28. Another way of calculating this limit is by double integration of $\sin x \exp(-xs)$ on a suitable domain.

Example 14.19. Using the absolute Newton integral we can show that the Lebesgue integrals

$$\int_0^1 \frac{dt}{t^\alpha}, \qquad \int_1^\infty \frac{dt}{t^\beta}$$

exist if and only if $\alpha < 1$ and $\beta > 1$, respectively. (See Problem 14.15.)

The functions from the preceding example can be used as integrable dominants in Theorem 13.41. First we introduce the following notation for the order of magnitude due to Landau.

Definition 14.20 (Order of magnitude; asymptotic functions).
Let f, g be complex valued functions defined in some open interval of the
real line. We write

$$f(x) = O(g(x)) \text{ as } x \to a \iff \frac{f(x)}{g(x)} \text{ is bounded in } (a - \delta, a + \delta), \ \delta > 0,$$

$$f(x) = o(g(x)) \text{ as } x \to a \iff \lim_{x \to a} \frac{f(x)}{g(x)} = 0,$$

$$f(x) \asymp g(x) \text{ as } x \to a \iff f(x) = O(g(x)) \text{ and } g(x) = O(f(x)) \text{ as } x \to a.$$

It is clear how to modify these definitions in the case that $x \to a+$, $x \to a-$,
$x \to \infty$ or $x \to -\infty$. We observe that

$$\lim_{x \to a} (f(x)/g(x)) \in \mathbb{R} \implies f(x) = O(g(x)) \text{ as } x \to a;$$

$$\lim_{x \to a} (f(x)/g(x)) \in \mathbb{R} \setminus \{0\} \implies f(x) \asymp g(x) \text{ as } x \to a;$$

$$f(x) = o(g(x)) \text{ as } x \to a \implies f(x) = O(g(x)) \text{ as } x \to a.$$

Theorem 14.21 (Comparison test). *Let* $f \colon (a, b) \to \mathbb{C}$ *be integrable on
each interval* (a, c), *where* $a < c < b$.

(i) *If* $f(x) = O(g(x))$ *as* $x \to b-$ *and* g *is integrable on* (a, b), *then so
is* $f(x)$.

(ii) *If* $f(x) \asymp g(x)$ *as* $x \to b-$ *and* g *is integrable on each interval* (a, c),
where $a < c < b$, *then* f *is integrable on* (a, b) *if and only if* g *is.*

Example 14.22. Consider the function $f(x) = 1/\sqrt{x(1 - x)}$ on the interval $(0, 1)$. Then f is Lebesgue integrable on each interval $[u, v]$, $0 < u <
v < 1$ (continuity), and

$$f(x) = O(1/\sqrt{x}) \text{ as } x \to 0+, \qquad f(x) = O(1/\sqrt{1 - x}) \text{ as } x \to 1 - .$$

Since $g(x) = 1/\sqrt{x}$ and $h(x) = 1/\sqrt{1 - x}$ are integrable on $(0, 1)$, f is
integrable on $(0, 1)$.

14.1.2 Calculus for complex valued functions of a real variable

In calculating Lebesgue integrals via the Newton integral we need to find primitives (or generalized primitives) to complex valued functions of a real variable. This is a reverse process to calculating derivatives of such functions. Let
$f(t) = u(t) + iv(t)$ with u, v real valued be defined on some interval (a, b). We
define the derivative of f at $t \in (a, b)$ by

$$f'(t) := u'(t) + iv'(t),$$

provided both u, v are differentiable at t. It is not difficult to check that differentiation of such functions is a linear operation which obeys the product and quotient rules

$$(fg)' = f'g + fg', \qquad \left(\frac{f}{h}\right)' = \frac{f'h - fh'}{h^2}, \ h \neq 0,$$

and the chain rule

$$(f \circ k)' = (f' \circ k)\, k'.$$

If f is a differentiable function and $n \neq 0$ is an integer, then

$$(f^n)' = nf^{n-1} f';$$

this can be confirmed by mathematical induction for a positive n; for a negative n we first prove that

$$(f^{-1})' = -f'/f^2 \tag{14.13}$$

and then obtain

$$(f^n)' = [(f^{-1})^{-n}]' = (f^{-1})'(-n)(f^{-1})^{-n-1} = nf^{n-1} f'.$$

If possible, do not separate the real and imaginary parts, as some computations may turn out to be quite involved. As a sample we verify (14.13) for $f = u + iv$ with u and v real valued, where such separation is inevitable:

$$\begin{aligned}
\frac{d}{dt}\frac{1}{f} &= \frac{d}{dt}\frac{1}{u + iv} = \frac{d}{dt}\left(\frac{u}{u^2 + v^2} - i\frac{v}{u^2 + v^2}\right)\\
&= \frac{u'(u^2 + v^2) - u(2uu' + 2vv')}{(u^2 + v^2)^2} - i\frac{v'(u^2 + v^2) - v(2uu' + 2vv')}{(u^2 + v^2)^2}\\
&= -\frac{(u' + iv')(u - iv)^2}{(u^2 + v^2)^2} = -\frac{u' + iv'}{(u + iv)^2} = -\frac{f'}{f^2}.
\end{aligned}$$

If $w \in \mathbb{C}$, then

$$\frac{d}{dt}\exp(wt) = w\exp(wt).$$

Recall that $\exp((a + ib)t) = e^{at}(\cos bt + i\sin bt)$, where $a, b \in \mathbb{R}$.

All these rules can be converted to rules for finding primitives. We can use the time honoured notation for primitives, called indefinite integrals in calculus,

$$\int f(t)\, dt = F(t) + C,$$

where F is a particular primitive of f and C is an arbitrary complex number (any two primitives to the same function differ by a constant on an interval). On appropriate intervals we have

$$\int f(t)^n f'(t)\,dt = \frac{f(t)^{n+1}}{n+1} + C, \quad n \in \mathbb{Z} \setminus \{-1\},$$

$$\int \exp(wt)\,dt = \frac{\exp(wt)}{w} + C, \quad w \in \mathbb{C} \setminus \{0\},$$

$$\int e^{at} \cos bt\,dt = \mathsf{Re} \int \exp((a+ib)t)\,dt = \frac{e^{at}}{a^2+b^2}(a\cos bt + b\sin bt) + C,$$

$$\int e^{at} \sin bt\,dt = \mathsf{Im} \int \exp((a+ib)t)\,dt = \frac{e^{at}}{a^2+b^2}(a\sin bt - b\cos bt) + C.$$

A useful technique is the so called *indefinite integration by parts*,

$$\int f'g = fg - \int fg'.$$

14.1.3 *Nonabsolute integrals*

It is interesting to know that there exist integration processes for real valued functions on the real line which include both the Lebesgue and Newton integrals. One such integral was developed by Oskar Perron in 1914, and carries his name. Unlike the Lebesgue integral, the Perron integral has the property that if $\lim_{t \to b}(\mathcal{P})\int_a^t f$ exists, then f is Perron integrable on $[a,b]$, and

$$\lim_{t \to b}(\mathcal{P})\int_a^t f = (\mathcal{P})\int_a^b f.$$

Unlike the Lebesgue integral, the Perron integral is a nonabsolute itegral. Every Newton or Lebesgue integrable function is Perron integrable, and the integrals are consistent. Every Perron integrable function is Lebesgue measurable, and equal a.e. to the derivative of its indefinite Perron integral. A function is Lebesgue integrable if and only if it is absolutely Perron integrable; the two integrals are then equal. For the theory of the Perron integral see Gordon [20]. Another approach to this integral in terms of functions absolutely continuous in the generalized sense is due to Denjoy (about 1912)—see again Gordon [20].

In the 1950s Jaroslav Kurzweil introduced an innocent looking change in the definition of the Riemann integral and obtained an integral with the range and power of Lebesgue integral. This idea was taken up independently by Ralph Henstock who developed the integral and showed its relation to the Lebesgue integral. The integral is usually called the *Kurzweil–Henstock integral* or the *generalized Riemann integral*. For a detailed discussion see Bartle [7], Gordon [20] and Kurtz and Swartz [30]. It is rather surprising that the Perron, Denjoy and Kurzweil-Henstock integrals are one and the same integral, given by different definitions.

14.1.4   *The Kurzweil–Henstock integral*

A *tagged partition* π of a closed bounded interval $[a, b]$ is determined by the partition points $a = t_0 < t_1 < \cdots < t_{n-1} < t_n = b$, and by *tags* $\tau_k \in [t_{k-1}, t_k]$ $(k = 1, \ldots, n)$. Let $\delta \colon [a, b] \to \mathbb{R}$ be a strictly positive function. We say that a tagged partition π is δ-*fine* if each subinterval of π satisfies $[t_{k-1}, t_k] \subset (\tau_k - \delta(\tau_k), \tau_k + \delta(\tau_k))$.

If f is a bounded complex valued function defined on $[a, b]$, we define the *Riemann sum* relative to a tagged partition π by

$$S(f; \pi) = \sum_{k=1}^{n} f(\tau_k)(t_k - t_{k-1}).$$

The function f is *Kurzweil–Henstock integrable* (or *generalized Riemann integrable*) if there is a complex number I such that for each $\varepsilon > 0$ there exists a strictly positive function δ on $[a, b]$ such that

$$|S(f; \pi) - I| < \varepsilon$$

whenever π is a δ-fine partition of $[a, b]$. The existence of δ-fine partitions for a given positive function δ follows from Cousin's lemma (Lemma B.23) given in Appendix B. The number I is called the *Kurzweil–Henstock integral* (or the *generalized Riemann integral*) of f on $[a, b]$. An elementary introduction to the generalized Riemann integral can be found in the Bartle and Sherbet's textbook [8].

Every Kurzweil–Henstock integrable function is Lebesgue measurable. It is not difficult to see that if the function δ in the preceding definition is restricted to a constant, we obtain the ordinary Riemann integral.

Further, the Kurzweil–Henstock integral integrates all derivatives, and subsumes both the Lebesgue and Newton integral on the line. The following result, proved in [7, 20, 30], shows its relation to the Lebesgue integral.

Theorem 14.23. *A function $f \colon [a, b] \to \mathbb{C}$ is Lebesgue integrable if and only if both f and $|f|$ are Kurzweil–Henstock integrable. The two integrals are then equal.*

The Kurzweil–Henstock integral obeys a version of the monotonic convergence theorem and Lebesgue's dominated convergence theorem. The usefulness of the Kurzweil–Henstock integral is in its simple definition and its great power, shown in applications to differential equations.

Riemann integrable functions are characterized in the following theorem.

Theorem 14.24. *A complex valued function f defined on $[a, b]$ is Riemann integrable if and only if it is bounded and continuous almost everywhere in $[a, b]$.*

Proof. Exercise. $\square$

An alternative approach to the Riemann integral (in $\mathbb{R}^k$) is given in Problem 16.5.

14.2 Lebesgue–Stieltjes integral on the real line

Let g be a right continuous increasing function on $\mathbb{R}$. We recall that there is a unique measure μ_g on the σ-algebra $\mathcal{B}$ of Borel sets such that

$$\mu_g((a,b]) = g(b) - g(a),$$

the so called the Lebesgue–Stieltjes measure generated by g. A construction of μ_g is described in Chapter 15 on construction of measures. We shall write $\mu = \mu_g$ when g is understood. Recall that, for any point $a \in \mathbb{R}$,

$$\mu(\{a\}) = g(a) - g(a-).$$

One point sets have in general nonzero measure, and so, unlike in the case of the Lebesgue integral, we must distinguish between integrals over the intervals (a,b), $(a,b]$, $[a,b)$ and $[a,b]$, which in general are all different. The integral relative to the Lebesgue–Stieltjes measure is called the *Lebesgue–Stieltjes integral*. For clarity we sometimes write $(\mathcal{LS}) \int_S f \, d\mu$ and $(\mathcal{L}) \int f \, dm$ for the Lebesgue–Stieltjes and Lebesgue integrals, respectively.

Example 14.25. Let g be a right continuous increasing function on $\mathbb{R}$, and let μ be the associated Lebesgue–Stieltjes measure. If $f(t) = c$ is a constant function on $\mathbb{R}$, then

$$\int_{(a,b)} f \, d\mu = c(g(b-) - g(a)), \qquad \int_{(a,b]} f \, d\mu = c(g(b) - g(a)),$$

$$\int_{[a,b)} f \, d\mu = c(g(b-) - g(a-)), \qquad \int_{[a,b]} f \, d\mu = c(g(b) - g(a-)).$$

Example 14.26. Let g be defined on $\mathbb{R}$ by

$$g = \sum_{n=1}^{\infty} (1 - (\tfrac{1}{2})^n) \chi_{[n-1,n)}.$$

Then g is a right continuous increasing function on $\mathbb{R}$. Let $\mu = \mu_g$ be the Lebesgue–Stieltjes measure associated with g, so that μ is defined on $\mathcal{B}$, and $\mu((a,b]) = g(b) - g(a)$ for every interval $(a,b]$. Let f be μ-integrable on $(n-1,n]$. Since

$$\mu((n-1,n)) = g(n-) - g(n-1) = g(n-1) - g(n-1) = 0$$

and

$$\mu(\{n\}) = g(n) - g(n-) = 1 - (\tfrac{1}{2})^{n+1} - 1 + (\tfrac{1}{2})^n = (\tfrac{1}{2})^{n+1},$$

we have

$$\int_{(n-1,n]} f \, d\mu = \int_{(n-1,n)} f \, d\mu + \int_{\{n\}} f \, d\mu = \int_{\{n\}} f \, d\mu = f(n)(\tfrac{1}{2})^{n+1}.$$

Note that $\mu((-\infty,0)) = 0$. A necessary and sufficient condition for a function $f \colon \mathbb{R} \to \mathbb{C}$ to be μ-integrable on $\mathbb{R}$ is that the series

$$\sum_{n=0}^{\infty} |f(n)|(\tfrac{1}{2})^n$$

converges. This follows from the extension lemma (Problem 13.29) and the identity

$$\int_{-\infty}^{k} |f| \, d\mu = \sum_{n=0}^{k} |f(n)|(\tfrac{1}{2})^{n+1}.$$

We then have

$$\int_{\mathbb{R}} f \, d\mu = \sum_{n=0}^{\infty} f(n)\mu(\{n\}).$$

We observe that, in general, the Lebesgue–Stieltjes measure is not translation invariant. In fact, it can be proved that any nonzero positive measure on $\mathcal{B}$ which is translation invariant and finite on compact sets is a multiple of Lebesgue measure by a positive constant.

The next result explores the calculation of Lebesgue–Stieltjes integrals as Lebesgue integrals in the case that g is differentiable. Then g is continuous; if μ is Lebesgue–Stieltjes measure associated with g, then $\mu(\{a\}) = 0$ for all one point sets. In the following theorem we postulate the existence of one sided derivatives at the end points of the interval.

Theorem 14.27. *Let g be a differentiable function with $g'(x) \geq 0$ for all $x \in [a,b]$, and let μ be the Lebesgue–Stieltjes measure associated with g. If f is continuous on $[a,b]$, then f is μ-integrable on $[a,b]$, and*

$$(\mathcal{LS}) \int_{[a,b]} f \, d\mu = (\mathcal{L}) \int_{[a,b]} fg' \, dm.$$

Proof.   Set

$$F(x) = (\mathcal{LS}) \int_{[a,x]} f \, d\mu.$$

We show that $F'(x) = f(x)g'(x)$ for all $x \in [a,b]$. First assume that $x \in [a,b)$. The case $g'(x) = 0$ is left as an exercise.

Let $g'(x) \neq 0$. Then there is $\delta > 0$ such that $g(x + h) \neq g(x)$ if $0 < h \leq \delta$; so $\mu(I_h) = g(x + h) - g(x) > 0$ for each interval $I_h = [x, x + h]$. Write $\varphi(h) = g'(x) - \mu(I_h)/h$. From the definition of derivative, $\varphi(h) \to 0$ as $h \to 0^+$. In the following calculation x is kept fixed and h varies in the interval $(0, \delta]$:

$$
\left| \frac{F(x + h) - F(x)}{h} - f(x)g'(x) \right|
$$

$$
= \left| \frac{1}{h} \int_{I_h} f(t)\, d\mu(t) - f(x)\frac{\mu(I_h)}{h} - f(x)\varphi(h) \right|
$$

$$
= \left| \frac{1}{h} \int_{I_h} (f(t) - f(x))\, d\mu(t) - f(x)\varphi(h) \right|
$$

$$
\leq \sup_{t \in I_h} |f(t) - f(x)| \frac{\mu(I_h)}{h} + |f(x)| \sup_{0 \leq h \leq \delta} |\varphi(h)|.
$$

The expression on the right has limit 0 as $\delta \to 0+$. This shows that the right derivative of F at x is $f(x)g'(x)$. A similar argument is used to show that the left derivative of F at x is also $f(x)g'(x)$.

Since f is continuous, so is $|f|$. By the above argument applied to $|f|$ in place of f,

$$
\frac{d}{dx} \int_{[a,x]} |f|\, d\mu = |f(x)|g'(x) = |f(x)g'(x)|.
$$

This shows that the function $G(x) = \int_{[a,x]} |f|\, d\mu$ is a primitive to $|fg'|$, and fg' is absolutely Newton integrable. By Theorem 14.13, fg' is Lebesgue integrable, and

$$
\int_{[a,b]} f\, d\mu = F(b) - F(a) = \int_{[a,b]} fg'\, dm.
$$

$\square$

We now turn our attention to the calculation of Lebesgue–Stieltjes integrals with a general g as Lebesgue integrals. We give only a sketch of such an approach.

Let g be a right continuous increasing function, and let

$$
A = \lim_{t \to -\infty} g(t), \quad B = \lim_{t \to \infty} g(t).
$$

With g we associate a function h (sometimes called the Banach indicatrix) by

$$
h(x) = \inf g^{-1}([x, \infty)); \tag{14.14}
$$

$h(x)$ is then defined only if the set $g^{-1}([x, \infty))$ is nonempty and bounded. We note that h is defined on the interval $J = (A, B)$, and that

$$h(x) \in (a, b] \iff x \in (g(a), g(b)].$$

This implies that

$$\chi_{(a,b]} \circ h = \chi_{(g(a),g(b)]}.$$

If f is a simple function of the form

$$f = \sum_{i=1}^{n} c_i \chi_{(a_i, b_i]}, \tag{14.15}$$

then

$$\int_{\mathbb{R}} f \, d\mu = \int_{J} f \circ h \, dm. \tag{14.16}$$

It can be proved that the simple functions of the form (14.15) are norm dense in $L^1(\mu)$. Then the formula (14.16) holds for arbitrary functions in the sense that if one side exists, so does the other, and they are equal.

Example 14.28. Define a function g on $\mathbb{R}$ by

$$g = \sum_{n=1}^{\infty} n \chi_{[n-1, n)}.$$

Then g is right continuous and increasing. With g we associate the function h by (14.14). Then

$$h(x) = \sum_{n=1}^{\infty} n \chi_{(n, n+1]}.$$

Let μ be the Lebesgue–Stieltjes measure generated by g. If f is μ-integrable, then

$$\int_{(0,\infty)} f \, d\mu = \int_{(1,\infty)} f \circ h \, dm = \sum_{n=1}^{\infty} \int_{(n, n+1]} f \circ h \, dm = \sum_{n=1}^{\infty} f(n),$$

as $f \circ h$ is constant on $(n, n+1]$, and

$$\int_{(n, n+1]} f \circ h \, dm = f(n)(g(n+2) - g(n+1)) = f(n)(n+2-n-1) = f(n).$$

If $f(x) = x^{-2}$, then

$$\int_{(0,\infty)} \frac{1}{x^2} \, d\mu = \sum_{n=1}^{\infty} \frac{1}{n^2} = \frac{\pi^2}{6}.$$

14.3 Problems for Chapter 14

1. Show that the function defined by $f(t) = t\sin(1/t)$ for $t \neq 0$ and $f(0) = 0$ is uniformly continuous in $[0, 1]$, absolutely continuous on every interval $[\varepsilon, 1]$ for $0 < \varepsilon < 1$, but is not absolutely continuous on $[0, 1]$.

2. *The Cantor singular function* (also known as the *devil's staircase*). Refer to the construction of the Cantor set C on the interval $[0, 1]$ (Problem 12.27). For each $x \in (\frac{1}{3}, \frac{2}{3})$ define $f(x) = \frac{1}{2}$. For each $x \in (\frac{1}{9}, \frac{2}{9})$ set $f(x) = \frac{1}{4}$ and for each $x \in (\frac{7}{9}, \frac{8}{9})$ set $f(x) = \frac{3}{4}$. Set $f(x)$ to be $1/2^n$, $3/2^n$, $5/2^n, \ldots$, for x on the various intervals removed from C_{n-1}. Then define $f(x) = \sup\{f(t) : t \in [0, 1] \setminus C \text{ and } t < x\}$.

(i) Show that f is constant on each interval complementary to the Cantor set C.

(ii) Show that f is increasing and maps $[0, 1]$ onto $[0, 1]$.

(iii) Show that f is continuous on $[0, 1]$.

(iv) Give a rough sketch of f.

(v) Show that f is an example of a continuous function whose derivative is equal to 0 almost everywhere and which is not constant.

(vi) Show that f is an example of an increasing continuous function which is not absolutely continuous.

3. Let F be a complex valued function which has a bounded derivative in (a, b). Show that F is absolutely continuous in (a, b).

4. Let f be a complex valued quasicontinuous function on an interval (a, b) (bounded or unbounded), and let f be Lebesgue integrable over (a, b). Show that f is Newton integrable on (a, b), and the two integrals are equal.

5. *Differentiation of integrals on unbounded intervals.* If $f\colon (a, b) \to \mathbb{C}$ is Lebesgue integrable on an interval (a, b), where $-\infty \leq a < b \leq \infty$, show that the function

$$F(x) = \int_c^x f(t)\,dt$$

defined for some $c \in (a, b)$, is absolutely continuous on (a, b), that $F'(x) = f(x)$ for almost all x in (a, b) and that

$$\int_a^b f(t)\,dt = F(b-) - F(a+).$$

6. *Integration of derivatives on unbounded intervals.* Let $F\colon [a, b] \to \mathbb{C}$ be absolutely continuous on an interval (a, b), where $-\infty \leq a < b \leq \infty$. Show that the derivative $F'(x) = f(x)$ exists for almost all x in (a, b), and that

$$\int_a^b f(t)\,dt = F(b-) - F(a+).$$

7. Show that the function $f(x) = x$ is absolutely Newton integrable on the interval $[-1, 1]$ by constructing primitives for f and $|f|$.

8. Let $f: [0, 1] \to \mathbb{C}$ be defined by $f(t) = t^2$ if $t \in [0, 1]$ is irrational and $f(t) = 0$ if $t \in [0, 1]$ is rational. Show that f is (absolutely) Newton integrable on $[0, 1]$ and evaluate the Newton integral $(\mathcal{N}) \int_0^1 f$.

9. Let $f: [0, 1] \to \mathbb{C}$ be defined by $f(t) = t^2$ if $t \in [0, 1] \setminus C$ where C is the Cantor set, and $f(t) = 0$ if $t \in C$. Show that f is Lebesgue but not Newton integrable on $[0, 1]$ and evaluate the Lebesgue integral $(\mathcal{L}) \int_0^1 f$.

10. Considering the function $f(t) = e^{(-a+ib)t}$, where $a > 0$ and $b \in \mathbb{R}$, evaluate with justification

$$\int_0^\infty e^{-at} \cos bt \ dt, \quad \int_0^\infty e^{-at} \sin bt \ dt.$$

11. Let $f_n(x) = (n+i)/(nx^2+i)$ for $n \in \mathbb{N}$ and $x > 1$. Show that the sequence (f_n) has an integrable dominant on $(1, \infty)$, and find $\lim_{n\to\infty} \int_1^\infty f_n(x) \, dx$.

12. Prove the existence of and evaluate the Lebesgue integrals $\int_0^1 (t + i)^{-2} \, dt$ and $\int_1^\infty (t + i)^{-2} \, dt$.

13. As $x \to 0$, show that $\sin x \asymp x$, $\cos x \asymp 1$, $\cos x - 1 \asymp x^2$, $\cot x \asymp 1/x$, $\arcsin x \asymp x$, $\arccos x \asymp 1$, $e^x - 1 \asymp x$, $\log(1 + x) \asymp x$.

14. Show that $f(x) = \operatorname{arccot} x$ is not Lebesgue integrable on $(1, \infty)$ ($\operatorname{arccot} x \asymp 1/x$ as $x \to \infty$).

15. Use the Newton integral to prove the following:

(i) $f(t) = 1/t^\alpha$ is Lebesgue integrable on $(0, 1)$ if and only if $\alpha < 1$.

(ii) $f(t) = 1/t^\beta$ is Lebesgue integrable on $(1, \infty)$ if and only if $\beta > 1$.

16. Show that $\log x = O(1/x^\alpha)$ as $x \to 0+$ for any $\alpha > 0$ and $\log x = O(x^\beta)$ as $x \to \infty$ for any $\beta > 0$.

17. Show that the function $f(x) = \log x/(x^2 - 1)$ is Lebesgue integrable on $(0, \infty)$.

18. Show that if $f: [a, b] \to \mathbb{C}$ is continuous on the closed bounded interval $[a, b]$, then both f and $|f|$ have primitives on $[a, b]$, that is, f is absolutely Newton integrable on (a, b).

19. Prove these properties of the Newton integral (supplying hypotheses where needed):

(i) Let f be Newton integrable on (a, b) and let $g = f$ nearly everywhere in (a, b). Then g is Newton integrable in (a, b) and the two integrals over (a, b) are equal.

(ii) $(\mathcal{N}) \displaystyle\int_a^b (\alpha f + \beta g) = \alpha\,(\mathcal{N}) \displaystyle\int_a^b f + \beta\,(\mathcal{N}) \displaystyle\int_a^b g.$

(iii) If $a < c < b$, then $(\mathcal{N}) \displaystyle\int_a^b f = (\mathcal{N}) \displaystyle\int_a^c f + (\mathcal{N}) \displaystyle\int_c^b f.$

(iv) If $f \leq g$ on (a, b), then $(\mathcal{N}) \displaystyle\int_a^b f \leq (\mathcal{N}) \displaystyle\int_a^b g.$

(v) $\left| (\mathcal{N}) \displaystyle\int_a^b f \right| \leq (\mathcal{N}) \displaystyle\int_a^b |f|.$

20. Suppose that $(\mathcal{N})\int_a^c f$ exists whenever $a < c < b$, and that the limit $L = \lim_{c \to b-} (\mathcal{N})\int_a^c f$ exists in $\mathbb{C}$. Show that f is Newton integrable on (a, b) and that $(\mathcal{N})\int_a^b f = L$. Is this true for the Lebesgue integral?

21. *Integration by parts for the Newton integral.* Let $f, g : (a, b) \to \mathbb{C}$ be continuous functions. Show that primitives F, G of f, g exist on (a, b). If the one sided limits $(FG)(a+)$ and $(FG)(b-)$ exist and Fg is Newton integrable on (a, b), show that

$$(\mathcal{N}) \int_a^b Fg = (FG)(b-) - (FG)(a+) - (\mathcal{N}) \int_a^b fG.$$

22. *Change of variable theorem for the Newton integral.* Let φ be a diffeomorphism from (α, β) onto (a, b). Show that

$$(\mathcal{N}) \int_a^b f = (\mathcal{N}) \int_\alpha^\beta (f \circ \varphi) |\varphi'|$$

whenever one of the integrals exists.

23. Discuss the existence of the following Lebesgue integrals, and, if possible, evaluate them.

(i) $\displaystyle\int_0^1 \log t \, dt$

(ii) $\displaystyle\int_1^\infty t^p \, dt$

(iii) $\displaystyle\int_0^1 t^q \, dt$

(iv) $\displaystyle\int_0^1 t \log t \, dt$

(v) $\displaystyle\int_0^1 \frac{\cos t - 1}{\sqrt{t}} \, dt$

(vi) $\displaystyle\int_0^\infty t \sin e^{-t} \, dt$

(vii) $\displaystyle\int_0^1 \frac{t^2 \, dt}{\sqrt{1 - t^4}}$

(viii) $\displaystyle\int_0^\infty \frac{dt}{\sqrt{t^3 + t^2}}$

(ix) $\displaystyle\int_0^\infty \frac{dt}{\sqrt{t^3 + t}}$

(x) $\displaystyle\int_0^{\pi/2} \frac{\cos x \, dx}{(1 - \sin x)^{2/3}}$

(xi) $\displaystyle\int_1^e \frac{dx}{x\sqrt{\log x}}$

(xii) $\displaystyle\int_1^\infty t^{-3/2} \sin t \, dt$

24. Using appropriate theorems, find the limit

$$\lim_{n \to \infty} \int_0^n \frac{e^{-t^2/n} \, dt}{t^2 + 1}.$$

Suggestion. Express all integrals over the same domain.

25. Show that the Lebesgue integral

$$\int_2^\infty \frac{dx}{x^\alpha \log^\beta x}$$

exists if and only if (i) $\alpha > 1$ or (ii) $\alpha = 1$ and $\beta > 1$.

26. Verify that the function g defined on $\mathbb{R}$ by

$$g(x) = x + 1 \text{ if } x < 0, \quad g(x) = x^3 + 4 \text{ if } x \geq 0,$$

is increasing and right continuous on $\mathbb{R}$. Let μ be the Lebesgue-Stieltjes measure induced by g, that is, satisfying $\mu((a, b]) = g(b) - g(a)$ for all $a < b$ in $\mathbb{R}$. Show that $f(x) = x^2 + 2$ is μ-integrable on $[-1, 1]$, and calculate the Lebesgue-Stieltjes integrals $\int_A f \, d\mu$ for A equal to

$$[-1, 0), \quad [-1, 0], \quad (0, 1], \quad [0, 1].$$

27. Let g be defined by

$$g = \sum_{n=1}^\infty \frac{1}{n} \chi_{[1/(n+1), 1/n)} + \chi_{[1, \infty)}.$$

(i) Sketch the graph of g and confirm that g is right continuous and increasing.

(ii) If μ is the Lebesgue–Stieltjes measure generated by g, find

$$\mu((-\infty, 0]), \quad \mu((\tfrac{1}{n+1}, \tfrac{1}{n})), \quad \mu((\tfrac{1}{2}, \infty)), \quad \mu(\{\tfrac{1}{n}\})$$

for $n = 2, 3, \ldots$.

(iii) If a function $f: \mathbb{R} \to \mathbb{C}$ is such that the series $\sum_{n=2}^\infty |f(1/n)|(n(n-1))^{-1}$ converges, show that $f \in L^1(\mu)$ and

$$\int_\mathbb{R} f \, d\mu = \sum_{n=2}^\infty \frac{f(\tfrac{1}{n})}{n(n-1)}.$$

Suggestion. For (iii) apply the theorem on the integration of series term-by-term.

28. For each $s > 0$ set

$$G(s) = \int_0^\infty x^{-1} \sin x \, e^{-sx} \, dx = \lim_{u \to \infty} \int_0^u x^{-1} \sin x \, e^{-sx} \, dx$$

checking the existence of the Lebesgue integral over $(0, \infty)$. Give a careful justification of the following argument:

$$G'(s) = -\int_0^\infty \sin x \, e^{-sx} \, dx = (1 + s^2)^{-1} \implies G(s) = -\arctan s + \tfrac{1}{2}\pi,$$

and

$$\lim_{x \to \infty} \int_0^u x^{-1} \sin x \, dx = \lim_{u \to \infty} \lim_{s \to 0+} \int_0^u x^{-1} \sin x \, e^{-sx} \, dx$$

$$= \lim_{s \to 0+} \lim_{u \to \infty} \int_0^u x^{-1} \sin x \, e^{-sx} \, dx = \lim_{s \to 0+} G(x) = \tfrac{1}{2}\pi.$$

Chapter 15

Construction of Measures

Lebesgue originally defined the integral for real valued functions on the real line. Some 14 years later, Carathéodory (a German mathematician of Greek origin) extended Lebesgue's ideas in a paper published in 1914. One of Carathéodory's principal innovations was the introduction of outer measures. Further contributions to measure and integration theory were given by Radon, Hahn, Saks, Nikodým and many others. The abstract approach to measure theory was used by Kolmogorov in 1930 to give an axiomatic foundation to probability theory.

Constantin Carathéodory (1873–1950)
Hans Hahn (1879–1934)
Johann Radon (1887–1956)
Andrey Kolmogorov (1903–1987)

15.1 Outer measures

We describe a general construction of measures which yields the Lebesgue and Lebesgue–Stieltjes measures on $\mathbb{R}^k$ as well as a variety of other measures on various spaces. The first step is to obtain a so-called outer measure, which is countably subadditive rather than countably additive. As always, S is a nonempty set.

Definition 15.1. A family $\mathcal{K}$ of subsets of S is called a *σ-covering family* of S if $\emptyset \in \mathcal{K}$ and every set $A \subset S$ has a countable cover by sets in $\mathcal{K}$. A function $\rho \colon \mathcal{K} \to [0, \infty]$ is called a *gauge* on $\mathcal{K}$ if
$$\rho(\emptyset) = 0.$$
A pair $(\mathcal{K}, \rho)$, where $\mathcal{K}$ is a σ-covering family and ρ a gauge on $\mathcal{K}$, will be called a *generating pair*.

Example 15.2. The family $\mathcal{O}$ consisting of $\emptyset$ and all bounded open intervals (a, b) is a σ-covering family of $\mathbb{R}$.

Example 15.3. The family $\mathcal{O}_k$ consisting of $\emptyset$ and all bounded open k-dimensional cells

$$(a_1, b_1) \times (a_2, b_2) \times \cdots \times (a_k, b_k)$$

is a σ-covering family of $\mathbb{R}^k$.

Definition 15.4. A function $\lambda \colon \mathcal{P}(S) \to [0, \infty]$ is an *outer measure* on S if

(i) $\lambda(\emptyset) = 0$;

(ii) $\lambda(A) \leq \sum_{n=1}^{\infty} \lambda(A_n)$ whenever $A \subset \bigcup_{n=1}^{\infty} A_n$.

We observe that $\lambda(A) \leq \lambda(B)$ if $A \subset B$. Clearly, every measure is an outer measure.

Theorem 15.5. *Let $(\mathcal{K}, \rho)$ be a generating pair. For each set $A \subset S$ define $\lambda(A)$ by*

$$\lambda(A) = \inf_{\{E_n\}} \sum_{n=1}^{\infty} \rho(E_n), \tag{15.1}$$

where the infimum is taken over all countable covers $\{E_n\}$ of A by sets in $\mathcal{K}$. Then λ is an outer measure on S.

Proof.   (i) We have $\lambda(\emptyset) = 0$ as $\{\emptyset, \emptyset, \dots\}$ is a countable cover for $\emptyset$.

(ii) Let (A_n) be a countable cover for A, and let $\varepsilon > 0$. From the definition of infimum, for each n there is a sequence $(E_{nk})_{k=1}^{\infty}$ of sets in $\mathcal{K}$ covering A_n such that

$$\sum_{k=1}^{\infty} \rho(E_{nk}) \leq \lambda(A_n) + \frac{\varepsilon}{2^n}.$$

Then (E_{nk}) is a countable cover for $A \subset \bigcup_{n=1}^{\infty} A_n$, and

$$\lambda(A) \leq \sum_{n,k=1}^{\infty} \rho(E_{nk}) = \sum_{n=1}^{\infty} \sum_{k=1}^{\infty} \rho(E_{nk}) \leq \sum_{n=1}^{\infty} (\lambda(A_n) + \varepsilon 2^{-n}) = \sum_{n=1}^{\infty} \lambda(A_n) + \varepsilon.$$

Since ε was arbitrary, we have

$$\lambda(A) \leq \sum_{n=1}^{\infty} \lambda(A_n).$$

$\square$

15.2   Measures from outer measures

In this section λ denotes an outer measure on S. In the preceding section we saw one method for constructing outer measures from generating pairs. We show below that an outer measure always yields a measure on a certain σ-algebra of subsets of S. The sets E we want to single out should have the property that they split every set A in such a way that λ is additive on the disjoint union $A = (A \cap E) \cup (A \cap E^c)$. It then turns out that λ is countably additive on that subfamily.

Definition 15.6 (Carathéodory). A set $E \subset S$ is called *measurable* with respect to an outer measure λ (or λ-*measurable*) if

$$\lambda(A) = \lambda(A \cap E) + \lambda(A \cap E^c)$$

for all $A \subset S$ (Figure 15.1)

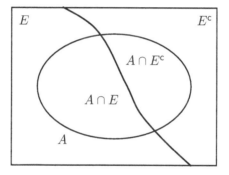

Figure 15.1

Since λ is subadditive, we always have the inequality $\lambda(A) \leq \lambda(A \cap E) + \lambda(A \cap E^c)$. This means that to check λ-measurability we only need to verify the inequality

$$\lambda(A) \geq \lambda(A \cap E) + \lambda(A \cap E^c).$$

Theorem 15.7 (Carathéodory's extension theorem). *The family $\mathcal{A}$ of all λ-measurable subsets of S is a σ-algebra. The restriction of λ to $\mathcal{A}$ is a complete positive measure.*

Proof.   It is left as an exercise to prove that $\emptyset$ and S are in $\mathcal{A}$, and that the complement of any set in $\mathcal{A}$ is itself in $\mathcal{A}$.

Next we show that if $E, F \in \mathcal{A}$, then $E \cap F \in \mathcal{A}$. We combine the following three identities with an arbitrary $A \subset S$:

$$\lambda(A \cap F) = \lambda(A \cap F \cap E) + \lambda(A \cap F \cap E^c),$$
$$\lambda(A) = \lambda(A \cap F) + \lambda(A \cap F^c),$$
$$\lambda(A \cap (E \cap F)^c) = \lambda(A \cap F \cap E^c) + \lambda(A \cap F^c)$$

to get

$$\lambda(A) = \lambda(A \cap E \cap F) + \lambda(A \cap (E \cap F)^c),$$

which shows that $E \cap F$ is λ-measurable.

Since $\mathcal{A}$ is closed under finite intersections and under complementation, it is closed also under finite unions.

Let (E_n) be a sequence of disjoint sets in $\mathcal{A}$, and let $S_n = E_1 \cup \cdots \cup E_n$. Then $S_n \in \mathcal{A}$, and we show that

$$\lambda(A \cap S_n) = \sum_{k=1}^{n} \lambda(A \cap E_k) \quad \text{for all } A \subset S. \tag{15.2}$$

This is proved by induction on n. For $n = 1$, (15.2) is an obvious identity. Assume that (15.2) is true for n. Then

$$\lambda(A \cap S_{n+1}) = \lambda(A \cap S_{n+1} \cap S_n) + \lambda(A \cap S_{n+1} \cap S_n^c)$$
$$= \lambda(A \cap S_n) + \lambda(A \cap E_{n+1})$$
$$= \sum_{k=1}^{n} \lambda(A \cap E_k) + \lambda(A \cap E_{n+1}) = \sum_{k=1}^{n+1} \lambda(A \cap E_k).$$

In the proof we used the measurability of S_n.

Let $T = \bigcup_{k=1}^{\infty} E_k$. We prove that, for any set $A \subset S$,

$$\lambda(A \cap T) = \sum_{k=1}^{\infty} \lambda(A \cap E_k). \tag{15.3}$$

Indeed, by the monotonicity of λ and (15.2),

$$\lambda(A \cap T) \geq \lambda(A \cap S_n) = \sum_{k=1}^{n} (A \cap E_k).$$

Taking the limit as $n \to \infty$, we get $\lambda(A \cap T) \geq \sum_{k=1}^{\infty} \lambda(A \cap E_k)$. The reverse inequality follows from the countable subadditivity of λ.

Now we can show that a countable disjoint union of sets E_n in $\mathcal{A}$ is again in $\mathcal{A}$. Indeed, for any set $A \subset S$,

$$\lambda(A) = \lambda(A \cap S_n) + \lambda(A \cap S_n^c) \geq \sum_{k=1}^{n} \lambda(A \cap E_k) + \lambda(A \cap T^c).$$

Taking the limit as $n \to \infty$ and using (15.3), we get

$$\lambda(A) \geq \lambda(A \cap T) + \lambda(A \cap T^c).$$

This shows that T is λ-measurable.

We have to show that λ is countably additive on $\mathcal{A}$. This follows from (15.3) when we set $A = S$.

To see that the restriction of λ to the σ-algebra $\mathcal{A}$ of λ-measurable sets is a complete measure, consider a set E of measure zero and a set $B \subset E$. For an arbitrary set $A \subset S$ we have

$$\lambda(A) = \lambda(E) + \lambda(A) \geq \lambda(A \cap B) + \lambda(A \cap B^c)$$

as $E \supset A \cap B$ and $A \supset A \cap B^c$; this shows that B is measurable. $\quad\square$

Let $(\mathcal{K}, \rho)$ be a generating pair on S and let λ be the outer measure on S induced by ρ via (15.1). The σ-algebra of all λ-measurable sets in S will be denoted by

$$\mathcal{A} = \mathcal{A}(\mathcal{K}, \rho).$$

For any $E \in \mathcal{K}$, $\{E, \emptyset, \emptyset, \dots\}$ is a countable cover for E. Hence

$$\lambda(E) \leq \rho(E) \text{ for all } E \in \mathcal{K}.$$

However, it is not always true that the sets in $\mathcal{K}$ are λ-measurable or that $\lambda(A) = \rho(A)$ for $A \in \mathcal{K}$. It is true in one significant case, when $\mathcal{K}$ is a *semiring* of sets (Definition 15.13) and the gauge ρ is countably additive (Definition 15.16). This result is described in the next section (Theorem 15.20).

Example 15.8. The *Lebesgue measure* on the real line can be constructed from the generating pair $(\mathcal{N}, \rho)$, where $\mathcal{N}$ is the family of subsets of $\mathbb{R}$ consisting of $\emptyset$ and all bounded semiclosed intervals $(a, b]$, and a gauge ρ on $\mathcal{N}$ is defined by

$$\rho(\emptyset) = 0 \text{ and } \rho((a, b]) = b - a.$$

Let λ be the outer measure on S generated by this pair. Then every interval $(a, b] \in \mathcal{N}$ is λ-measurable, and

$$\lambda((a, b]) = \inf_{(a_n, b_n]} \sum_{n=1}^{\infty} (b_n - a_n) = b - a,$$

where the infimum is taken over all countable covers of $(a, b]$ by elements of $\mathcal{N}$. The result is true because $\mathcal{N}$ is a semiring, and the gauge ρ is countably additive on $\mathcal{N}$ (Section 15.3). The σ-algebra $\Lambda = \mathcal{A}(\mathcal{N}, \rho)$ of all λ-measurable sets is the σ-algebra of Lebesgue measurable sets in $\mathbb{R}$, and λ is the Lebesgue measure on Λ.

Example 15.9. The *Lebesgue-Stieltjes measure* on $\mathbb{R}$ is a generalization of the Lebesgue measure. Given a real valued right-continuous increasing function g on $\mathbb{R}$, we define the Stieltjes gauge ρ_g by

$$\rho_g(\emptyset) = 0, \qquad \rho_g((a,b]) = g(b) - g(a)$$

for every interval $(a,b]$. The Lebesgue-Stieltjes measure is then constructed from the generating pair $(\mathcal{N}, \rho_g)$, where $\mathcal{N}$ is the σ-covering family from the preceding example. The outer measure λ_g generated by this pair coincides with ρ_g on all intervals $(a,b]$, which can be shown to be λ_g-measurable. This follows from the fact that $\mathcal{N}$ is a semiring of sets and the Stieltjes gauge ρ_g is countably additive. See Section 15.3 for details.

Example 15.10. The *Lebesgue measure* on $\mathbb{R}^k$ constructed from the generating pair $(\mathcal{N}_k, \rho)$ where $\mathcal{N}_k$ is the family consisting of $\emptyset$ and all bounded cells in $\mathbb{R}^k$ of the form

$$A = (a_1, b_1] \times (a_2, b_2] \times \cdots \times (a_k, b_k],$$

and gauge ρ on $\mathcal{N}_k$ defined as the Euclidean volume: $\rho(\emptyset) = 0$ and

$$\rho(A) = \mathsf{vol}\,(A) = (b_1 - a_1)(b_2 - a_2) \cdots (b_k - a_k).$$

If λ is the outer measure on S generated by this pair, then every cell of the above form is λ-measurable and λ agrees with ρ on $\mathcal{N}_k$. Again, the reason for this desirable behavior is that $\mathcal{N}_k$ is a semiring of sets, while the Euclidean volume is countably additive on $\mathcal{N}_k$.

Example 15.11. *Carathéodory's construction in a metric space.* Let (S, d) be a metric space, $\mathcal{K}$ a σ-covering family of S, and ρ any gauge on $\mathcal{K}$. For any real number $c > 0$ let $\mathcal{K}_c$ be the family of all sets in $\mathcal{K}$ with diameter not exceeding c. (Recall that $\mathrm{diam}(A) = \sup\{d(x, y) : x, y \in A\}$.) Let μ_c be the outer measure on S generated by the pair $(\mathcal{K}_c, \rho)$, and let

$$\mu(A) = \lim_{c \to 0+} \mu_c(A) \qquad \text{for all } A \subset S.$$

Then μ is an outer measure on S called the *Carathéodory measure* on S associated with $(\mathcal{K}, \rho)$. It turns out that all Borel sets in S are μ-measurable. (Problem 15.12.)

Example 15.12. *m-dimensional Hausdorff measure* in $\mathbb{R}^k$. Let α_m be the Lebesgue measure of the closed unit ball $\{u \in \mathbb{R}^m : \|u\|_2 \le 1\}$ in the Euclidean space $\mathbb{R}^m$. We apply Carathéodory's construction with $S = \mathbb{R}^k$, $\mathcal{K}$ the set of all subsets of $\mathbb{R}^k$ and

$$\rho(A) = \frac{\alpha_m}{2^m}(\mathrm{diam}(A))^m.$$

The resulting measure is called the *m-dimensional Hausdorff measure* over $\mathbb{R}^k$, denoted by $\mathcal{H}^m$. The Hausdorff measure $\mathcal{H}^m$ is used to define the area of m-dimensional manifolds in $\mathbb{R}^k$, and the *Hausdorff dimension* of a nonempty subset A of $\mathbb{R}^k$ by $h(A) = \inf\{m \geq 0 : \mathcal{H}^m(A) < \infty\}$.

15.3 Extension of measures

The main results of this section are the Hahn extension Theorem 15.20 and the Hahn uniqueness Theorem 15.23. Usually, their formulations are given for set algebras; however we chose to replace algebras by semirings of sets. The reason is that for the customary formulation we have to construct a set algebra, usually from a semiring, extend a given finitely additive gauge from a semiring to an algebra, and verify that the extended gauge remains finitely additive; the choice of a semiring from the outset avoids a necessity for this intermediate construction.

Definition 15.13. A family $\mathcal{K}$ of subsets of S is a *semiring* if

   (i) $\emptyset \in \mathcal{K}$;

   (ii)   if $A, B \in \mathcal{K}$, then $A \cap B \in \mathcal{K}$;

   (iii)   if $A, B \in \mathcal{K}$, then $A \setminus B$ can be expressed as a finite disjoint union of elements of $\mathcal{K}$.

Example 15.14. Let $\mathcal{N}$ be the family of subsets of $\mathbb{R}$ consisting of $\emptyset$ and all bounded semiclosed intervals $(a, b]$. The elements of $\mathcal{N}$ are sometimes called *nails* (from the resemblance of ( ] to a fingernail). The family $\mathcal{N}$ is a σ-covering family for $\mathbb{R}$ and a semiring (Problem 15.5).

Example 15.15. Let $\mathcal{N}_k$ be a k-dimensional analogue of the set $\mathcal{N}$ from the preceding example; $\mathcal{N}_k$ then consists of $\emptyset$ and all bounded cells in $\mathbb{R}^k$ of the form

$$A = (a_1, b_1] \times (a_2, b_2] \times \cdots \times (a_k, b_k]. \tag{15.4}$$

Then $\mathcal{N}_k$ is a σ-covering family for $\mathbb{R}^k$ and a semiring of sets in $\mathbb{R}^k$ (Problem 15.5).

Definition 15.16. Let $(\mathcal{K}, \rho)$ be a generating pair. The gauge ρ is said to be *finitely additive* on $\mathcal{K}$ if, for every finite disjoint family $A_1, \ldots, A_p$ of sets in $\mathcal{K}$ whose union $A = \bigcup_{n=1}^{p} A_n$ is also an element of $\mathcal{K}$, we have $\rho(A) = \sum_{n=1}^{p} \rho(A_n)$; ρ is *countably additive* on $\mathcal{K}$ if, for every disjoint sequence (A_n) of sets in $\mathcal{K}$ whose union $A = \bigcup_{n=1}^{\infty} A_n$ is also an element of $\mathcal{K}$, we have $\rho(A) = \sum_{n=1}^{\infty} \rho(A_n)$.

Example 15.17. Let $\mathcal{N}$ be the family of nails defined in Example 15.14 and let g be a real valued right-continuous increasing function on $\mathbb{R}$. If ρ is defined by

$$\rho(\emptyset) = 0 \ \text{ and } \ \rho((a,b]) = g(b) - g(a),$$

then ρ is a gauge on $\mathcal{N}$ which is finitely additive; in fact, ρ is countably additive (see the next section). In particular, the interval length

$$m(\emptyset) = 0 \ \text{ and } \ m((a,b]) = b - a$$

is countably additive on $\mathcal{N}$.

Example 15.18. Let $\mathcal{N}_k$ be the semiring defined in Example 15.15, and let ρ be the Euclidean volume on $\mathbb{R}^k$, that is, let $\rho(\emptyset) = 0$ and let

$$\rho(A) = \mathsf{vol}\,(A) = (b_1 - a_1)(b_2 - a_2) \cdots (b_k - a_k)$$

for any cell A in (15.4). Then ρ is finitely additive; in fact, ρ is a countably additive gauge on the semiring $\mathcal{N}_k$ (see the next section).

In the next lemma we put together some properties of a semiring $\mathcal{K}$ and of finitely (countably) additive gauges on $\mathcal{K}$ which we need in our treatment of extensions of measures.

Lemma 15.19. *Let $\mathcal{K}$ be a semiring and ρ a finitely additive gauge on $\mathcal{K}$. Then:*

(i)   *If $A, A_1, \ldots, A_n \in \mathcal{K}$, then $A_1 \cup \cdots \cup A_n$, $A \cap (A_1 \cup \cdots \cup A_n)$ and $A \cap (A_1 \cup \cdots \cup A_n)^c$ can be expressed as finite disjoint unions of sets in $\mathcal{K}$.*

(ii)   *If $A, B \in \mathcal{K}$ and $A \subset B$, then $\rho(A) \leq \rho(B)$.*

(iii)   *If $A_n \in \mathcal{K}$ are disjoint and $\bigcup_{n=1}^{p} A_n \subset A$ with $A \in \mathcal{K}$ then $\sum_{n=1}^{p} \rho(A_n) \leq \rho(A)$.*

(iv)   *If $A_n \in \mathcal{K}$, $A \in \mathcal{K}$ and $A \subset \bigcup_{n=1}^{p} A_n$, then $\rho(A) \leq \sum_{n=1}^{p} \rho(A_n)$.*

(v)   *If $A_n \in \mathcal{K}$, $A \in \mathcal{K}$ and $A \subset \bigcup_{n=1}^{\infty} A_n$, then $\rho(A) \leq \sum_{n=1}^{\infty} \rho(A_n)$ provided ρ is countably additive.*

Proof.   Problem 15.6. $\square$

If Let $(\mathcal{K}, \rho)$ is a generating pair and $\mathcal{K} \subset \mathcal{A}$, where $\mathcal{A}$ is a σ-algebra, then a measure λ on $\mathcal{A}$ is called an *extension of the gauge ρ* provided

$$\lambda(E) = \rho(E) \ \text{ for all } \ E \in \mathcal{K}.$$

The following result is a version of Hahn's theorem formulated for semirings.

Theorem 15.20 (The Hahn extension theorem). *Let $(\mathcal{K}, \rho)$ be a generating pair of S with $\mathcal{K}$ a semiring and ρ countably additive on $\mathcal{K}$, let λ be the outer measure generated by $(\mathcal{K}, \rho)$, and let $\mathcal{A}$ be the σ-algebra of all λ-measurable sets. Then $\mathcal{K} \subset \mathcal{A}$, and λ is an extension of ρ.*

Proof. Let E be in $\mathcal{K}$. Since $\{E, \emptyset, \emptyset, \dots\}$ is a countable cover for E, $\lambda(E) \leq \rho(E)$.

Conversely, let (A_n) be any countable cover of $E \in \mathcal{K}$ by sets in $\mathcal{K}$. Then by Lemma 15.19 (v),

$$\rho(E) \leq \sum_{n=1}^{\infty} \rho(A_n).$$

Since (A_n) was an arbitrary countable cover of E by sets in $\mathcal{K}$, it follows that

$$\rho(E) \leq \lambda(E).$$

It remains to prove that every set $E \in \mathcal{K}$ is λ-measurable. First we prove

$$\rho(A) \geq \lambda(A \cap E) + \lambda(A \cap E^c) \quad \text{for all } A \in \mathcal{K}. \tag{15.5}$$

By the definition of a semiring,

$$A \cap E = B, \qquad A \cap E^c = C_1 \cup \cdots \cup C_k,$$

where $B, C_1, \dots, C_k$ are disjoint sets in $\mathcal{K}$. Then

$$
\begin{aligned}
\lambda(A \cap E) + \lambda(A \cap E^c) &= \lambda(B) + \lambda(C_1 \cup \cdots \cup C_k) \\
&\leq \lambda(B) + \lambda(C_1) + \cdots + \lambda(C_k) \\
&= \rho(B) + \rho(C_1) + \cdots + \rho(C_k) \\
&= \rho(B_1 \cup C_1 \cup \cdots \cup C_k) = \rho(A).
\end{aligned}
$$

This proves (15.5). Let A be an arbitrary subset of S, and let $\varepsilon > 0$. There is a countable cover (A_n) of A by sets in $\mathcal{K}$ such that

$$\sum_{n=1}^{\infty} \rho(A_n) \leq \lambda(A) + \varepsilon.$$

Then

$$A \cap E \subset \bigcup_{n=1}^{\infty} (A_n \cap E), \qquad A \cap E^c \subset \bigcup_{n=1}^{\infty} (A_n \cap E^c),$$

so that

$$\lambda(A \cap E) \leq \sum_{n=1}^{\infty} \lambda(A_n \cap E), \qquad \lambda(A \cap E^c) \leq \sum_{n=1}^{\infty} \lambda(A_n \cap E^c).$$

Using (15.5) and properties of outer measure, we get

$$\lambda(A \cap E) + \lambda(A \cap E^c) \le \sum_{n=1}^{\infty} \{\lambda(A_n \cap E) + \lambda(A_n \cap E_n^c)\}$$

$$\le \sum_{n=1}^{\infty} \rho(A_n) \le \lambda(A) + \varepsilon.$$

Since ε was arbitrary, we have

$$\lambda(A) \ge \lambda(A \cap E) + \lambda(A \cap E^c).$$ □

If $\mathcal{K}$ is a semiring of subsets of S, then a $\mathcal{K}$-*simple function* is defined as a linear combination of characteristic functions of elements of $\mathcal{K}$. If the conditions of the preceding theorem are fulfilled, then the λ-integrable functions can be approximated by $\mathcal{K}$-simple functions. This fact will be used to our advantage in the proof of Fubini's theorem in Section 16.2.

Theorem 15.21. *Let the conditions of the preceding theorem be fulfilled. Then the $\mathcal{K}$-simple functions are dense in $(L^1(\lambda), \|\cdot\|_{L^1(\lambda)})$.*

Proof.   Let $E \in \mathcal{A}$, $\lambda(E) < \infty$, and let $\varepsilon > 0$. There exists a sequence of disjoint sets $A_n \in \mathcal{K}$ such that

$$E \subset \bigcup_{n=1}^{\infty} A_n, \qquad \sum_{n=1}^{\infty} \rho(A_n) < \lambda(E) + \tfrac{1}{2}\varepsilon.$$

There exists $N \in \mathbb{N}$ such that $\sum_{n=N+1}^{\infty} \rho(A_n) < \tfrac{1}{2}\varepsilon$. Set $A = \bigcup_{n=1}^{\infty} A_n$ and $B = \bigcup_{n=1}^{N} A_n$. Then $\sum_{n=1}^{\infty} \rho(A_n) = \lambda(A)$, and

$$\|\chi_E - \chi_B\|_{L^1(\lambda)} \le \|\chi_E - \chi_A\|_{L^1(\lambda)} + \|\chi_A - \chi_B\|_{L^1(\lambda)}$$

$$\le (\lambda(E) - \lambda(A)) + \tfrac{1}{2}\varepsilon < \varepsilon.$$

Hence any simple function in $L^1(\lambda)$ can be approximated by a $\mathcal{K}$-simple function. The result follows as the simple functions are dense in $L^1(\lambda)$. □

Let $(\mathcal{K}, \rho)$ be a generating pair, and λ the measure constructed in Theorem 15.20. If μ is another measure on the σ-algebra $\mathcal{A} = \mathcal{A}(\mathcal{K}, \rho)$ such that $\mu(A) = \rho(A)$ for all $A \in \mathcal{K}$, then in general $\mu \ne \lambda$. The next result describes an important case when we always have $\mu = \lambda$.

Definition 15.22. Let ρ be a gauge on a semiring $\mathcal{K}$. We say that ρ is σ-*finite* if S can be expressed as a countable union of disjoint sets (A_n) in $\mathcal{K}$ with $\rho(A_n) < \infty$ for all n.

Theorem 15.23 (The Hahn uniqueness theorem). *Let* $(\mathcal{K}, \rho)$ *be a generating pair of S with $\mathcal{K}$ a semiring and ρ countably additive and σ-finite on S. Then ρ has a unique extension to a measure μ on the σ-algebra $\mathcal{A} = \mathcal{A}(\mathcal{K}, \rho)$.*

Proof. Let λ be the outer measure generated by $(\mathcal{K}, \rho)$. By Carathéodory's extension theorem, λ is a measure on $\mathcal{A}$. By the preceding theorem, λ is an extension of ρ. Suppose that μ is a measure on $\mathcal{A}$ such that

$$\mu(A) = \rho(A) \text{ for all } A \in \mathcal{K}.$$

Let E be any set in $\mathcal{A}$ and let (A_n) be a countable cover of E by sets in $\mathcal{K}$. Then

$$\mu(E) \leq \mu\left(\bigcup_{n=1}^{\infty} A_n\right) \leq \sum_{n=1}^{\infty} \mu(A_n) = \sum_{n=1}^{\infty} \rho(A_n).$$

By the definition of the outer measure λ,

$$\mu(E) \leq \lambda(E).$$

Suppose that E is contained in a set $A \in \mathcal{K}$ with $\rho(A) < \infty$. Replacing E by $A \cap E^c$ in the preceding inequality, we get $\mu(A \cap E^c) \leq \lambda(A \cap E^c)$. Then

$$\mu(E) = \mu(A) - \mu(A \cap E^c) = \rho(A) - \mu(A \cap E^c)$$
$$= \lambda(E) + \lambda(A \cap E^c) - \mu(A \cap E^c) \geq \lambda(E).$$

This proves that $\mu(E) = \lambda(E)$. In the general case, S can be expressed as a disjoint union of sets A_n in $\mathcal{K}$ with $\rho(A_n) < \infty$. Observe that μ and λ agree on finite disjoint unions of sets in $\mathcal{K}$; in particular, they agree on $A_n \cap E$. Then

$$\mu(E) = \sum_{n=1}^{\infty} \mu(A_n \cap E) = \sum_{n=1}^{\infty} \lambda(A_n \cap E) = \lambda(E).$$

$\square$

15.4 Lebesgue and Lebesgue–Stieltjes measure

The main result of this section is Theorem 15.27 which shows that the volume (or more generally a Stieltjes gauge) is countably additive; we can then apply the Hahn extension theorem of the preceding section.

A construction of the Lebesgue and Lebesgue-Stieltjes measures in $\mathbb{R}^k$ relies on properties of cells in $\mathbb{R}^k$, and in the case of the Lebesgue measure, also on properties of the Euclidean volume. Two properties of the volume are most relevant for this construction. One is that the volume is finitely additive, the other is the fact that the volume is regular in the sense that

$$\text{vol}\,(A) = \sup\,\{\text{vol}\,(K) : K \subset A,\ K \text{ is a closed cell}\},$$
$$\text{vol}\,(A) = \inf\,\{\text{vol}\,(G) : G \supset A,\ G \text{ is an open cell}\}. \qquad (15.6)$$

Faces of k-cells in $\mathbb{R}^k$ have zero k-dimensional Euclidean volume, so for the construction of the Lebesgue measure in $\mathbb{R}^k$ we could use open cells or closed cells instead of the semiclosed cells from the semiring $\mathcal{N}_k$. However, for the construction of the Lebesgue-Stieltjes measure this distinction is all important; we have to use semiclosed cells as the Stieltjes gauge (or Stieltjes 'volume') of the cell face is not necessarily zero. We also have to formulate property (15.6) in terms of semiclosed cells.

Definition 15.24. Let $\mathcal{N}_k$ be the semiring consisting of $\emptyset$ and the bounded semiclosed cells

$$A = (a_1, b_1] \times \cdots \times (a_k, b_k]$$

in $\mathbb{R}^k$. A gauge ρ on $\mathcal{N}_k$ is called a *Stieltjes gauge* if it has the following properties:

S1 ρ takes finite values on semiclosed cells and is finitely additive on $\mathcal{N}_k$.

S2 For each $A \in \mathcal{N}_k$ and each $\varepsilon > 0$ there is a cell C in $\mathcal{N}_k$ such that
$$A \subset C^\circ \text{ and } 0 \le \rho(C) - \rho(A) < \varepsilon.$$

S3 For each $A \in \mathcal{N}_k$ and each $\varepsilon > 0$ there is a cell B in $\mathcal{N}_k$ such that
$$\overline{B} \subset A \text{ and } 0 \le \rho(A) - \rho(B) < \varepsilon.$$

Example 15.25. The Euclidean volume on $\mathcal{N}_k$ is a Stieltjes gauge. Indeed, for any cell $A = (a_1, b_1] \times \cdots \times (a_k, b_k]$, $\text{vol}\,(A)$ is the product

$$\text{vol}\,(A) = m((a_1, b_1]) \cdots m((a_k, b_k]),$$

where the length $m((a_i, b_i]) = b_i - a_i$ is a Stieltjes gauge on one dimensional nails (verify). From this we deduce that volume is finitely additive on disjoint unions which form partitions of A, that is, cartesian products of one dimensional partitions. The finite additivity in the general case can be deduced from this.

Let $A = (a_1, b_1] \times \cdots \times (a_k, b_k]$ be a semiclosed cell and let $\varepsilon > 0$. For all sufficiently small $\eta \ge 0$ define

$$B_\eta = (a_1 + \eta, b_1] \times \cdots \times (a_k + \eta, b_k], \quad C_\eta = (a_1, b_1 + \eta] \times \cdots \times (a_k, b_k + \eta].$$

The functions $f(\eta) = \text{vol}\,(B_\eta)$ and $g(\eta) = \text{vol}\,(C_\eta)$ are continuous at 0 being products of k continuous functions of η. Since $f(\eta) \le f(0) = \text{vol}\,(A)$, $g(\eta) \ge g(0) = \text{vol}\,(A)$, we can choose $B = B_\eta$ and $C = C_\eta$ with a sufficiently small $\eta > 0$ to satisfy **S2** and **S3**.

Example 15.26. If g is a real valued right-continuous increasing function on $\mathbb{R}$, then the gauge defined by $\rho(\emptyset) = 0$ and $\rho((a, b]) = g(b) - g(a)$ is a Stieltjes gauge on the semiring of one dimensional nails (verify). More generally, let $g_1, g_2, \ldots, g_k$ be real valued right-continuous increasing functions on $\mathbb{R}$, and let $\rho_i((a_i, b_i]) = g_i(b_i) - g_i(a_i)$ for each i. If ρ is defined on the semiring $\mathcal{N}_k$ by $\rho(\emptyset) = 0$ and

$$\rho(A) = \rho_1((a_1, b_1]) \cdots \rho_k((a_k, b_k])$$

on any cell $A = (a_1, b_1] \times \cdots \times (a_k, b_k]$, then ρ is a Stieltjes gauge. Proof of this is left as an exercise (Problem 15.10). Not every Stieltjes gauge on $\mathcal{N}_k$ can be obtained in this way.

Theorem 15.27. *Any Stieltjes gauge ρ on the semiring $\mathcal{N}_k$ is countably additive.*

Proof. Suppose that $A \in \mathcal{N}_k$ and $A = \bigcup_{i=1}^{\infty} A_i$, where A_i are disjoint members of $\mathcal{N}_k$. By Lemma 15.19 (iii), $\rho(A) \geq \sum_{i=1}^{n} \rho(A_i)$ for all n. Consequently we have

$$\sum_{i=1}^{\infty} \rho(A_i) \leq \rho(A). \tag{15.7}$$

We prove the inequality reverse to (15.7). Let $\varepsilon > 0$. By **S3**, there is is a cell $B \in \mathcal{N}_k$ such that $\overline{B} \subset A$ and

$$\rho(A) < \rho(B) + \varepsilon.$$

By **S2**, for each n there is a cell $C_n \in \mathcal{N}_k$ such that $A_n \subset C_n^\circ$ and $\rho(C_n) < \rho(A_n) + 2^{-n}\varepsilon$. Then $\{C_n^\circ\}$ is an open cover for the compact set $\overline{B}$. By the Heine–Borel theorem for $\mathbb{R}^k$ there is a finite subcover; hence there is p such that $B \subset \overline{B} \subset \bigcup_{n=1}^{p} C_n$. By Lemma 15.19,

$$\rho(B) \leq \sum_{n=1}^{p} \rho(C_n) \leq \sum_{n=1}^{\infty} \rho(C_n) < \sum_{n=1}^{\infty} \rho(A_n) + \varepsilon.$$

So

$$\rho(A) < \rho(B) + \varepsilon < \sum_{n=1}^{\infty} \rho(A_n) + 2\varepsilon.$$

Since ε was arbitrary, the inequality $\rho(A) \leq \sum_{n=1}^{\infty} \rho(A_n)$ follows. $\square$

Example 15.28. A specialization of Example 15.26 shows that $\rho((a, b]) = g(b) - g(a)$ is a Stieltjes gauge on $\mathcal{N}$ if g is a right continuous increasing function on $\mathbb{R}$. By the preceding theorem, ρ is countably additive on $\mathcal{N}$.

In particular, the length gauge $\rho((a, b]) = b - a$ is countably additive on $\mathcal{N}$. It is interesting to observe that any Stieltjes gauge on $\mathcal{N}$ is generated by some real valued right-continuous increasing function g so that $\rho((a, b]) = g(b) - g(a)$ for every interval $(a, b]$ (Problem 15.8). There is no such simple characterization of Stieltjes gauges in $\mathbb{R}^k$ for $k > 1$.

Theorem 15.29. *Let ρ be a Stieltjes gauge on the semiring $\mathcal{N}_k$. Then there is a unique measure μ on the σ-algebra $\mathcal{B}_k$ of all Borel sets in $\mathbb{R}^k$ such that for all $A \in \mathcal{N}_k$,*

$$\mu(A) = \rho(A). \tag{15.8}$$

Note. This measure is called the *Lebesgue–Stieltjes measure* on $\mathbb{R}^k$ generated by ρ. In the case that ρ is the Euclidean volume in $\mathbb{R}^k$, the measure μ is called the *Lebesgue measure* in $\mathbb{R}^k$. It is the unique measure on $\mathcal{B}_k$ with

$$\mu(A) = \mathsf{vol}\,(A)$$

for all semiclosed cells $A \in \mathcal{N}_k$.

Proof. Let λ be the outer measure generated by the pair $(\mathcal{N}_k, \rho)$, and let $\mathcal{A}$ be the σ-algebra of all λ-measurable sets in $\mathbb{R}^k$. By Theorem 15.27, ρ is a countably additive gauge on $\mathcal{N}$. Also, ρ is σ-finite on $\mathcal{N}_k$ as ρ takes only finite values on semiclosed cells and $\mathbb{R}^k$ can be covered by a countable family of these cells. By Theorem 15.20 and 15.23, there is a a unique measure μ on $\mathcal{A}$ such that (15.8) is satisfied.

The σ-algebra $\mathcal{A}$ contains all Borel subsets of $\mathbb{R}^k$ as it contains the semiclosed cells, which are known to generate the Borel σ-algebra $\mathcal{B}_k$ in $\mathbb{R}^k$. The restriction of μ to $\mathcal{N}_k$ satisfies the conditions of the theorem. □

In the case that the Stieltjes gauge ρ is the Euclidean volume, we may investigate the relation between $\mathcal{B}_k$ and $\mathcal{A}$ from the preceding theorem. We know that the Lebesgue measure $\mu = \mu_k$ is complete on $\mathcal{A}$ (Carathéodory's extension theorem). So $\mathcal{A}$ must contain all Lebesgue measurable sets in $\mathbb{R}^k$, which were defined as elements of the measure completion Λ^k of $\mathcal{B}_k$ with respect to μ. We then have

$$\mathcal{B}_k \subset \Lambda^k \subset \mathcal{A}.$$

In fact, we have $\mathcal{A} = \Lambda^k$, as the next result shows.

Theorem 15.30. *Let vol be the Euclidean volume on $\mathbb{R}^k$, and λ the outer measure generated by the pair $(\mathcal{N}_k, \mathsf{vol})$. The σ-algebra $\mathcal{A}$ of all λ-measurable sets in $\mathbb{R}^k$ coincides with the measure completion Λ^k of $\mathcal{B}_k$ with respect to the Lebesgue measure μ.*

Proof. We have to show that, for each set $E \in \mathcal{A}$, there are Borel sets A, B such that $A \subset E \subset B$ and $\lambda(B \cap A^c) = 0$.

Assume first that $E \in \mathcal{A}$ satisfies $\lambda(E) < \infty$. Given $\varepsilon > 0$, there is a countable cover (A_n) of E by sets in $\mathcal{N}_k$ with

$$\sum_{n=1}^{\infty} \text{vol}\,(A_n) < \lambda(E) + \varepsilon.$$

The union $D = \bigcup_{n=1}^{\infty} A_n$ is a Borel set in $\mathbb{R}^k$ with $D \supset E$ and

$$\lambda(D) \leq \sum_{n=1}^{\infty} \text{vol}\,(A_n) < \lambda(E) + \varepsilon.$$

Since λ is a measure on $\mathcal{A}$, we have $\lambda(D) = \lambda(E) + \lambda(D \cap E^c)$, so that $\lambda(D \cap E^c) < \varepsilon$.

Choosing in succession ε to be $1, \frac{1}{2}, \frac{1}{3}, \ldots$ we see that for each n there is a Borel set D_n with $E \subset D_n$ and $\lambda(D_n \cap E^c) < 1/n$. Then the set $B = \bigcap_{n=1}^{\infty} D_n$ is Borel, and

$$E \subset B, \qquad \lambda(B \cap E^c) = 0. \tag{15.9}$$

If $\lambda(E) = \infty$, we can choose a sequence C_n of disjoint Borel sets of finite measure with the union $\mathbb{R}^k$, apply the preceding result to each $C_n \cap E$, and then take the union. We conclude that there is a Borel set B satisfying (15.9). Details are left as an exercise.

Let E be an arbitrary element of $\mathcal{A}$. By the preceding result there are Borel sets B and C such that $E \subset B$ and $\lambda(B \cap E^c) = 0$, and $E^c \subset C$ and $\lambda(C \cap E) = 0$. Set $A = C^c$. Then A, B are Borel sets with

$$A \subset E \subset B, \qquad \lambda(B \cap A^c) = 0. \qquad \square$$

Definition 15.31. Let S be a Hausdorff topological space and μ a measure on the σ-algebra $\mathcal{B} = \mathcal{B}(S)$ of all Borel subsets of S. We say that μ is *regular* if

(i) $\mu(K) < \infty$ for every compact set K in S,

(ii) for every $E \in \mathcal{B}$,

$$\mu(E) = \inf\,\{\mu(G) : G \supset E, \ G \text{ open}\},$$

(iii) for every $E \in \mathcal{B}$,

$$\mu(E) = \sup\,\{\mu(K) : K \subset E, \ K \text{ compact}\}.$$

Definition 15.32. A Hausdorff topological space is said to be *σ-compact* if it is a countable union of compact sets. The space $\mathbb{R}^k$ is σ-compact as we can write $\mathbb{R}^k = \bigcup_{n=1}^{\infty} K_n$, where

$$K_n = [-n, n]^k = [-n, n] \times \cdots \times [-n, n]. \tag{15.10}$$

Theorem 15.33. *Any Lebesgue–Stieltjes measure on $\mathcal{B}_k$ is regular.*

Proof. Let ρ be a Stieltjes gauge on $\mathcal{N}_k$ and μ the Lebesgue-Stieltjes measure on $\mathcal{B}_k$ generated by ρ.

(i) Let K a compact set in $\mathbb{R}^k$. Then K is contained in some cell $A \in \mathcal{N}_k$, so that $\mu(K) \leq \rho(A) < \infty$.

(ii) Let $E \in \mathcal{B}_k$ and let $\varepsilon > 0$. We show that there exists an open set G such that

$$G \supset E \text{ and } \mu(G \backslash E) \leq \varepsilon.$$

There is a countable cover (A_n) of E by elements of $\mathcal{N}_k$ such that

$$\sum_{n=1}^{\infty} \rho(A_n) \leq \mu(E) + \tfrac{1}{2}\varepsilon.$$

By **S2**, for each n there is $C_n \in \mathcal{N}_k$ such that $A_n \subset C_n^\circ$ and $\rho(C_n) < \rho(A_n) + (\frac{1}{2})^{n+1}\varepsilon$. Then the set $G = \bigcup_{n=1}^{\infty} C_n^\circ$ is open with $G \supset E$ and

$$\mu(G) \leq \sum_{n=1}^{\infty} \rho(C_n^\circ) \leq \sum_{n=1}^{\infty} \rho(C_n) \leq \sum_{n=1}^{\infty} \left(\rho(A_n) + (\tfrac{1}{2})^{n+1}\varepsilon \right) \leq \mu(E) + \varepsilon.$$

If $\mu(E) < \infty$, then $\mu(G \backslash E) = \mu(G) - \mu(E) < \varepsilon$. If $\mu(E) = \infty$, we use the σ-compactness of $\mathbb{R}^k$. Let K_n be the cells defined in (15.10). For each n, $K_n \cap E$ is a Borel set of finite measure, and by the preceding part of this proof there is an open set $G_n \supset K_n \cap E$ such that $\mu(G_n \backslash (K_n \cap E)) < (\frac{1}{2})^n \varepsilon$. The set $G = \bigcup_{n=1}^{\infty} G_n$ is then open with $G \supset E$, and $G \backslash E \subset \bigcup_{n=1}^{\infty} G_n \backslash (K_n \cap E)$. Hence

$$\mu(G \backslash E) \leq \sum_{n=1}^{\infty} \mu(G_n \backslash (K_n \cap E)) \leq \sum_{n=1}^{\infty} (\tfrac{1}{2})^n \varepsilon = \varepsilon.$$

The result we have just proved implies property (ii) of Definition 15.31.

(iii) Let $E \in \mathcal{B}_k$ and let $\varepsilon > 0$. We show that there is a closed set $F \subset E$ with $\mu(E \backslash F) < \varepsilon$. By part (ii) of this proof applied to E^c in place of E, there is an open set $G \supset E^c$ with $\mu(G \cap E) = \mu(G \backslash E^c) < \varepsilon$. The set $F = G^c$ is closed, $F \subset E$ and $\mu(E \backslash F) = \mu(E \cap G) < \varepsilon$.

Let K_n be the cells (15.10). Then $F \cap K_n \nearrow F$ with the sets $F \cap K_n$ compact. So $\mu(F \cap K_n) \to \mu(F)$. If $\mu(E) < \infty$, then $\mu(E) < \mu(F) + \varepsilon$ and, for a sufficiently large n, $\mu(F) < \mu(F \cap K_n) + \varepsilon$. Property (iii) of Definition 15.31 then follows. If $\mu(E) = \infty$, then $\mu(E) = \mu(F) + \mu(E \backslash F)$ implies that $\mu(F) = \infty$ and the result again follows. □

15.5 Signed and complex measures

Up to now we have dealt only with measures that take values in the interval $[0, \infty]$, that is, with *positive measures*. In many applications we encounter measures that take positive and negative values, or even complex values. We make the following definition.

Definition 15.34. Let Σ be a σ-algebra of subsets of a nonempty set S. A *signed measure* on Σ is a countably additive set function on Σ taking values in $\mathbb{R} \cup \{\infty\}$ or in $\mathbb{R} \cup \{-\infty\}$ and satisfying $\mu(\emptyset) = 0$. A *complex measure* is a countably additive set function $\mu \colon \Sigma \to \mathbb{C}$.

From the countable additivity of a complex measure we have $\mu(\emptyset) = \sum_{n=1}^{\infty} \mu(\emptyset)$, which implies $\mu(\emptyset) = 0$. The difference $\mu = \mu_1 - \mu_2$ of two positive measures μ_i, $i = 1, 2$, is a signed measure provided one of the measures is finite. We show that the converse is also true, that is, any signed measure μ can be expressed as the difference of two positive measures, one of them finite. Our proof follows Swartz [42].

Theorem 15.35 (Jordan decomposition). *Let μ be a signed measure on a σ-algebra Σ. Then there exists a unique decomposition of μ of the form*

$$\mu = \mu^+ - \mu^-,$$

where μ^+, μ^- are positive measures, one of which is finite. The measures μ^+ and μ^- are mimimal in the sense that $\mu^+ \leq \lambda$ and $\mu^- \leq \nu$ for any positive measures λ, ν satisfying $\mu = \lambda - \nu$.

Proof. We may assume that $\mu \colon \Sigma \to \mathbb{R} \cup \{\infty\}$, for in the case that $\mu \colon \Sigma \to \mathbb{R} \cup \{-\infty\}$, we would consider $-\mu$ instead of μ.

For any $E \in \Sigma$ define

$$\mu^+(E) = 0 \vee \sup \{\mu(A) : A \in \Sigma, A \subset E\}.$$

Then μ^+ is a set function on Σ taking values in $[0, \infty]$ and satisfying $\mu^+(\emptyset) = 0$. We prove that μ^+ is countably additive, that is, a positive measure. Let $E_i \in \Sigma$, $i \in \mathbb{N}$, be disjoint and let $E = \bigcup_{i=1}^{\infty} E_i$. If $A \subset E$ and $A \in \Sigma$, then

$$\mu(A) = \sum_{i=1}^{\infty} \mu(A \cap E_i) \leq \sum_{i=1}^{\infty} \mu^+(E_i), \quad \text{and} \quad \mu^+(E) \leq \sum_{i=1}^{\infty} \mu^+(E_i).$$

Suppose that $\mu^+(E_i) < \infty$ for all $i \in \mathbb{N}$, and let $\varepsilon > 0$ be given. For each $i \in \mathbb{N}$ there exists $A_i \subset E_i$, $A_i \in \Sigma$, such that $\mu(A_i) > \mu^+(E_i) - (\frac{1}{2})^i \varepsilon$. Then

$$\mu^+(E) \geq \mu\left(\bigcup_{i=1}^{\infty} A_i\right) = \sum_{i=1}^{\infty} \mu(A_i) > \sum_{i=1}^{\infty} \mu^+(E_i) - \varepsilon.$$

Since $\varepsilon > 0$ was arbitrary, we get $\mu^+(E) \geq \sum_{i=1}^{\infty} \mu^+(E_i)$. Hence $\mu^+(E) = \sum_{i=1}^{\infty} \mu^+(E_i)$.

Suppose that $\mu^+(E_k) = \infty$ for some $k \in \mathbb{N}$. Given $n \in \mathbb{N}$, there exists $A \subset E_k \subset E$, $A \in \Sigma$, such that $\mu(A) > n$. Then $n < \mu(A) \leq \mu^+(E)$, which proves that $\mu^+(E) = \infty$. Again we get $\mu^+(E) = \infty = \sum_{i=1}^{\infty} \mu^+(E_i)$. This proves that μ^+ is a positive (possibly infinite) measure.

The rest of the proof is devoted to showing that the positive measure μ^- defined by $\mu^-(E) = (-\mu)^+(E)$ for each $E \in \Sigma$ is finite, and that $\mu = \mu^+ - \mu^-$.

First we show that $\mu^+(E) = \infty$ for some $E \in \Sigma$ implies $\mu(E) = \infty$. Set $E_0 = E$. Then there exists $A_1 \subset E_0$, $A_1 \in \Sigma$, such that $\mu(A_1) > 1$. Since μ^+ is additive, either $\mu^+(A_1)$ or $\mu^+(E_0 \setminus A_1)$ is equal to ∞. Denote the one with infinite μ^+ measure by E_1. Further, there exists $A_2 \subset E_1$, $A_2 \in \Sigma$, with $\mu(A_2) > 2$. Again, one of the sets A_2 and $E_1 \setminus A_2$ has infinite μ^+ measure. Denote the one with infinite measure by E_2. Proceeding in this way, we construct two sequences of sets in Σ, (E_n) and (A_n), such that $\mu^+(E_n) = \infty$, $A_n \subset E_{n-1}$ and $\mu(A_n) > n$.

We distinguish two cases.

Suppose first that $E_n = E_{n-1} \setminus A_n$ for infinitely many $n \in \mathbb{N}$. Then there exists a subsequence (A_{k_n}) of disjoint sets A_{k_n} such that for $A = \bigcup_{n=1}^{\infty} A_{k_n} \subset E$,

$$\mu(A) = \mu\left(\bigcup_{n=1}^{\infty} A_{k_n}\right) = \sum_{n=1}^{\infty} \mu(A_{k_n}) \geq \sum_{n=1}^{\infty} k_n = \infty;$$

then $\mu(E) = \mu(A) + \mu(E \setminus A) = \infty$.

Next suppose that there exists an index N such that $E_n = A_n$ for all $n \geq N$. Then the sequence $(A_n)_{n=N}^{\infty}$ is contracting, and for $B = \bigcap_{n=N}^{\infty} A_n \subset E$,

$$\mu(B) = \mu\left(\bigcap_{n=N}^{\infty} A_n\right) = \lim_{n \to \infty} \mu(A_n) = \infty,$$

which again gives $\mu(E) = \infty$.

For each $E \in \Sigma$ define a positive measure μ^- by

$$\mu^-(E) = (-\mu)^+(E), \quad E \in \Sigma.$$

Then μ^- is finite: If $\mu^-(E)$ were equal to ∞, by the preceding argument we would have $-\mu(E) = \infty$, that is, $\mu(E) = -\infty$, which is excluded.

If $\mu^+(E) = \infty$, then $\mu(E) = \infty = \mu^+(E) - \mu^-(E)$. If $\mu^+(E)$ is finite, then

$$
\begin{aligned}
\mu^+(E) - \mu(E) &= \sup\{\mu(A) - \mu(E) : A \subset E,\ A \in \Sigma\} \\
&= \sup\{-\mu(E \setminus A) : A \subset E,\ A \in \Sigma\} \\
&= \sup\{-\mu(B) : B \subset E,\ B \in \Sigma\} = \mu^-(E),
\end{aligned}
$$

which gives $\mu(E) = \mu^+(E) - \mu^-(E)$.

Finally we prove the uniqueness property of μ^+ and μ^-. Suppose that $\mu = \lambda - \nu$ for positive measures λ and ν. Let $E \in \Sigma$. Then $\mu(A) = \lambda(A) - \nu(A) \leq \lambda(A) \leq \lambda(E)$ for any $A \subset E$, $A \in \Sigma$. Hence $\mu^+(E) \leq \lambda(E)$. If $\lambda(E) = \infty$, then $\mu(E) = \infty$, and $\mu^-(E) = 0 \leq \nu(E)$. If $\lambda(E) < \infty$, then $-\mu(E) = \nu(E) - \lambda(E)$, and $\mu^-(E) = (-\mu)^+(E) \leq \nu(E)$ by the preceding argument applied to $-\mu$. $\qquad\square$

The Jordan decomposition can be used to obtain the so-called *Hahn decomposition* of the set S relative to a given signed measure. See Problem 15.31.

Definition 15.36. The measures μ^+ and μ^- associated with a signed measure μ are called the *positive variation* of μ and the *negative variation* of μ, respectively. We also define the *total variation* $|\mu|$ of μ by

$$
|\mu| = \mu^+ + \mu^-.
$$

Corollary 15.37. *If μ is a finite signed measure, then μ^+, μ^- and $|\mu|$ are finite positive measures.*

Proof. In the proof of Theorem 15.35 we established that $\mu^+(E) = \infty$ implies $\mu(E) = \infty$, and $\mu^-(E) = \infty$ implies $\mu(E) = -\infty$. This means that if μ is finite, then so are μ^+ and μ^-; hence $|\mu| = \mu^+ - \mu^-$ is also finite. $\qquad\square$

The positive, negative and total variations of a signed measure μ satisfy the inequalities

$$
-\mu^-(A) \leq \mu(A) \leq \mu^+(A), \quad |\mu(A)| \leq |\mu|(A), \quad A \in \Sigma.
$$

From the second inequality and Corollary 15.37 it follows that any finite signed measure μ is in fact bounded:

$$
|\mu(A)| \leq |\mu|(A) \leq |\mu|(S) = \text{const} < \infty.
$$

Let μ be a complex measure on Σ. Then $\mu_1 = \operatorname{Re}\mu$ and $\mu_2 = \operatorname{Im}\mu$ are finite signed measures on Σ, and $\mu = \mu_1 + i\mu_2$. For each $E \in \Sigma$ we define $|\mu|(E)$ by

$$|\mu|(E) = \sup_{\{A_n\}} \sum_{n=1}^{\infty} |\mu(A_n)|$$

over all measurable partitions $\{A_n\}$ of E. We call $|\mu|$ the *total variation* of a complex measure μ. It can be proved that $|\mu|$ is a positive measure on Σ (Problem 15.25). Also, if μ is a signed measure, this expression agrees with the previous definition of $|\mu|$ (Problem 15.26). As in the case of a signed measure, $|\mu|$ is a (finite) positive measure. Indeed, if $\mu = \mu_1 + i\mu_2$, then for any $A \in \Sigma$,

$$|\mu(A)| = |\mu_1(A) + i\mu_2(A)| \le |\mu_1(A)| + |\mu_2(A)| \le |\mu_1|(A) + |\mu_2|(A) = \nu(A),$$

where $\nu = |\mu_1| + |\mu_2|$ is a finite positive measure. For each measurable partition $\{A_n\}$ of $E \in \Sigma$,

$$\sum_{n=1}^{\infty} |\mu(A_n)| \le \sum_{n=1}^{\infty} \nu(A_n) = \nu(E).$$

Again, this shows that any complex measure μ is bounded: $|\mu(A)| \le |\mu|(A) \le |\mu|(S)$.

It is useful to define integration with respect to a complex (or signed) measure. Let μ be a complex measure on a σ-algebra Σ and let $|\mu|$ be its total variation; then $|\mu|$ is a finite positive measure. A Σ-simple function $f = \sum_{i=1}^{k} c_i \chi_{A_i}$ is always integrable with respect to the complex measure μ since the expression

$$\int_S f \, d\mu = \sum_{i=1}^{k} c_i \mu(A_i)$$

is always defined (sums of products of complex numbers). We introduce a seminorm on the set of all simple functions by setting

$$\|f\|_{L^1} = \int_S |f| \, d|\mu|. \tag{15.11}$$

From the integration with respect to positive measures we know that $\|f\|_{L^1} = 0$ if and only if $f = 0$ $|\mu|$-a.e. It is not difficult to check that for any simple function f,

$$\left| \int_S f \, d\mu \right| \le \int_S |f| \, d|\mu|. \tag{15.12}$$

Definition 15.38. A function $f: S \to \mathbb{C}$ is *integrable with respect to a complex measure* μ if there exists a Cauchy sequence (f_n) of Σ-simple functions (with respect to the seminorm defined by (15.11)) such that

$$f_n \to f \quad |\mu|\text{-a.e.}$$

Such a sequence is called an *approximating sequence* for f. The integral of f is defined by

$$\int_S f \, d\mu = \lim_{n \to \infty} \int_S f_n \, d\mu.$$

We need to check that the complex sequence $(\int_S f_n \, d\mu)$ is Cauchy:

$$\left| \int_S f_n \, d\mu - \int_S f_m \, d\mu \right| = \left| \int_S (f_n - f_m) \, d\mu \right| \leq \int_S |f_n - f_m| \, d|\mu| = \|f_n - f_m\|_{L^1}.$$

The independence of the limit of the choice of the approximating sequence can be deduced from the corresponding result for positive measures (see Section 13.3). The inequality (15.12) holds also for general integrable functions. It is interesting to observe that, as sets of functions, $L^1(\mu)$ and $L^1(|\mu|)$ are the same, however the integrals with respect to μ and $|\mu|$ are different.

15.5.1 *Absolute continuity of measures*

We discuss a typical situation in which complex measures arise. Let (S, Σ, μ) be a measure space (with a positive measure μ), and let $f \in L^1(\mu)$. We define

$$\nu(E) = \int_E f \, d\mu, \quad E \in \Sigma. \tag{15.13}$$

Then ν is a complex valued set function on Σ which is countably additive, that is, a complex measure. Another set function is defined by

$$\rho(E) = \int_E |f| \, d\mu, \quad E \in \Sigma;$$

this time ρ is a finite positive measure, and $|\nu(E)| \leq \rho(E)$ for all $E \in \Sigma$. There is a close relation between these two measures: ρ is the total variation of ν, that is, $\rho = |\nu|$.

Theorem 15.39. *Let μ be a positive measure and let $f \in L^1(\mu)$. If ν is the complex measure defined by* (15.13), *then*

$$|\nu|(E) = \int_E |f| \, d\mu, \quad E \in \Sigma. \tag{15.14}$$

Proof.   A proof of the inequality $|\nu|(E) \le \int_E |f|\, d\mu$ is left as exercise (Problem 15.27). For the proof of the converse inequality we may assume that f is measurable and define h by

$$h(t) = \begin{cases} \dfrac{|f(t)|}{f(t)} & \text{if } f(t) \ne 0, \\ 1 & \text{if } f(t) = 0. \end{cases}$$

Then h is measurable, $|h| = 1$ and $hf = |f|$. There exists a sequence (h_n) of simple functions such that $h_n \to h$ and $|h_n| \le |h| = 1$. We have

$$h_n f \to hf = |f| \quad \text{and} \quad |h_n f| \le |f|.$$

By Lebesgue's dominated convergence theorem, for any $E \in \Sigma$,

$$\int_E h_n f \, d\mu \to \int_E |f| \, d\mu.$$

Then for any $\varepsilon > 0$ there exists N such that, for $g = h_N$,

$$\left| \int_E gf \, d\mu - \int_E |f| \, d\mu \right| < \varepsilon.$$

Without a loss of generality we may assume that $g = \sum_{i=1}^n c_i \chi_{A_i}$, where $\{A_i\}$ is a measurable partition of E. Then

$$\int_E |f| \, d\mu - \varepsilon < \left| \int_E gf \, d\mu \right| = \left| \int_E \sum_{i=1}^n c_i \chi_{A_i} f \, d\mu \right| = \left| \sum_{i=1}^n c_i \int_{A_i} f \, d\mu \right|$$

$$= \left| \sum_{i=1}^n c_i \nu(A_i) \right| \le \sum_{i=1}^n |c_i| |\nu(A_i)| \le \sum_{i=1}^n |\nu(A_i)| \le |\nu|(E).$$

Since $\varepsilon > 0$ was arbitrary, we have $\int_E |f| \, d\mu \le |\nu|(E)$. Thus (15.14) holds.   □

If f is real valued, then ν is a finite signed measure, and we have

$$\nu^+(E) = \int_E f^+ \, d\mu, \quad \nu^-(E) = \int_E f^- \, d\mu, \quad E \in \Sigma. \tag{15.15}$$

(Problem 15.28.)

Let μ be a positive measure, $f \in L^1(\mu)$, and let ν be defined by (15.13). According to Problem 13.27, $\lim_{\mu(E) \to 0} \int_E |f| \, d\mu = 0$. This means that

$$\mu(E) \to 0 \implies |\nu|(E) \to 0. \tag{15.16}$$

Surprisingly enough, (15.16) is equivalent to

$$\mu(E) = 0 \implies |\nu|(E) = 0 \tag{15.17}$$

(Problem 15.29).

Definition 15.40. Let μ be a positive measure and ν a complex measure. We say that ν is *absolutely continuous with respect to* μ, written $\nu \ll \mu$, if the condition (15.17) is satisfied. For a complex measure μ, $\nu \ll \mu$ means $\nu \ll |\mu|$.

If μ is a positive measure and ν is a signed measure, then

$$\nu \ll \mu \iff \nu^+ \ll \mu \text{ and } \nu^- \ll \mu.$$

Any complex measure ν can be written in the form $\nu = \nu_1 - \nu_2 + i\nu_3 - i\nu_4$, where ν_i are finite positive measures. Then

$$\nu \ll \mu \iff \nu_k \ll \mu, \ k \in \{1, 2, 3, 4\}.$$

If ν is given by (15.13), then $\nu \ll \mu$. The Radon–Nikodým theorem gives the converse of this fact provided μ is σ-finite.

15.5.2 *The Radon–Nikodým theorem*

The Radon–Nikodým theorem is regarded as one of the most important results in measure theory and integration.

Theorem 15.41 (Radon–Nikodým). *Let ν be a complex measure on Σ, let μ be a σ-finite positive measure on Σ, and let $\nu \ll \mu$. Then there is a μ-integrable function h, unique up to equality μ-a.e., such that*

$$\nu(A) = \int_A h \, d\mu, \quad A \in \Sigma. \tag{15.18}$$

For any $f \in L^1(\nu)$ the function fh is is $L^1(\mu)$, and

$$\int_S f \, d\nu = \int_S fh \, d\mu. \tag{15.19}$$

Proof. (i) Assume first that ν and μ are finite positive measures. Then $\lambda = \nu + \mu$ is a positive finite measure, and

$$\int_S f \, d\lambda = \int_S f \, d\nu + \int_S f \, d\mu$$

for any λ-integrable function f on S. Define

$$\varphi(f) = \int_S f \, d\nu, \quad f \in L^2(\lambda);$$

then φ is linear and bounded on $L^2(\lambda)$ as

$$|\varphi(f)| \leq \int_S |f|\, d\nu \leq \int_S |f|\, d\lambda \leq \lambda(S)^{1/2}\left(\int_S |f|^2\, d\lambda\right)^{1/2} = c\|f\|_{L^2(\lambda)}.$$

By the Riesz representation theorem for functionals on the Hilbert space $L^2(\lambda)$ (Theorem 9.32) there exists $g \in L^2(\lambda)$ such that $\varphi(f) = \langle f, \bar{g}\rangle$ for all $f \in L^2(\lambda)$, that is,

$$\int_S f\, d\nu = \int_S fg\, d\lambda, \quad f \in L^2(\lambda). \tag{15.20}$$

We may assume that g is measurable; then $\operatorname{Im} g$ is also measurable.

We show that $g(t) \in [0,1)$ for λ-almost all $t \in S$. Let $B = \{t \in S : \operatorname{Im} g(t) > 0\} = (\operatorname{Im} g)^{-1}(0,\infty)$. Then

$$\nu(B) = \varphi(\chi_B) = \int_B \operatorname{Re} g\, d\lambda + i\int_B \operatorname{Im} g\, d\lambda,$$

where $\nu(B)$ is real. Hence $\int_B \operatorname{Im} g\, d\lambda = 0$, and $\operatorname{Im} g = 0$ λ-a.e. on B. A similar argument shows that $\operatorname{Im} g = 0$ λ-a.e. on $C = \{t \in S : \operatorname{Im} g(t) < 0\} = (\operatorname{Im} g)^{-1}(-\infty,0)$. Hence $\operatorname{Im} g = 0$ λ-a.e. on S, and g is real λ-a.e.

Let $D = \{t \in S : g(t) < 0\}$. If $\lambda(D) > 0$, we would have $\nu(D) = \varphi(\chi_D) < 0$, which is a contradiction. Hence $\lambda(D) = 0$. Finally, let $E = \{t \in S : g(t) \geq 1\}$. From (15.20) we obtain $\int_S f\, d\nu = \int_S fg\, d\nu + \int_S fg\, d\mu$, that is,

$$\int_S f(1-g)d\nu = \int_S fg\, d\mu, \quad f \in L^2(\lambda). \tag{15.21}$$

We can set $f = \chi_E$ in (15.21) as $\chi_E \in L^2(\lambda)$. This yields

$$0 \leq \mu(E) = \int_E d\mu \leq \int_E g\, d\mu = \int_E (1-g)\, d\nu \leq 0.$$

Therefore $\mu(E) = 0$, and since $\nu \ll \mu$, we also have $\nu(E) = 0$. Replacing g by $\tilde{g} = g\chi_{(D\cup E)^c}$, we achieve $0 \leq \tilde{g} < 1$ everywhere on S, while $\tilde{g} = g$ a.e. with respect to both μ and ν. Without a loss of generality we will write g instead of $\tilde{g}$.

If f is a bounded measurable function on S, then $(1 + g + \cdots + g^n)f$ is in $L^2(\lambda)$ for any $n \in \mathbb{N}$. Then by (15.21),

$$\int_S (1 - g^{n+1})f\, d\nu = \int_S g(1-g)^{-1}(1-g^{n+1})f\, d\mu. \tag{15.22}$$

Since $(1 - g^{n+1})f \to f$ and $|(1 - g^{n+1})f| = (1 - g^{n+1})|f| \leq |f|$, we may apply Lebesgue's dominated convergence theorem on both sides to obtain

$$\int_S f\, d\nu = \int_S fh\, d\mu, \tag{15.23}$$

where $h = g(1-g)^{-1} \in L^1(\mu)$. If f is ν-integrable, measurable and unbounded, there is a sequence (f_n) of simple functions such that $f_n \to f$ and $|f_n| \leq |f|$. We apply the preceding argument to f_n and then use Lebesgue's dominated convergence theorem.

(ii) Assume that ν is a finite positive measure and μ is positive and σ-finite. The proof of this case is left as an exercise.

(iii) Let ν be a signed measure and μ positive and σ-finite. Then ν has a Jordan decomposition $\nu = \nu^+ - \nu^-$, where ν^+, ν^- are finite positive measures. The result then follows from part (ii) of this proof.

(iv) Let ν be a complex measure and μ positive and σ-finite. Then $\nu = \nu_1 + i\nu_2$, where ν_1, ν_2 are signed measures. The result then follows from part (iii) of this proof. $\qquad\square$

The condition that μ is σ-finite cannot be omitted from the hypotheses of the Radon–Nikodým theorem. This is seen from the following example.

Example 15.42. Let $S = [0,1]$ and let $\mathcal{B}$ be the σ-algebra of all Borel subsets of S. If ν is the counting measure on S and m the Lebesgue measure on S, then $m \ll \nu$, but the Radon-Nikodým theorem fails (Problem 15.30).

15.6 Problems for Chapter 15

In this section m stands for the Lebesgue measure on the space $\mathbb{R}^k$, and Λ^k for the σ-algebra of all Lebesgue measurable sets in $\mathbb{R}^k$.

1. Let $\mathcal{K}$ be the family consisting of $\emptyset$ and all bounded open intervals (a, b) in $\mathbb{R}$. Define ρ by

$$\rho(\emptyset) = 0 \quad \text{and} \quad \rho((a,b)) = (b-a)^2.$$

Describe the outer measure λ generated by $(\mathcal{K}, \rho)$.

2. Give an example of a generating pair $(\mathcal{K}, \rho)$ which induces an outer measure λ such that $\lambda(A) < \rho(A)$ for some $A \in \mathcal{K}$. (Preceding problem.)

3. Let S be a countable set, and $\mathcal{K}$ the family of subsets of S consisting of $\emptyset$ and all the one point sets. Let $\rho(\emptyset) = 0$, $\rho(S) = \infty$ and $\rho(\{a\}) = 1$ for any $a \in S$. Show that $(\mathcal{K}, \rho)$ is a generating pair. Describe the outer measure λ generated by $(\mathcal{K}, \rho)$.

4. Let $\mathcal{K}$ be a σ-algebra and ρ a positive measure on $\mathcal{K}$, and λ be the induced outer measure. Show that:

(i) $\lambda(A) = \rho(A)$ for each $A \in \mathcal{K}$.

(ii) Each set $A \in \mathcal{K}$ is λ-measurable.

5. Prove the following:

(i) The set $\mathcal{N}$ of nails in $\mathbb{R}$ is a semiring.

(ii) The set $\mathcal{N}_k$ of nails in $\mathbb{R}^k$ is a semiring.

6. Prove the properties of semirings and finitely (countably) additive gauges stated in Lemma 15.19.

7. Let ρ be a finitely additive gauge on the semiring $\mathcal{N}$ of nails in $\mathbb{R}$. Show that ρ is countably additive on $\mathcal{N}$ if and only if it has the following property:

$$\rho((a, b_1]) < \infty \text{ and } b_n \searrow b \implies \rho((a, b_n]) \to \rho((a, b]).$$

8. Let ρ be a Stieltjes gauge on the semiring $\mathcal{N}$ of nails in $\mathbb{R}$. Show that there exists a real valued right-continuous increasing function g (unique up to an additive constant) such that $\rho((a, b]) = g(b) - g(a)$ for every interval $(a, b] \in \mathcal{N}$. *Suggestion.* Define g by setting $g(x) = \rho((0, x])$ if $x > 0$, and modifying this definition for $x \le 0$.

9. Let ρ be a finitely additive gauge on the family $\mathcal{N}_k$ consisting of $\emptyset$ and semi-closed cells in $\mathbb{R}^k$ taking only finite values. Prove the following equivalences:

(i) ρ is countably additive on $\mathcal{N}_k$.

(ii) If $A_n \searrow A$ in $\mathcal{N}$, then $\rho(A_n) \to \rho(A)$.

(iii) If $A_n \searrow \emptyset$ in $\mathcal{N}$, then $\rho(A_n) \to 0$.

(iv) ρ is a Stieltjes gauge.

10. Let $g_1, g_2, \ldots g_k$ be real valued right-continuous increasing functions on $\mathbb{R}$, and let ρ be defined on the semiring $\mathcal{N}_k$ by $\rho(\emptyset) = 0$ and

$$\rho(A) = [g_1(b_1) - g_1(a_1)][g_2(b_2) - g_2(a_2)] \cdots [g_k(b_k) - g_k(a_k)]$$

on any cell $A = (a_1, b_1] \times \cdots \times (a_k, b_k]$. Show that ρ is a Stieltjes gauge.

11. Show that for each unbounded set $A \in \Lambda^k$ of finite measure and each $\varepsilon > 0$ there exists a bounded measurable set B such that $B \subset A$ and $m(A \setminus B) < \varepsilon$.

12. Show that:

(i) $m(G) > 0$ for every nonempty open set $G \subset \mathbb{R}^k$.

(ii) $m(K) < \infty$ for every compact set $K \subset \mathbb{R}^k$.

13. Let $A \in \Lambda^k$ and let $\varepsilon > 0$. Independently of Theorem 15.33 show that:

(i) There is an open set $G \supset A$ such that $m(G \setminus A) < \varepsilon$.

(ii) $m(A) = \inf_{G \supset A} m(G)$ over all open sets G in $\mathbb{R}^k$.

14. Let $A \in \Lambda^k$ be a set of finite measure, and let $\varepsilon > 0$. Independently of Theorem 15.33 show that:

(i) There is a compact set $K \subset A$ such that $m(A \setminus K) < \varepsilon$.

(ii) $m(A) = \sup\limits_{K \subset A} m(K)$ over all compact sets K in $\mathbb{R}^k$.

Suggestion: Problem 15.13 and 15.11.

15. Let f be Lebesgue integrable and nonnegative on the interval $[a, b]$. Prove that the set

$$\Omega = \{(x, y) \in \mathbb{R}^2 : a \le x \le b,\ 0 \le y \le f(x)\}$$

is Lebesgue measurable, and $m(\Omega) = \int_{[a,b]} f\,dm$.

16. Let A be a subset of $\mathbb{R}^k$ such that for every $\varepsilon > 0$ there exists a closed set F and an open set G such that $F \subset A \subset G$ and $m(G \setminus F) < \varepsilon$. Prove that $A \in \Lambda^k$.

Suggestion. For each $k \in \mathbb{N}$ let F_k, G_k be sets such that $F_k \subset A \subset G_k$ and $m(G_k \setminus F_k) < 1/k$, with F_k closed and G_k open. Let $F = \bigcup_{k=1}^{\infty} F_k$ and $G = \bigcap_{k=1}^{\infty} G_k$, and show that $m(G \setminus F) = 0$.

17. Let A be a bounded subset of $\mathbb{R}^k$ such that for every $\varepsilon > 0$ there exists a closed set F with $F \subset A$ and $\lambda(A \setminus F) < \varepsilon$ (λ is the Lebesgue outer measure in $\mathbb{R}^k$). Prove that $A \in \Lambda^k$.

18. If $A \subset \mathbb{R}^k$ is such that $A \cap K \in \Lambda^k$ for every compact set K, show that $A \in \Lambda^k$.

19. Let A be a subset of $\mathbb{R}^k$ such that for every $\varepsilon > 0$ there exists an open set G such that $G \supset A$ and $\lambda(G \setminus A) < \varepsilon$, where λ is the Lebesgue outer measure in $\mathbb{R}^k$. Prove that $A \in \Lambda^k$.

20. Find the Lebesgue measure of the following sets in $\mathbb{R}^2$.

(i) $A = \{(x, y) : 0 < y < |x|^{-1/2}$ if $0 < |x| < 1$ and $0 < y < |x|^{-2}$ if $|x| \ge 1\}$.

(ii) $B = \{(x, y) : 0 < y < |\log x|$ if $0 < |x| < 1\}$.

(iii) $C = \{(x, y) : 0 < y < 1$ if $|x| < 1$ and $0 < y < |x|^{-1}$ if $|x| \ge 1\}$.

21. Prove *Carathéodory's criterion*: Let λ be an outer measure on a metric space (S, d). Then every Borel set is λ-measurable if and only if $\lambda(A \cup B) \ge \lambda(A) + \lambda(B)$ for any two sets A, B in S a positive distance apart.

Suggestion. To show that Borel sets are λ-measurable if the inequality holds: Let E be an open proper subset of S, and let A be a nonempty subset of E. For each $n \in \mathbb{N}$ define $A_n = \{x \in S : d(x, E^c) \ge 1/n\}$. Considering the sets $B_{2n} = A_{2n} \cap A'_{2n-1}$ and $B_{2n+1} = A_{2n+1} \cap A'_{2n}$, prove that $\lim_{n \to \infty} \lambda(A_n) = \lambda(A)$. Hence conclude that E is measurable.

22. Let μ be the Carathéodory measure on a metric space (S, d) as constructed in Example 15.11. Prove that μ is an outer measure on S. Using the preceding problem, show that all Borel sets in S are λ-measurable.

23. Let A be the interval $A = [0, 1]$ in $\mathbb{R}$. Prove that the one-dimensional

Hausdorff measure $\mathcal{H}^1(A)$ of A is equal to 1.

24. (Harder.) Show that the Hausdorff dimension $h(C) = \inf\{m \geq 0 : \mathcal{H}^m(C) < \infty\}$ of the Cantor set is $\log 2/\log 3$.

25. Prove that the total variation $|\mu|$ of a complex measure μ is a positive measure.

26. Let μ be a signed measure, and let $|\mu| = \mu^+ + \mu^-$ be the total variation of μ. Prove that, for any $E \in \Sigma$,

$$|\mu|(E) = \sup_{\{A_n\}} \sum_{n=1}^{\infty} |\mu(A_n)|,$$

where the supremum is taken over all measurable partitions $\{A_n\}$ of E.

27. If $\nu(E) = \int_E f\, d\mu$ with $f \in L^1(\mu)$, prove that $|\nu|(E) \leq \int_E |f|\, d\mu$ for any $E \in \Sigma$.

28. Prove equation (15.15) for the positive and negative variation of a signed measure ν defined by $\nu(E) = \int_E f\, d\mu$, $E \in \Sigma$, where $f \in L^1(\mu)$.

29. Prove that equation (15.16) follows from (15.17).

Suggestion. If not, there exists a sequence (A_n) in Σ and $\varepsilon > 0$ such that $\mu(A_n) < 2^{-n}$ and $|\nu|(A_n) \geq \varepsilon$. Let $B_n = \bigcup_{k=n}^{\infty} A_k$. Then $B_n \searrow B$ for some $B \in \Sigma$. Show that $\mu(B) = 0$ and $|\nu|(B) \geq \varepsilon$.

30. Prove the assertion of Example 15.42.

31. *Hahn decomposition.* Let μ be a signed measure on a measurable space (S, Σ). Show that there exists a Σ-measurable partition $S = P \dot\cup N$ of S such that

$$A \in \Sigma \text{ and } A \subset P \implies \mu(A) \geq 0, \quad B \in \Sigma \text{ and } B \subset N \implies \mu(B) \leq 0.$$

Suggestion. Assume that $\mu: \Sigma \to \mathbb{R}\cup\{-\infty\}$. Construct a sequence (E_n) in Σ such that for each n, $\mu^+(S) - (\frac{1}{2})^n < \mu^+(E_n) < \infty$, and show that $\mu^-(E_n) < (\frac{1}{2})^n$. Set

$$P = \limsup_{n \to \infty} E_n = \bigcap_{n=1}^{\infty} \bigcup_{k=n}^{\infty} E_k, \quad N = \liminf_{n \to \infty} E_n = \bigcup_{n=1}^{\infty} \bigcap_{k=1n}^{\infty} E_k.$$

Adapt the construction in the case that $\mu: \Sigma \to \mathbb{R} \cup \{\infty\}$.

Chapter 16

Product Measures. Integration on $\mathbb{R}^k$

16.1   Product measures

In this section, (S, Σ, μ) and (T, Ω, ν) are two σ-finite measure spaces. A set $A \times B$, where $A \in \Sigma$ and $B \in \Omega$, is called a *measurable rectangle* in $S \times T$. We show that the Cartesian product $S \times T$ can be made into a measure space in which the product measure is constructed as an extension of $\rho(A \times B) = \mu(A)\nu(B)$.

Definition 16.1. The smallest σ-algebra Π in $S \times T$ which contains all measurable rectangles is called the *product σ-algebra*

$$\Pi = \Sigma \otimes \Omega.$$

Theorem 16.2. *Let (S, Σ, μ) and (T, Ω, ν) be σ-finite measure spaces. There exists a unique measure π on the product σ-algebra $\Pi = \Sigma \otimes \Omega$ such that*

$$\pi(A \times B) = \mu(A)\nu(B)$$

for all measurable rectangles $A \times B$.

Proof.   Let $\mathcal{K}$ be the family of all measurable rectangles $A \times B$ with $\mu(A) < \infty$ and $\nu(B) < \infty$. Let $\rho : \mathcal{K} \to [0, \infty)$ be defined by

$$\rho(A \times B) = \mu(A)\nu(B).$$

Then $(\mathcal{K}, \rho)$ is a generating pair for $S \times T$ (verify).

(i) $\mathcal{K}$ *is a semiring.* This follows from the identities

$$(A_1 \times B_1) \cap (A_2 \times B_2) = (A_1 \cap A_2) \times (B_1 \cap B_2),$$

$$(A_1 \times B_1) \setminus (A_2 \times B_2) = [(A_1 \setminus A_2) \times B_1)] \cup [(A_1 \cap A_2) \times (B_1 \setminus B_2)]$$

$$= U \cup V;$$

note that the union $U \cup V$ is disjoint (see Figure 16.1).

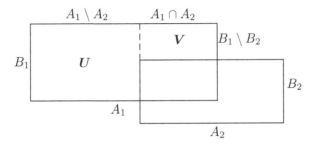

Figure 16.1

(ii) ρ *is finitely additive on* $\mathcal{K}$. Let the measurable rectangle $A \times B$ be partitioned as the union of n disjoint measurable rectangles $A_i \times B_i$,

$$A \times B = \bigcup_{i=1}^{n} (A_i \times B_i).$$

Note that $A = \bigcup_i A_i$ and $B = \bigcup_i B_i$ (the unions not necessarily disjoint). Let all sets A_i and B_i be of finite measure. Write $f = \chi_{A \times B}$ and $f_i = \chi_{A_i \times B_i}$, $i = 1, \ldots, n$. Then $f = f_1 + \cdots + f_n$ (disjointness of the $A_i \times B_i$), and

$$\int_T f(s, \cdot)\, d\nu = \sum_{i=1}^{n} \int_T f_i(s, \cdot)\, d\nu,$$

that is,

$$\nu(B)\chi_A = \sum_{i=1}^{n} \nu(B_i)\chi_{A_i}.$$

Hence

$$\int_S \nu(B)\chi_A\, d\mu = \sum_{i=1}^{n} \int_S \nu(B_i)\chi_{A_i}\, d\mu$$

and

$$\rho(A \times B) = \mu(A)\nu(B) = \sum_{i=1}^{n} \mu(A_i)\nu(B_i) = \sum_{i=1}^{n} \rho(A_i \times B_i).$$

If any of the sets A_i, B_i has measure ∞, then $\mu(A)\nu(B) = \infty$, so that $\rho(A \times B) = \sum_{i=1}^{n} \rho(A_i \times B_i)$ is again true.

(iii) ρ *is countably additive on* $\mathcal{K}$. For the proof of countable additivity suppose that

$$(A_n \times B_n) \nearrow (A \times B)$$

with all sets of finite measure. Set

$$f_n = \chi_{A_n \times B_n}, \qquad f = \chi_{A \times B}.$$

Then $f_n \nearrow f$. For each $s \in S$, $f_n(s, \cdot) \nearrow f(s, \cdot)$ on T. By the monotone convergence theorem with respect to ν,

$$\int_T f_n(s, \cdot)\, d\nu \nearrow \int_T f(s, \cdot)\, d\nu,$$

that is,

$$\nu(B_n)\chi_{A_n} \nearrow \nu(B)\chi_A \text{ on } S.$$

By the monotone convergence with respect to μ,

$$\int_S \nu(B_n)\chi_{A_n}\, d\mu \to \int_S \nu(B)\chi_A\, d\mu,$$

that is,

$$\mu(A_n)\nu(B_n) \to \mu(A)\nu(B) \quad \text{or} \quad \rho(A_n \times B_n) \to \rho(A \times B).$$

Since ρ is finitely additive, this is equivalent to the countable additivity of ρ.

By Theorems 15.20 and 15.23, there is a unique measure π on the σ-algebra Π generated by $(\mathcal{K}, \rho)$. $\qquad\square$

The measure π whose existence and uniqueness is guaranteed by Theorem 16.2 is called the *product measure* of μ and ν and is denoted by

$$\pi = \mu \otimes \nu.$$

The following fact is often useful: If $\mathcal{R}$ is the family consisting of all finite disjoint unions of measurable rectangles, then $\mathcal{R}$ is an algebra of sets. Recall that an algebra of sets contains $\emptyset$, is closed under taking of finite unions, and is closed under complementation. This is true for $\mathcal{R}$ since $\mathcal{R}$ is a semiring and $S \times T \in \mathcal{R}$.

Let $\mathcal{B}^k$ denote the σ-algebra of all Borel measurable sets in $\mathbb{R}^k$, and Λ^k the σ-algebra of all Lebesgue measurable sets in $\mathbb{R}^k$. Then we can show that

$$\mathcal{B}^k \otimes \mathcal{B}^s = \mathcal{B}^{k+s}, \qquad \overline{\Lambda^k \otimes \Lambda^s} = \Lambda^{k+s}$$

where the bar denotes the measure completion with respect to the Lebesgue measure. Let m_k be the Lebesgue measure on $\mathcal{B}^k$ and λ_k its extension (measure completion) to Λ^k. Then

$$m_k \otimes m_s = m_{k+s}, \qquad \overline{\lambda_k \otimes \lambda_s} = \lambda_{k+s},$$

where the bar denotes measure completion.

16.2   Fubini's and Tonelli's theorem

The following two theorems deal with the relation between integration with respect to the product measure and iterated integrals with respect to measures on the factor spaces. Let f be a complex valued function defined on the product space $S \times T$. The existence of the iterated integral

$$\int_S \left(\int_T f(s,t)\, d\nu(t) \right) d\mu(s)$$

assumes the following:

(i)   For almost all $s \in S$ the function $t \mapsto f(s,t)$ is in $L^1(\nu)$.

(ii)   The function $s \mapsto \int_T f(s,t)\, d\nu(t)$ is defined μ-a.e. in S and is in $L^1(\mu)$.

In the above equations the symbol $d\mu(s)$ indicates that we integrate the function with respect to the variable s. A similar convention applies to $d\nu(t)$. We will use this as a standard notation from now on. In the proof of Fubini's theorem we use the following characterization of the integral which can be deduced from Theorems 13.35 and 15.21. Recall that if $\mathcal{K}$ is a semiring, then a linear combination of characteristic functions of elements of $\mathcal{K}$ is called a $\mathcal{K}$-simple function.

Lemma 16.3. *Let $(\mathcal{K}, \rho)$ be a generating pair with $\mathcal{K}$ a semiring and ρ a countably additive gauge on $\mathcal{K}$. Let (S, Σ, μ) be a measure space generated by $(\mathcal{K}, \rho)$. A function $f\colon S \to \mathbb{C}$ is μ-integrable if and only if there exists a norm-absolutely convergent series $\sum_n f_n$ of μ-integrable $\mathcal{K}$-simple functions such that $f(t) = \sum_{n=1}^{\infty} f_n(t)$ for all $t \in S$ for which the numerical series converges absolutely.*

We recall that the family of all measurable rectangles $A \times B$ in $S \times T$ is a semiring. The preceding lemma will be applied to functions defined on the cartesian product $S \times T$. The following theorem is due to Guido Fubini.

Theorem 16.4 (Fubini's theorem). *Let (S, Σ, μ) and (T, Ω, ν) be σ-finite measure spaces. If $f \in L^1(\mu \otimes \nu)$, then*

$$\int_{S \times T} f(s,t)\, d(\mu \otimes \nu) = \int_S \left(\int_T f(s,t)\, d\nu(t) \right) d\mu(s)$$

$$= \int_T \left(\int_S f(s,t)\, d\mu(s) \right) d\nu(t).$$

Proof.   Let $\mathcal{R}$ be the set of all measurable rectangles in $S \times T$ with finite $\mu \otimes \nu$ measure. We observe that the theorem holds for $\mathcal{R}$-simple $(\mu \otimes \nu)$-integrable functions.

If f is an arbitrary $L^1(\mu \otimes \nu)$ function, then by Lemma 16.3 there exists a series $\sum_n f_n$ of $\mathcal{R}$-simple functions on $S \times T$ with

$$\sum_{n=1}^{\infty} \int_{S \times T} |f_n| \, d(\mu \otimes \nu) < \infty \tag{16.1}$$

and $f(s,t) = \sum_{n=1}^{\infty} f_n(s,t)$ for all points (s,t) for which the numerical series absolutely converges. By the monotone convergence theorem,

$$\sum_{n=1}^{\infty} \int_{S \times T} |f_n| \, d(\mu \otimes \nu) = \sum_{n=1}^{\infty} \int_S \left(\int_T |f_n(s,t)| \, d\nu(t) \right) d\mu(s)$$

$$= \int_S \left(\sum_{n=1}^{\infty} \int_T |f_n(s,t)| \, d\nu(t) \right) d\mu(s) < \infty, \tag{16.2}$$

and there exists a set E of μ-measure zero such that $\sum_n |f_n(s,t)| < \infty$ for all $s \in E^c$. For any $s \in E^c$ the monotone convergence theorem implies that

$$\sum_{n=1}^{\infty} \int_T |f_n(s,t)| \, d\nu(t) = \int_T \sum_{n=1}^{\infty} |f_n(s,t)| \, d\nu(t),$$

so that the numerical series $\sum_n f_n(s,t)$ is absolutely convergent in $\mathbb{C}$ for ν-almost all $t \in T$. Keep $s \in E^c$ fixed; we can then apply Lebesgue's dominated convergence theorem to conclude that $t \mapsto f(s,t)$ is ν-integrable and that $\int_T f(s,t) \, d\nu(t) = \sum_n \int_T f_n(s,t) \, d\nu(t)$. Moreover, for $s \in E^c$,

$$\left| \int_T f_n(s,t) \, d\nu(t) \right| \leq \sum_{n=1}^{\infty} \int_T |f_n(s,t)| \, d\nu(t),$$

where the function on the right is μ-integrable by (16.2). Lebesgue's dominated convergence theorem implies that $s \mapsto \int_T f(s,t) \, d\nu(t)$ is μ-integrable with

$$\int_S \left(\int_T f(s,t) \, d\nu(s) \right) d\mu(t) = \sum_{n=1}^{\infty} \int_S \left(\int_T f_n(s,t) \, d\nu(t) \right) d\mu(s)$$

$$= \int_{S \times T} f(s,t) \, d(\mu \otimes \nu)$$

in view of (16.1).

The other equality in Fubini's theorem is proved analogously. $\square$

Fubini's theorem can only be applied if we can establish that the function f is $(\mu \otimes \nu)$-integrable, which can be often difficult. Luckily, we have the following theorem due to Leonida Tonelli to fall back on to, where only the existence of repeated integrals of nonnegative functions is required.

Theorem 16.5 (Tonelli's theorem). *Let (S, Σ, μ) and (T, Ω, ν) be σ-finite measure spaces and let $f \colon S \times T \to \mathbb{C}$ be a $(\mu \otimes \nu)$-measurable function. If one of the integrals*

$$\int_S \left(\int_T |f(s,t)| \, d\nu(t) \right) d\mu(s), \quad \int_T \left(\int_S |f(s,t)| \, d\mu(s) \right) d\nu(t)$$

exists (as a real number), then $f \in L^1(\mu \otimes \nu)$, and Fubini's theorem applies.

Proof. If the measures μ and ν are finite, the result follows from Fubini's theorem.

If μ and ν are σ-finite, there exists an expanding sequence (E_n) in $\Sigma \otimes \Omega$ such that $E_n \nearrow S \times T$ and $(\mu \otimes \nu)(E_n) < \infty$. For each $n \in \mathbb{N}$ set $f_n = (|f| \wedge n)\chi_{E_n}$. Fubini's theorem applies to each f_n. The rest of the proof follows by an application of the monotone convergence theorem to the sequence (f_n); the details of the proof are left as an exercise. $\square$

16.3   Lebesgue integral on $\mathbb{R}^k$ and change of variables

In this section we consider the measure space $(\mathbb{R}^k, \Lambda(\mathbb{R}^k), m)$, where m is the k-dimensional Lebesgue measure, and more generally $(\mathbb{R}^k, \mathcal{B}(\mathbb{R}^k), \mu)$, where μ is the Lebesgue–Stieltjes measure. Fubini's theorem enables us to reduce a k-dimensional integral to k repeated one dimensional integrals, for instance,

$$\int_A f \, dm = \int_{A_1} \int_{A_2} \cdots \int_{A_k} f(x_1, x_2, \ldots, x_k) \, dx_k \, \ldots \, dx_2 \, dx_1,$$

where $A = A_1 \times A_2 \times \cdots \times A_k$ is the Cartesian product of Lebesgue measurable sets A_k in $\mathbb{R}$, and where the integration is first executed in x_k over A_k, followed by integration in x_{k-1} over A_{k-1}, etc. Further, the Euclidean space $\mathbb{R}^k$ has a rich geometric structure, and we have many special results for the Lebesgue and Lebesgue–Stieltjes integrals.

In general measure spaces the integration is based on simple functions, which are linear combinations of characteristic functions of measurable sets. Measurable sets in $\mathbb{R}^k$ can be quite exotic, with even sets of finite measure possibly unbounded and scattered in $\mathbb{R}^k$. If the measure is derived from a generating pair $(\mathcal{K}, \rho)$, where $\mathcal{K}$ is a semiring and ρ is countably additive, $\mathcal{K}$-simple functions can be used in the construction of the integral (Theorem 15.21). In $\mathbb{R}^k$, $\mathcal{N}_k$-simple functions are a special case of step functions.

Definition 16.6. A function $f \colon \mathbb{R}^k \to \mathbb{C}$ is called a *step function* if there is a semiclosed k-cell K partitioned into a finite union of disjoint semiclosed k-cells E_i $(i = 1, \ldots, n)$ such that f is constant on the interior of each cell E_i and zero outside K.

The following result is a special case of Theorem 15.21 in which we use step functions which are finite linear combinations of characteristic functions of semiclosed k-cells. Arbitrary step functions can be used if ρ is the Euclidean volume.

Theorem 16.7. *Let ρ be a Lebesgue–Stieltjes gauge on the semiring $\mathcal{N}_k$ of semiclosed k-cells in $\mathbb{R}^k$, and let μ be the Lebesgue–Stieltjes measure generated by ρ. Then the step functions are dense in the normed space $(L^1(\mu), \|\cdot\|_{L^1(\mu)})$.*

For any function $f \colon \mathbb{R}^k \to \mathbb{C}$ we define the *support of f*, written $\operatorname{supp} f$, as the closure in $\mathbb{R}^k$ of the set $\{x \in \mathbb{R}^k : f(x) \neq 0\}$. If f is a function with compact support, then there exists a compact k-cell K such that f vanishes outside K. For Lebesgue integrable functions we get the following useful result.

Theorem 16.8. *Continuous functions with compact support are dense in the normed space $(L^1(\mathbb{R}^k), \|\cdot\|_{L^1})$ of Lebesgue integrable functions.*

Proof. In view of the preceding theorem it is enough to verify that characteristic functions of compact k-cells can be approximated by continuous functions with compact support in $(L^1(\mathbb{R}^k), \|\cdot\|_{L^1})$.

Let K be a compact k-cell and let $\varepsilon > 0$. There is a compact k-cell L such that $K \subset L^0$ and $\operatorname{vol}(L) < \operatorname{vol}(K) + \varepsilon$. Then there exists a continuous (piecewise linear) function g on $\mathbb{R}^k$ such that $0 \le g \le 1$, $g(x) = 1$ for all $x \in K$ and $g(x) = 0$ for all $x \notin L$. Then

$$\|\chi_K - g\|_{L^1} = \int_L (g - \chi_K)\, dm \le \int_{L \setminus K} dm = m(L \setminus K) = \operatorname{vol}(L) - \operatorname{vol}(K) < \varepsilon.$$

The proof is now complete. $\qquad\square$

We now turn our attention to the change of variables in the Lebesgue integral on $\mathbb{R}^k$. Let Δ be an open subset of $\mathbb{R}^k$ and let $u \colon \Delta \to \mathbb{R}^k$ be a mapping such that the components of $u = (u_1, \ldots, u_n)$ have continuous partial derivatives of order one. The *Jacobian matrix* of u is the function $Ju: \Delta \to \mathbb{R}^{k \times k}$ defined by

$$Ju(t) = \begin{bmatrix} \dfrac{\partial u_1}{\partial t_1} & \cdots & \dfrac{\partial u_1}{\partial t_k} \\ \cdots & \cdots & \cdots \\ \dfrac{\partial u_k}{\partial t_1} & \cdots & \dfrac{\partial u_k}{\partial t_k} \end{bmatrix}.$$

If $u: A \to \mathbb{R}^k$ and $v: B \to \mathbb{R}^k$ are two continuously differentiable composable functions $(v(B) \subset A)$, then we have the generalized *chain rule for Jacobian matrices*:

$$J(u \circ v) = (Ju \circ v)Jv.$$

The determinant of Ju is called the *Jacobian* of u. The chain rule extends also to the Jacobians: $\det J(u \circ v) = \det(Ju \circ v)\det(Jv)$. Let Δ, Ω be open subsets of $\mathbb{R}^k$ and let $u: \Delta \to \Omega$ be a bijection such that both $u = (u_1, \ldots, u_n)$ and $u^{-1} = (v_1, \ldots, v_n)$ have continuous partial derivatives of order one. Then u is called a *diffeomorphism* of Δ onto Ω.

The next result is a k-dimensional substitution, or change of variables theorem for the Lebesgue integral.

Theorem 16.9 (Change of variables). *Let $f: \Omega \to \mathbb{C}$ be a Lebesgue measurable function on an open set $\Omega \subset \mathbb{R}^k$, and let $u: \Delta \to \Omega$ be a diffeomorphism between open subsets of $\mathbb{R}^k$. Then*

$$\int_\Omega f\, dm = \int_\Delta (f \circ u)\,|\det Ju|\, dm. \tag{16.3}$$

whenever one of the integrals exists.

Proof. We give only a sketch of the proof as the details are quite technical. A complete proof can be found in [15] and [31].

Suppose that the integral on the left exists, which means that f is Lebesgue integrable on Ω. Then (16.3) can be proved by an argument based on Problem 16.18. Conversely, assume that the integral on the right exists, and let v be the inverse function of u, so that $u \circ v = \mathrm{id}$. Then the function $F = (f \circ u)|\det Ju|$ is Lebesgue integrable, and

$$(F \circ v)\,|\det Jv| = (f \circ u \circ v)\,|\det(Ju \circ v)|\,|\det Jv|$$
$$= f\,|\det(Ju \circ v)\det Jv|$$
$$= f\,|\det J(u \circ v)| = f.$$

Apply (16.3) to F in place of f, v in place of u and Δ in place of Ω. We obtain

$$\int_\Delta F \, dm = \int_\Omega (F \circ v) \, |\det Jv| \, dm,$$

which is (16.3) with the left and right sides interchanged. □

The Lebesgue integral can be also written in notation convenient for the change of variables, that is,

$$\int_S f \, dm = \int_S f(x) \, dx.$$

Then equation (16.3) can be written more explicitly as

$$\int_\Omega f(x) \, dx = \int_\Delta f(u(t)) \, |\det Ju(t)| \, dt. \tag{16.4}$$

For integration on the real line the formula takes the following convenient form.

Theorem 16.10. *Let* $u \colon (a, b) \to \mathbb{R}$ *be a continuously differentiable mapping on* (a, b) *with either* $u'(t) > 0$ *for all* t *or* $u'(t) < 0$ *for all* t, *and let* f *be a complex valued Lebesgue measurable function on the set* $\Omega := u((a, b))$. *Then*

$$\int_{u(a)}^{u(b)} f(x) \, dx = \int_a^b f(u(t)) u'(t) \, dt \tag{16.5}$$

provided one of the integrals exists.

Proof. The hypotheses on u ensure that u is either strictly increasing or strictly decreasing on (a, b), and that the inverse mapping v of u has continuous derivative. Hence u is a diffeomorphism of $\Delta := (a, b)$ onto the image Ω of (a, b) under u, which is again an interval. According to Theorem 16.9, $\int_\Omega f(x) \, dx = \int_\Delta f(u(t)) |u'(t)| \, dt$ provided one of the integrals exists. Observe:

$$\text{if } u' > 0, \quad \text{then } \Omega = (u(a), u(b)) \quad \text{and } |u'| = u',$$
$$\text{if } u' < 0, \quad \text{then } \Omega = (u(b), u(a)) \quad \text{and } |u'| = -u'.$$

This results in the equation (16.5), applying the convention $\int_a^c = -\int_c^d$. □

It is important to bear in mind that the substitution operates from $\mathbb{R}$ to $\mathbb{R}$; the temptation to substitute $u(t) = \sin t + 3i$ in $\int_0^{\pi/2} (\sin t + 3i)^{-2} \cos t \, dt$ must be resisted. The right substitution is $u(t) = \sin t$.

Example 16.11. Calculate this slightly extravagant Lebesgue integral if it exists:

$$I := \int_1^{\exp(\pi/2)} (2\log t + i) \exp(\log^2 t)(\cos(\log t) + i\sin(\log t)) t^{-1}\, dt.$$

We try the substitution $x = u(t) = \log t$. Then $u'(t) = t^{-1} > 0$ on the interval $(1, \exp(\tfrac{1}{2}\pi))$, while $u(1) = 0$, $u(\exp(\tfrac{1}{2}\pi)) = \tfrac{1}{2}\pi$. The function $f(x) = (2x + i)\exp(x^2 + ix)$ is continuous, and therefore Lebesgue measurable, on $(0, \tfrac{1}{2}\pi)$. We can easily verify that the integrand in I above is equal to $f(u(t))u'(t)$. The Lebesgue integral $\int_0^{\pi/2} f(x)\, dx$ exists since f is continuous on the compact interval $[0, \tfrac{1}{2}\pi]$. We note that f has a primitive $F(x) = \exp(x^2 + ix)$. By the compatibility of the Lebesgue and Newton integral we obtain

$$I = \int_0^{\pi/2} (2x + i)\exp(x^2 + ix)\, dx = F(\tfrac{1}{2}\pi-) - F(0+) = i\exp(\tfrac{1}{4}\pi^2) - 1.$$

Example 16.12. Calculate the Lebesgue integral

$$\int_0^\infty \frac{dx}{\sqrt{\exp(2x) - 1}}$$

if it exists. We use the substitution $x = u(t) = -\log t$ for t in the interval $(0, 1)$. Then $u'(t) = -t^{-1} < 0$ and $u(0+) = \infty$, $u(1) = 0$; u is a diffeomorphism from $(0, 1)$ to $(0, \infty)$. Substituting $x = u(t) = -\log t$ into the function $f(x) = (\exp(2x) - 1)^{-1/2}$ and using the equation $\int_0^\infty f(x)\, dx = \int_1^0 f(u(t))u'(t)\, dt$, we get

$$\int_0^\infty \frac{dx}{\sqrt{\exp(2x) - 1}} = -\int_1^0 \frac{dt}{\sqrt{1 - t^2}} = \int_0^1 \frac{dt}{\sqrt{1 - t^2}} = \big[\arcsin t\big]_0^1 = \tfrac{1}{2}\pi,$$

where we used the absolute Newton integrability of $(1 - t^2)^{-1/2}$ on $(0, 1)$.

Example 16.13. We consider polar coordinates in the Euclidean plane $\mathbb{R}^2$:

$$x_1 = r\cos\theta, \qquad x_2 = r\sin\theta.$$

Calculate the Jacobian of the mapping $(x_1, x_2) = u(r, \theta) = (r\cos\theta, r\sin\theta)$:

$$\det(Ju(r, \theta)) = \begin{vmatrix} \cos\theta & -r\sin\theta \\ \sin\theta & r\cos\theta \end{vmatrix} = r(\cos^2\theta + \sin^2\theta) = r.$$

Let Δ be the rectangle $\Delta = \{0 < r < R,\ \theta_1 < \theta < \theta_2\}$ in the (r, θ)-plane, and $\Omega = u(\Delta)$ the resulting angular region in the (x_1, x_2)-plane. Then

$$\int_\Omega f(x_1, x_2)\, dx_1\, dx_2 = \int_\Delta f(r\cos\theta, r\sin\theta)\, r\, dr\, d\theta.$$

$$= \int_0^R r \left(\int_{\theta_1}^{\theta_2} f(r\cos\theta, r\sin\theta) \, d\theta \right) dr$$

$$= \int_{\theta_1}^{\theta_2} \left(\int_0^R f(r\cos\theta, r\sin\theta) \, r \, dr \right) d\theta.$$

In particular, choose $f(x_1, x_2) = x_1^2 + x_2^2$, $R = 1$, $\theta_1 = 0$ and $\theta_2 = \frac{1}{2}\pi$. Then Ω is the quarter of the open unit disc in the first quadrant, and

$$\int_\Omega (x_1^2 + x_2^2) \, dx_1 \, dx_2 = \int_0^{\pi/2} \left(\int_0^1 r^2 \, r \, dr \right) d\theta = \frac{1}{4} \int_0^{\pi/2} d\theta = \frac{1}{4} \frac{\pi}{2} = \frac{\pi}{8}.$$

Example 16.14. In $\mathbb{R}^3$ we have the spherical coordinates expressed by equations

$$x_1 = r\sin\varphi\cos\theta, \quad x_2 = r\sin\varphi\sin\theta, \quad x_3 = r\cos\varphi.$$

Calculate the Jacobian of this transformation and write down a similar formula for the change of variable as in the preceding example. Observe that φ is the Euclidean angle between $(0,0,1)$ and (x_1, x_2, x_3), and θ is the angle between $(1,0,0)$ and $(x_2, x_2, 0)$ (sketch a diagram).

16.4   Problems for Chapter 16

1.   Let $\mathcal{B}^k$ denote the σ-algebra of all Borel measurable sets in $\mathbb{R}^k$, and Λ^k the σ-algebra of all Lebesgue measurable sets in $\mathbb{R}^k$. Show that

$$\mathcal{B}^k \otimes \mathcal{B}^s = \mathcal{B}^{k+s}, \qquad \overline{\Lambda^k \otimes \Lambda^s} = \Lambda^{k+s},$$

where the bar denotes the measure completion with respect to the Lebesgue measure.

2.   Show that the boundary of a k-cell in $\mathbb{R}^k$ has Lebesgue measure 0. (This is in general false for Lebesgue-Stieltjes measures.)

3.   Let $p > 1$. Show that the set of all step functions is dense in the space $(L^p(\mathbb{R}^k), \|\cdot\|_{L^p})$.

4.   Let $p > 1$. Show that the set of all continuous functions on $\mathbb{R}^k$ with compact support is dense in the normed space $(L^p(\mathbb{R}^k), \|\cdot\|_{L^p})$.

5.   Let I be a closed bounded k-cell in $\mathbb{R}^k$ and f a bounded real valued function on I. Then f is said to be *Riemann integrable* if, for each $\varepsilon > 0$, there is a pair of step functions g, h such that $g \leq f \leq h$ on I and $\int_I (h - g) < \varepsilon$ (the elementary integral). The *Riemann integral* is then defined as $(\mathcal{R}) \int_I f(t) \, dt = \sup\{\int_I g : g \leq f, \ g \text{ a step function}\}$.

(i) If f is Riemann integrable on I, show that f is Lebesgue integrable on I, and that the two integrals are equal.

(ii) Show that a function f is Riemann integrable on I if and only if f is bounded and continuous a.e. in I.

(iii) Show that every bounded monotonic function f on $[a, b]$ is Riemann integrable.

6. If $E \in \Sigma \otimes \Omega$, $s \in S$, and $t \in T$, define
$$E_s = \{t \in T : (s, t) \in E\}, \quad E^t = \{s \in S : (s, t) \in E\}.$$
If $E \in \Sigma \otimes \Omega$, prove that $E_s \in \Omega$ and $E^t \in \Sigma$.

7. If $f : S \times T \to \mathbb{C}$, $s \in S$ and $t \in T$, define
$$f_s(t) = f(s, t), \quad t \in T, \quad f^t(s) = f(s, t), \quad s \in S.$$
If $f : S \times T \to \mathbb{C}$ is $(\Sigma \otimes \Omega)$-measurable, prove that f_s is Ω-measurable and f^t is Σ-measurable.

8. Convert each repeated Lebesgue integral to the double integral of the form $\iint_{A \times B} f(x, y) \, dx \otimes dy$ with suitable A, B and f, and evaluate the integrals by changing the order of integation. The emphasis is on justification using appropriate theorems of Lebesgue integration theory.

(i) $\displaystyle \int_0^1 \left(\int_{2x}^2 \exp(y^2) \, dy \right) dx,$ (ii) $\displaystyle \int_0^2 \left(\int_{x^2}^1 x \cos y^2 \, dy \right) dx,$

(iii) $\displaystyle \int_0^1 \left(\int_x^1 y^{-1} \sin y \cos(xy^{-1}) \, dy \right) dx,$ (iv) $\displaystyle \int_0^8 \left(\int_{\sqrt[3]{x}}^2 \frac{x}{\sqrt{16 + y^7}} \, dy \right) dx,$

(v) $\displaystyle \int_0^{\pi/2} \left(\int_y^{\pi/2} \frac{\sin x}{\sqrt{xy}} \, dx \right) dy,$ (vi) $\displaystyle \int_1^\infty \left(\int_{\arcsin(1/x)}^{\pi/2} \frac{\sqrt{\cos y}}{x^2} \, dy \right) dx.$

(Answers: (i) $\frac{1}{4}e^4 - \frac{1}{4}$, (ii) $\frac{1}{4} \sin 16$, (iii) $\sin 1(1 - \cos 1)$, (iv) $\frac{8}{7}$, (v) 2, (vi) $\frac{2}{3}$.)

9. Apply Fubini's theorem to the double integral
$$\iint_\Omega x^y \, dx \, dy$$
where $\Omega = [0, 1] \times [a, b]$, $0 < a < b$, to prove that
$$\int_0^1 \frac{x^b - x^a}{\log x} \, dx = \log \frac{1 + b}{1 + a}.$$

10. Using the change of variable and giving careful justification, evaluate the Lebesgue integrals

(i) $\displaystyle \int_{-\pi/2}^0 \log(\sin t + 1) \cos t \, dt,$ (ii) $\displaystyle \int_e^\infty \frac{dt}{t(\log t + i)^2},$

(iii) $\displaystyle\int_{-\infty}^{0} e^t (e^t + 2i)^2 \, dt,$ (iv) $\displaystyle\int_{-1}^{-1+\pi^2/4} \frac{\cos\sqrt{x+1}}{\sqrt{x+1}} \, dx,$

(v) $\displaystyle\int_{-\infty}^{0} \frac{e^x + i}{e^x + 1} \, dx,$ (vi) $\displaystyle\int_{0}^{1} \frac{t \, dt}{(1 - i\sqrt{1-t^2})^2 \sqrt{1-t^2}}.$

11. Using substitution discuss the existence of the Lebesgue integral over $(0, \infty)$.

$$\int_{e}^{\infty} \frac{dt}{t(\log t + i)}.$$

12. Calculate the Lebesgue measure of the set

$$A = \{(x, y, z) \in \mathbb{R}^3 : x^2 + y^2 < \min\{1, 1/z\}, z > 0\}.$$

13. Giving detailed justification, calculate the integral

$$\iint_{(0,\infty)\times(0,\infty)} e^{-x^2 - y^2} \, dx \, dy$$

and hence determine the value of

$$\int_{0}^{\infty} e^{-x^2} \, dx.$$

14. Calculate

$$\iint_{\Delta} \frac{3s \, dt \, ds}{\sqrt{1 + (t+s)^3}}$$

where $\Delta = \{(t, s) : t > 0, \ s > 0, \ t + s < a\}$, where $a > 0$.
Suggestion. Change of variables $x = t + s$, $y = t - s$, then Fubini. (Answer: $\sqrt{1 + a^3} - 1$.)

15. Giving complete arguments, calculate

$$\iint_{x>0,\, y>0} \frac{dx \, dy}{(1+y)(1+x^2 y)}$$

and deduce that

$$\int_{0}^{\infty} \frac{\log x}{x^2 - 1} \, dx = \frac{\pi^2}{4}.$$

16. Let Δ be the region in the first quadrant $t \geq 0$, $s \geq 0$ bounded by the curves $s - t = 0$, $s^2 - t^2 = 1$, $ts = a$, $ts = b$, where $0 < a < b$. Calculate the integral

$$\iint_{\Delta} (s^2 - t^2)^{ts} (t^2 + s^2) \, dt \, ds$$

by using a change of variables that transforms Δ into a rectangle.

Suggestion. Substitution $x = s^2 - t^2$ and $y = ts$. Use Fubini's theorem to convert a double integral to a repeated integral. (Answer: $\frac{1}{2} \log[(1+b)/(1+a)]$.)

17.   Use Tonelli's theorem to decide whether the following functions are integrable on the square $Q = [0, 1] \times [0, 1]$.

(i) $f(x, y) = \dfrac{x^2 - y^2}{(x^2 + y^2)^2}$.

(ii) $f(x, y) = x^{-3/2}$ if $0 < y < x$ and $f(x, y) = 0$ otherwise.

(iii) $f(x, y) = (1 - xy)^{-p}$, where $p > 0$.

18.   (Harder.) Let $T \colon \mathbb{R}^k \to \mathbb{R}^k$ be an invertible linear transformation, and let $f \colon \mathbb{R}^k \to \mathbb{C}$ be Lebesgue measurable. Prove the following:

(i) $f \circ T$ is Lebesgue measurable.

(ii)   If f is Lebesgue integrable, then

$$\int_{\mathbb{R}^k} f \, dm = |\det T| \int_{\mathbb{R}^k} (f \circ T) \, dm.$$

(iii)   If $A \subset \mathbb{R}^k$ is Lebesgue measurable, then $T(A)$ is Lebesgue measurable and

$$m(T(A)) = |\det T| \, m(A).$$

Suggestion. (ii) It is enough to consider only three types of linear transformations described by their effect on the elements of the standard basis $e_1, \ldots, e_k$: (1) $e_j \mapsto c e_j$ for one j; (2) $e_j \leftrightarrow e_k$; (3) $e_j \mapsto e_j + c e_k$ $(j \neq k)$. Use Fubini's theorem.

Chapter 17

Lebesgue Spaces

Functional analysis and measure and integration theory come together to produce some of the most important examples of Banach spaces, the Lebesgue L^p spaces. In order to obtain a norm $\|f\|_{L^p} = (\int_S |f|^p \, d\mu)^{1/p}$, we must have $p \geq 1$; the triangle inequality fails for $0 < p < 1$. There is exactly one Hilbert space among the L^p spaces, namely L^2. Surprisingly, L^2 is a prototype of all Hilbert spaces in the sense that each Hilbert space is isometrically isomorphic to some L^2 space.

One of the most interesting problems in L^p space theory is the representation of bounded linear functionals. A representation theorem for $L^2(0, 1)$ was the first to appear; it was discovered independently by Fréchet and Riesz in 1907. In 1910, Riesz obtained a representation theorem for $L^p(0, 1)$, $1 < p < \infty$. Extensions to general measure spaces had to wait until the 1930s. The case $p = 1$ was obtained in 1919 by Steinhaus.

17.1 Lebesgue spaces $L^p(\mu)$

In this section, (S, Σ, μ) is an arbitrary measure space. By p we will denote a real number, usually satisfying the inequality $p \geq 1$. If $p > 1$, we define $q = p/(p-1)$ to be the *conjugate index* of p; then

$$\frac{1}{p} + \frac{1}{q} = 1.$$

For real numbers $a, b \geq 0$ we have the inequalities

$$ab \leq \frac{a^p}{p} + \frac{b^q}{q} \tag{17.1}$$

and

$$(a + b)^p \leq 2^p(a^p + b^q). \tag{17.2}$$

For a proof of (17.1) see Appendix C. Inequality (17.2) follows from

$$(a + b)^p \leq (2 \max(a, b))^p = 2^p \max(a^p, b^p) \leq 2^p(a^p + b^p).$$

Definition 17.1. Let $p \geq 1$. The space $L^p(\mu) = L^p(S, \Sigma, \mu)$ consists of all functions $f \colon S \to \mathbb{C}$ such that

(i) f is μ-measurable,

(ii) $|f|^p$ is μ-integrable.

For $p = 1$ this definition is consistent with the earlier definition of $L^1(\mu)$.

For any $f \in L^p(\mu)$ we define

$$\|f\|_{L^p} = \left(\int_S |f|^p \, d\mu \right)^{1/p}.$$

For $p = 1$, this definition is again consistent with the earlier definition of $\|f\|$. We will show below that $\|\ \|_{L^p}$ is a norm on $L^p(\mu)$ if the equality μ-a.e. is adopted as the new equality. We observe that

$$\|f\|_{L^p} = 0 \iff f = 0 \ \mu\text{-a.e.}$$

Example 17.2. Consider the measure space $(\mathbb{N}, \mathcal{P}(\mathbb{N}), \nu)$, where ν is the counting measure on $\mathbb{N}$. Every function $f \colon \mathbb{N} \to \mathbb{C}$ is ν-measurable as $f^{-1}(G) \in \mathcal{P}(\mathbb{N})$ for each set $G \subset \mathbb{C}$. Further, given $p \geq 1$, $|f|^p$ is ν-integrable if and only if the series $\sum_{n=1}^{\infty} |f(n)|^p$ converges; in this case

$$\|f\|_{L^p} = \left(\int_S |f|^p \, d\mu \right)^{1/p} = \left(\sum_{n=1}^{\infty} |f(n)|^p \right)^{1/p}.$$

This shows that the space $L^p(\nu)$ can be identified with ℓ^p under the correspondence

$$f \leftrightarrow (f(1), f(2), f(3), \ldots).$$

Theorem 17.3 (Hölder's inequality). *Let $p > 1$, and let $q = p/(p-1)$ be the conjugate index of p. If $f \in L^p(\mu)$ and $h \in L^q(\mu)$, then $fh \in L^1(\mu)$, and*

$$\|fh\|_{L^1} \leq \|f\|_{L^p} \|h\|_{L^q}. \qquad (17.3)$$

Proof.   Let $f \in L^p(\mu)$ and $h \in L^q(\mu)$. Then fh is measurable μ-a.e. being the product of two functions measurable μ-a.e., and

$$|fh| \leq \frac{1}{p}|f|^p + \frac{1}{q}|h|^q \in L^1(\mu);$$

so $fh \in L^1(\mu)$ by Theorem 13.41. If $\|f\|_{L^p} = 0$ or $\|h\|_{L^q} = 0$, then $fh = 0$ μ-a.e. and Hölder's inequality (17.3) holds.

Assume that $\|f\|_{L^p} > 0$ and $\|h\|_{L^q} > 0$ and define

$$a(t) = \frac{f(t)}{\|f\|_{L^p}}, \qquad b(t) = \frac{h(t)}{\|h\|_{L^q}}.$$

Then $|ab| \leq p^{-1}|a|^p + q^{-1}|b|^q$, and

$$\int_S |ab|\, d\mu \leq \frac{1}{p} \int_S |a|^p\, d\mu + \frac{1}{q} \int_S |b|^q\, d\mu \leq \frac{1}{p} + \frac{1}{q} = 1,$$

while

$$\int_S |ab|\, d\mu = \frac{\|fh\|}{\|f\|_{L^p}\|h\|_{L^q}}.$$

$\square$

Theorem 17.4 (Minkowski's theorem). *Let $p \geq 1$. If $f, g \in L^p(\mu)$, then $f + g \in L^p(\mu)$, and*

$$\|f + g\|_{L^p} \leq \|f\|_{L^p} + \|g\|_{L^p}.$$

Proof. The case $p = 1$ has been settled in Chapter 13. Suppose that $p > 1$. The function $f + g$ is μ-measurable, and $|f + g|^p$ is integrable as

$$|f + g|^p \leq 2^p(|f|^p + |g|^p)$$

in view of (17.2). So $f + g \in L^p(\mu)$. Note that

$$|f + g|^p \leq (|f| + |g|)\,|f + g|^{p-1}. \tag{17.4}$$

Since $(p-1)q = p$, $|f+g|^{p-1} \in L^q(\mu)$; we can then apply Hölder's inequality to (17.4) to get

$$\int_S |f + g|^p\, d\mu \leq (\|f\|_{L^p} + \|g\|_{L^p})(\|f + g\|_{L^p})^{p/q}.$$

If $A = \|f + g\|_{L^p} = 0$, then the inequality holds with 0 on either side. If $A \neq 0$, we can divide by $A^{p/q}$ to get Minkowski's inequality. $\square$

Minkowski's inequality is the triangle inequality for the L^p-norm in $L^p(\mu)$. We leave the verification of **N1** – **N3** as an exercise. The space $L^p(\mu)$ then becomes a normed space with the equality μ-a.e. as the new equality in $L^p(\mu)$. Theorem 17.7 will enable us to conclude that $L^p(\mu)$ is complete.

Example 17.5. Let S be any nonempty set, $p \geq 1$ and ν the counting measure on S. We define $\ell^p(S)$ to be

$$\ell^p(S) = L^p(S, \mathcal{P}(S), \nu).$$

Observe that a function f is in $\ell^p(S)$ if and only if it there is a countable set $A \subset S$ such that $f(t) = 0$ for all $f \in A^c$, and

$$\int_S |f|^p \, d\nu = \sum_{t \in A} |f(s)|^p$$

(where the sum is independent of the order of the elements in A). For $S = \mathbb{N}$ we get $\ell^p(\mathbb{N}) = \ell^p$. The norm on $\ell^p(S)$ is defined by

$$\|f\|_p = \left(\sum_{t \in S} |f(t)|^p \right)^{1/p}$$

(the sum is written over all of S, but only countably many terms are nonzero). The Hölder and Minkowski inequalities take the form

$$\sum_{t \in S} |f(t)h(t)| \leq \left(\sum_{t \in S} |f(t)|^p \right)^{1/p} \left(\sum_{t \in S} |h(t)|^q \right)^{1/q}, \quad p > 1, \ \frac{1}{p} + \frac{1}{q} = 1$$

and

$$\left(\sum_{t \in S} |f(t) + g(t)|^p \right)^{1/p} \leq \left(\sum_{t \in S} |f(t)|^p \right)^{1/p} + \left(\sum_{t \in S} |g(t)|^p \right)^{1/p}.$$

If $S = \mathbb{N}$, we obtain the ordinary ℓ^p spaces. For $S = \{1, 2, \ldots, n\}$ the $\ell^p(S)$ spaces reduce to $\mathbb{R}^n$ or $\mathbb{C}^n$ equipped with the ℓ^p norm

$$\|x\|_p = \left(\sum_{k=1}^{n} |\xi_k - \eta_k|^p \right)^{1/p}.$$

For $p = 2$ we obtain the Euclidean spaces.

Theorem 17.6 (Lebesgue's L^p dominated convergence theorem).
Let (f_n) be a sequence of functions measurable μ-a.e. with $f_n \to f$ μ-a.e. Assume that there is $g \in L^p(\mu)$ with $g \geq 0$ and $|f_n| \leq g$ μ-a.e. for all n. Then $f_n, f \in L^p(\mu)$ and $\|f_n - f\|_{L^p} \to 0$.

Proof. Problem 17.1.   □

Let f be a simple function in S. Recall that f is μ-integrable if the set $\{t \in S : f(t) \neq 0\}$, which is measurable, is of finite measure. We observe that every μ-integrable simple f is in $L^p(\mu)$, $p \geq 1$. The set $E(\mu)$ of all integrable simple functions can be equipped with the L^p-norm; the next theorem shows that the completion of this space coincides with $L^p(\mu)$.

Theorem 17.7. *Let $p \geq 1$. Then $f \in L^p(\mu)$ if and only if there is a sequence (f_n) of integrable simple functions such that*

(i) (f_n) *is Cauchy in the L^p-norm,*

(ii) $f_n \to f$ *μ-a.e.*

Proof. First assume that $f \in L^p(\mu)$. Then f is measurable μ-a.e. and there is a sequence (f_n) of simple functions such that

$$f_n \to f \ \mu\text{-a.e.} \quad \text{and} \quad |f_n| \le |f| \ \text{for all} \ n.$$

By Lebesgue's L^p-dominated convergence theorem, $\|f_n - f\|_{L^p} \to 0$. So (f_n) is L^p-Cauchy.

Conversely assume that $f_n \to f$ μ-a.e. with (f_n) an L^p-Cauchy sequence. Let $\varepsilon > 0$. There is $N \in \mathbb{N}$ such that

$$\|f_m - f_n\|_{L^p} < \varepsilon \ \text{if} \ m, n \ge N.$$

Choose $m \ge N$ and keep it fixed. Then

$$\liminf_{n \to \infty} \int_S |f_m - f_n|^p \, d\mu \le \varepsilon^p.$$

By Fatou's lemma,

$$\int_S |f_m - f|^p \, d\mu = \int_S \liminf_{n \to \infty} |f_m - f_n|^p \, d\mu \le \liminf_{n \to \infty} \int_S |f_m - f_n|^p \, d\mu \le \varepsilon^p;$$

so $\|f_m - f\|_{L^p} \le \varepsilon$ if $m \ge N$. Thus $f_m - f \in L^p(\mu)$, and

$$f = f_m + (f - f_m) \in L^p(\mu).$$

$\square$

Theorem 17.8. *$L^p(\mu)$ is the completion of $E(\mu)$ under the L^p-norm. Therefore $L^p(\mu)$ is a Banach space.*

Proof. The result follows from the preceding theorem. $\square$

17.2 Lebesgue space $L^2(\mu)$

The only self-conjugate index among $p \ge 1$ is $p = 2$. The corresponding Lebesgue space $L^2(\mu)$ is a Hilbert space.

Theorem 17.9. *The space $L^2(\mu)$ is a Hilbert space with inner product*

$$\langle f, g \rangle = \int_S f\bar{g} \, d\mu.$$

Proof.   Let $f, g \in L^2(\mu)$. By Hölder's theorem, $f\bar{g} \in L^1(\mu)$, so that $\langle f, g \rangle$ is well defined. A verification of the inner product axioms is left as an exercise. The induced norm coincides with the L^2-norm:

$$\langle f, g \rangle^{1/2} = \left(\int_S f\bar{f}\, d\mu \right)^{1/2} = \left(\int_S |f|^2\, d\mu \right)^{1/2} = \|f\|_2.$$

The completeness of $L^2(\mu)$ follows from Theorem 17.8.   □

One of the most striking facts in the theory of Hilbert spaces is that every Hilbert space can be identified with some $\ell^2(S)$.

Theorem 17.10. *Let H be a nonzero Hilbert space. Then there is a nonempty set S and a linear isomorphism $T\colon H \to \ell^2(S)$ which preserves inner products (and so is an isometry).*

Proof.   By Theorem 8.19, H possesses an orthonormal basis S (a total orthonormal set in H). For any $x \in H$ define a function $f\colon S \to \mathbb{C}$ by

$$f(t) = \langle x, t \rangle, \quad \text{for all } t \in S.$$

We show that $f \in \ell^2(S) = L^2(S, \mathcal{P}(S), \nu)$, where ν is the counting measure on S.

Given $\alpha > 0$, there are only finitely many $t \in S$ for which $|\langle x, t \rangle| \geq \alpha$ (Bessel's inequality). Therefore the set A of all $t \in S$ with $\langle x, t \rangle \neq 0$ is countable. Hence

$$\int_S |f|^2\, d\nu = \sum_{t \in A} |\langle x, t \rangle|^2 \leq \|x\|^2.$$

Let the functions f, g be associated with $x, y \in H$, respectively. Then

$$f(t) = \langle x, t \rangle, \qquad g(t) = \langle y, t \rangle, \quad \text{for all } t \in S,$$

and

$$\int_S f\bar{g}\, d\mu = \sum_{t \in S} f(t)\overline{g(t)} = \sum_{t \in S} \langle x, t \rangle \overline{\langle y, t \rangle}.$$

Parseval's identity gives

$$\langle x, y \rangle = \sum_{t \in S} \langle x, t \rangle \overline{\langle y, t \rangle};$$

so

$$\int_S f\bar{g}\, d\mu = \langle x, y \rangle.$$

Define $Tx = f$, where $f(t) = \langle x, t \rangle$ for all $t \in S$. A proof of linearity of T is left as an exercise. We already showed that T preserves inner products. It remains to be proved that T is surjective.

Let $f \in \ell^2(S)$. Then f vanishes outside a countable set $A \subset S$, and

$$\sum_{t \in S} |f(t)|^2 < \infty.$$

Order A in a sequence, say $A = \{t_1, t_2, t_3, \ldots\}$, and define

$$x_n = \sum_{k=1}^{n} f(t_k) t_k.$$

Then (x_n) is a Cauchy sequence as

$$\|x_n - x_m\|^2 = \left\| \sum_{k=m+1}^{n} f(t_k) t_k \right\|^2 = \sum_{k=m+1}^{n} |f(t_k)|^2$$

(the last equality by Pythagoras's theorem). Since H is a complete inner product space, (x_n) converges to some element x of H,

$$x = \sum_{k=1}^{\infty} f(t_k) t_k = \sum_{t \in S} f(t) t.$$

Let B be the set of all $t \in S$ such that either $f(t) \neq 0$ or $\langle x, t \rangle \neq 0$. Then B is countable, and can be ordered in a sequence $s_1, s_2, s_3, \ldots$ For a fixed p and all $m \geq p$,

$$\left| \langle x, s_p \rangle - f(s_p) \right| = \left| \langle x, s_p \rangle - \sum_{n=1}^{m} f(s_n) \langle s_n, s_p \rangle \right|$$

$$= \left| \langle x - \sum_{n=1}^{m} f(s_n) s_n, s_p \rangle \right|$$

$$\leq \left\| x - \sum_{n=1}^{m} f(s_n) s_n \right\| \|s_p\| \to 0 \text{ as } m \to \infty.$$

So $f(s_p) = \langle x, s_p \rangle$, and $f(t) = \langle x, t \rangle$ for all $t \in S$. $\qquad \square$

17.3 Lebesgue space $L^\infty(\mu)$

In this section, (S, Σ, μ) is a σ-finite measure space. There are two reasons for this restriction. First, the definition of an essentially bounded function is more complicated for a general measure space than it is for a σ-finite space. Second, most applications of L^∞ spaces are to σ-finite spaces.

Definition 17.11. A function $f \colon S \to \mathbb{C}$ is *essentially bounded* if there is a set A of measure zero such that f is bounded on A^c. The space $L^\infty(\mu)$ consists of all functions measurable μ-a.e. and essentially bounded on S, and is equipped with the equality μ-a.e. as the new equality. The norm in $L^\infty(\mu)$ is defined by

$$\|f\|_{L^\infty} = \inf_{\mu(A)=0} \sup_{t \in A^c} |f(t)|.$$

In fact, the infimum is attained on some set A (Problem 17.5).

Theorem 17.12. $L^\infty(\mu)$ *is a Banach space.*

Proof.   Exercise.   □

Theorem 17.13. *Let* $f \in L^1(\mu)$ *and* $g \in L^\infty(\mu)$. *Then* $fg \in L^1(\mu)$, *and*

$$\left| \int_S fg \, d\mu \right| \le \int_S |fg| \, d\mu \le \|f\|\|g\|_{L^\infty}.$$

Proof.   Let g be bounded on A^c, where $\mu(A) = 0$. By the assumption, fg is measurable μ-a.e. and

$$|f(t)g(t)| \le \|g\|_{L^\infty} |f(t)| \text{ for all } t \in A^c;$$

so, if $\alpha = \|g\|_{L^\infty}$,

$$|fg| \le \alpha |f| \in L^1(\mu).$$

So fg is μ-integrable being dominated by an integrable function. The inequality follows from the monotonicity of the integral and from the inequality $|fg| \le \alpha |f|$.   □

17.4   Bounded linear functionals on $L^p(\mu)$

In this section we discuss a representation of a bounded linear functional on $L^p(\mu)$ in the form

$$\varphi(f) = \int_S fg \, d\mu, \text{ for all } f \in L^p(\mu),$$

and describe the duals of $L^p(\mu)$ for $1 \le p < \infty$. It turns out that the representations work for arbitrary measure spaces when $1 < p < \infty$, but that we must assume σ-finite spaces to get a corresponding result for $p = 1$. First we discuss the case of $p > 1$.

Theorem 17.14. *Let $1/p + 1/q = 1$, and let $g \in L^q(\mu)$. Let $\varphi \colon L^p(\mu) \to \mathbb{C}$ be defined by*

$$\varphi(f) = \int_S f g \, d\mu \quad \text{for all } f \in L^p(\mu).$$

Then φ is a bounded linear functional on $L^p(\mu)$ with

$$\|\varphi\| = \|g\|_{L^q}.$$

Proof. By Hölder's theorem, $fg \in L^1(\mu)$ if $f \in L^p(\mu)$ and $g \in L^q(\mu)$. So φ is well defined. The linearity of φ can be verified as an exercise. The boundedness of φ follows from Hölder's inequality:

$$|\varphi(f)| \le \int_S |fg| \, d\mu \le \|f\|_{L^p} \|g\|_{L^q}$$

for all $f \in L^p(\mu)$. This shows that $\|\varphi\| \le \|g\|_{L^q}$.

The function g is μ-measurable. To prove the equality of norms, define a function $h \colon S \to \mathbb{C}$ by

$$h(t) = \begin{cases} \dfrac{|g(t)|}{g(t)} & \text{if } g(t) \ne 0, \\ 1 & \text{if } g(t) = 0. \end{cases}$$

Then h is measurable μ-a.e. with

$$|h| = 1 \quad \text{and} \quad gh = |g|.$$

Choose

$$f = |g|^{q-1} h.$$

Then f is measurable μ-a.e. and

$$|f|^p = |g|^{(q-1)p} |h|^p = |g|^q \in L^p(\mu).$$

So $f \in L^p(\mu)$, and

$$\|f\|_{L^p} = \left(\|g\|_{L^q} \right)^{q-1}.$$

Therefore

$$\varphi(f) = \int_S f g \, d\mu = \int_S |g|^q \, d\mu = \left(\|g\|_{L^q} \right)^q,$$

and

$$\frac{\varphi(f)}{\|f\|_{L^p}} = \|g\|_{L^q}.$$

$\square$

Theorem 17.15. *Let S be a σ-finite measure space and let $g \in L^\infty(\mu)$. Define $\varphi \colon L^1(\mu) \to \mathbb{C}$ by*

$$\varphi(f) = \int_S fg \, d\mu \quad \text{for all } f \in L^1(\mu).$$

Then φ is a bounded linear functional on $L^1(\mu)$ with

$$\|\varphi\| = \|g\|_{L^\infty}.$$

Proof. The linearity of φ and the inequality $\|\varphi\| \leq \|g\|_{L^\infty}$ are left as an exercise.

To prove the equality of norms, we observe that the equality holds if $\|g\|_{L^\infty} = 0$. Let $\|g\|_{L^\infty} > 0$ and let $\varepsilon > 0$ be such that

$$\|g\|_{L^\infty} - \varepsilon > 0.$$

Changing g on a set of measure zero, we may assume that g is measurable. Define E to be the set of all $t \in S$ such that $|g(t)| \geq \|g\|_{L^\infty} - \varepsilon$. Then $E \in \Sigma$ and $\mu(E) \neq 0$ (the definition of $\|g\|_{L^\infty}$ as an infimum). Since S is σ-finite, $S = \bigcup_{n=1}^\infty S_n$ with $\mu(S_n) < \infty$ for all n. One of the sets $E \cap S_n$ has nonzero measure (otherwise $\mu(E) = 0$); denote this set by A. Define a function h as in the proof of the preceding theorem. So $|h| = 1$ and $gh = |g|$. Choose

$$f = \chi_A h.$$

Then

$$\frac{\varphi(f)}{\|f\|} = \frac{\int_S fg \, d\mu}{\int_S |f| \, d\mu} = \frac{\int_A |g| \, d\mu}{\mu(A)} \geq \frac{(\|g\|_{L^\infty} - \varepsilon)\mu(A)}{\mu(A)} = \|g\|_{L^\infty} - \varepsilon$$

(possible as $0 < \mu(A) < \infty$). Since ε was arbitrary, we have the equality $\|\varphi\| = \|g\|_{L^\infty}$. $\qquad\square$

Theorem 17.14 states that every function $g \in L^q(\mu)$ induces aF bounded linear functional on $L^p(\mu)$, where p and q are conjugate indices. It is our aim to show that every bounded linear functional can be represented in this way. For this we need the Radon–Nikodým theorem (Theorem 15.41).

Theorem 17.16 (Representation of functionals on $L^p(\mu)$). *Let φ be a bounded linear functional on $L^p(\mu)$, where $1 < p < \infty$. Then there is a unique (up to equality μ-a.e.) function $g \in L^q(\mu)$, where $q = p/(p-1)$ is the conjugate index of p, such that*

$$\varphi(f) = \int_S fg \, d\mu \quad \text{for all } f \in L^p(\mu).$$

Proof. (i) *The case* $\mu(S) < \infty$. Define $\nu: \Sigma \to \mathbb{C}$ by

$$\nu(A) = \varphi(\chi_A), \quad \text{for all } A \in \Sigma.$$

Then ν is a complex measure on Σ (check!) which is absolutely continuous relative to μ (check!), that is, $\nu \ll \mu$. Let g be a function satisfying $\nu(E) = \int_E g \, d\mu$ for each $E \in \Sigma$ whose existence is guaranteed by the Radon–Nikodým theorem (Theorem 15.41). Then for any simple function f,

$$\varphi(f) = \varphi\left(\sum_{k=1}^{s} c_k \chi_{A_k}\right) = \sum_{k=1}^{s} c_k \varphi(\chi_{A_k})$$

$$= \sum_{k=1}^{s} c_k \nu(A_k) = \sum_{k=1}^{s} c_k \int_{A_k} g \, d\mu$$

$$= \int_S \left(\sum_{k=1}^{s} c_k \chi_{A_k}\right) g \, d\mu = \int_S f g \, d\mu.$$

We want to show that $g \in L^q(\mu)$. By changing it on a set of measure zero, we may (and will) assume that g is measurable. For any n let $E_n = \{t \in S : |g(x)| \le n\}$, and let

$$g_n = g \chi_{E_n}.$$

Since μ is finite, the bounded measurable function g_n is in $L^q(\mu)$. For any simple function f the function $f \chi_{E_n}$ is also simple, and

$$\left|\int_S f g_n \, d\mu\right| = \left|\int_S f g \chi_{E_n} \, d\mu\right| = |\varphi(f \chi_{E_n})| \le \|\varphi\| \, \|f \chi_{E_n}\|_{L^p} \le \|\varphi\| \, \|f\|_{L^p}.$$

So $\|g_n\|_{L^q} \le \|\varphi\|$ by Theorem 17.14. Since $g_n \to g$ (and $|g_n|^q \nearrow |g|^q$), the monotone convergence theorem implies that $g \in L^q(\mu)$.

Next we show that $\varphi(f) = \int_S f g \, d\mu$ holds for all $f \in L^p(\mu)$. Given $f \in L^p(\mu)$, we can find a sequence (f_n) of simple functions such that $f_n \to f$ μ-a.e. and $|f_n| \le |f|$. Then $f_n g \to f g$ with $|f_n g| \le |f g|$ and $|f g| \in L^1(\mu)$ ($f g$ is the product of an $L^p(\mu)$ and an $L^q(\mu)$ function). By Lebesgue's dominated convergence theorem,

$$\int_S f g \, d\mu = \lim_{n \to \infty} \int_S f_n g \, d\mu = \lim_{n \to \infty} \varphi(f_n) = \varphi(f);$$

the last equality holds since φ is continuous and $\|f_n - f\|_{L^p} \to 0$ (by Lebesgue's dominated convergence theorem for $L^p(\mu)$).

Uniqueness of g is left as an exercise.

(ii) *The case of σ-finite S.* There is a sequence (S_n) of measurable sets of finite measure such that $S_n \nearrow S$. By part (i) of the proof, for each n there is $g_n \in L^q(\mu)$ which vanishes outside S_n such that

$$\varphi(f) = \int_S f g_n \, d\mu$$

for all $f \in L^p(\mu)$ vanishing outside S_n. Changing values on a set of measure zero, we may assume that, for each n, g_{n+1} coincides with g_n on S_n. Then we can define a function g on S by

$$g(t) = g_n(t) \text{ if } t \in S_n.$$

The function g is measurable μ-a.e. and $|g_n| \nearrow |g|$. By the monotone convergence theorem,

$$\int_S |g|^q \, d\mu = \lim_{n \to \infty} \int_S |g_n|^q \, d\mu \le \|\varphi\|^q.$$

So $g \in L^q(\mu)$. For any $f \in L^p(\mu)$, set

$$f_n = f \chi_{S_n}.$$

Then $f_n \to f$ μ-a.e. and $|f_n| \le |f|$. This can be used, together with Lebesgue's dominated convergence theorem, to obtain the desired result. (See the preceding part of the proof.)

A proof of uniqueness is left as an exercise.

(iii) *The case of an arbitrary measure space.* Let Σ_1 be the set of all σ-finite measurable sets in S. Then Σ_1 is a σ-ring; that is, Σ_1 contains $\emptyset$, countable unions of its members, and relative complements of its members. By part (ii) of the proof, for each $E \in \Sigma_1$, there is a function $g_E \in L^q(\mu)$ which vanishes outside E such that

$$\varphi(f) = \int_S f g_E \, d\mu$$

for all $f \in L^p(\mu)$ vanishing outside E. Define $\lambda \colon \Sigma_1 \to [0, \infty]$ by

$$\lambda(E) = \int_S |g_E|^q \, d\mu \text{ for all } E \in \Sigma_1.$$

If $A, B \in \Sigma_1$ and $A \subset B$, the uniqueness of g_E implies that $g_A = g_B$ μ-a.e. on A. So

$$A \subset B \implies \lambda(A) \le \lambda(B) \le \|\varphi\|^q.$$

Let

$$\alpha = \sup_{E \in \Sigma_1} \lambda(E).$$

Then there is a sequence (E_n) in Σ_1 with $\lambda(E_n) \to \alpha$. Define $A = \bigcup_{n=1}^{\infty} E_n$, so that $A \in \Sigma_1$. Moreover,

$$\lambda(E_n) \leq \lambda(A) \leq \alpha;$$

taking the limit as $n \to \infty$, we get $\lambda(A) = \alpha$.

Let $g = g_A$ on A and $g = 0$ on A^c. Then $g \in L^q(\mu)$. If $E \in \Sigma_1$ and $E \supset A$, then $g_E = g_A$ μ-a.e. on A. But

$$\int_{E \cap A^c} |g|^q \, d\mu = \int_E |g|^q \, d\mu - \int_A |g|^q \, d\mu = \lambda(E) - \lambda(A) \leq 0;$$

so the integral on the left is equal to zero, and $g_E = 0$ μ-a.e. on $E \cap A^c$. This shows that $g_E = g_A$ μ-a.e. on E.

If $f \in L^p(\mu)$, then there is a set $N \in \Sigma_1$ such that f vanishes outside N (Problem 13.28). So $E = A \cup N \in \Sigma_1$. Therefore $f = 0$ on E^c, and

$$\varphi(f) = \int_S f g_E \, d\mu = \int_S f g \, d\mu.$$

The equality $\|\varphi\| = \|g\|_{L^q}$ follows from Theorem 17.14. A proof of uniqueness is left as an exercise. $\square$

Theorem 17.17 (Representation of functionals on $L^1(\mu)$). *Let μ be a σ-finite measure. For each bounded linear functional φ on $L^1(\mu)$ there is a unique function $g \in L^\infty(\mu)$ such that*

$$\varphi(f) = \int_S f g \, d\mu \text{ for all } f \in L^1(\mu),$$

and $\|\varphi\| = \|g\|_{L^\infty}$.

Proof. Problem 17.8. $\square$

17.5 Problems for Chapter 17

1. *Lebesgue's dominated convergence theorem for L^p.* Let (f_n) be a sequence of functions measurable μ-a.e. such that $f_n \to f$ μ-a.e. and that $|f_n| \leq g$ for some function $g \in L^p(\mu)$, where $p \geq 1$. Prove that $f_n, f \in L^p(\mu)$, and that $\|f_n - f\|_{L^p} \to 0$.

2. If S is a finite measure space and $q \geq 1$, show that every bounded function measurable μ-a.e. on S is in $L^q(\mu)$.

3. Let S be a finite measure space and let $1 \leq r < s$. Show that $L^s(\mu) \subset L^r(\mu)$, and that

$$\|f\|_{L^r} \leq \mu(S)^{\frac{1}{r} - \frac{1}{s}} \|f\|_{L^s}$$

for all $f \in L^s(\mu)$.

4. If S is a nonempty set and $1 \leq r < s$, show that $\ell^r(S) \subset \ell^s(S)$, and that

$$\|f\|_s \leq \|f\|_r$$

for all $f \in \ell^r(S)$. If S is an infinite set, show that the inclusion is proper.

5. Let S be a σ-finite measure space and let $f \in L^\infty(\mu)$. Show that there is a measurable set A of measure zero such that

$$\|f\|_{L^\infty} = \sup_{t \in A^c} |f(t)|.$$

6. Let S be a finite measure space, $p \geq 1$, and let f be measurable. Show that $f \in L^p(\mu)$ if and only if

$$\sum_{n=1}^{\infty} n^p \mu(E_n) < \infty.$$

Formulate the preceding result for functions measurable μ-a.e.

7. Let $S = (0, \infty)$, let μ be Lebesgue measure on S, and let

$$f(t) = \frac{1}{\sqrt{t}(1 + |\log t|)}.$$

Find all $p \geq 1$ such that $f \in L^p(\mu)$.

8. Prove Theorem 17.17.

9. If $p \geq 1$, show that the space $L^p[0, 1]$ is separable. Prove that $L^\infty[0, 1]$ is not separable.

10. Show that $L^\infty(\mu) \cap L^1(\mu) \subset L^p(\mu)$ whenever $1 < p < \infty$.

11. Let μ be a finite measure and $f \in L^\infty(\mu)$. Prove that

$$\|f\|_\infty = \lim_{p \to \infty} \|f\|_{L^p}.$$

Appendix A

Sets and Mappings

A.1 Sets

We assume that the reader encountered the concept of a set before, and review only basic notation and set operations.

Sets will be denoted by upper case roman letters, such as $A, B, C, \ldots$, and elements of sets by lower case letters, such as $x, y, z \ldots$ The symbol $x \in A$ means that x is an element of the set A; $x \notin B$ means that x is not an element of B. We use the 'set builder' notation, for example

$$A = \{x \in \mathbb{R} : x^2 < 2\}$$

for the set of all real numbers x with the property that $x^2 < 2$.

Definition A.1. ($\Longrightarrow$, **and**, **or** are logical connectives.) If A, B are sets, we define:

(i)   *Subset*: $A \subset B$ means ($x \in A \implies x \in B$).

(ii) *Equality*: $A = B$ means ($A \subset B$ **and** $B \subset A$).

(iii) *Union*: $x \in A \cup B$ if and only if ($x \in A$ **or** $x \in B$).

(iv) *Intersection*: $x \in A \cap B$ if and only if ($x \in A$ **and** $x \in B$).

(v)   *Set complement*: $x \in A \setminus B$ if and only if ($x \in A$ **and** $x \notin B$).

(vi) *Power set*: $\mathcal{P}(A)$ is the family of all subsets of the set A (including the empty set and A).

The *empty set* $\emptyset$ is characterized by the fact that the statement '$x \in \emptyset$' is false for all x. If all sets we are dealing with are subsets of the same universal set S, the complement $S \setminus A$ is written as A^c.

We often have to deal with indexed families of sets. For example, $\mathcal{A} = \{A_\alpha : \alpha \in D\}$ is a family indexed by α, where D is the *index set*. The index set can be any convenient set, finite or infinite, countable or uncountable (see Appendix E). If $D = \{1, \ldots, n\}$ for some positive integer n, then we usually write $\{A_1, \ldots, A_n\}$ for the family $\mathcal{A}$. The definitions of union and intersection can be rewritten for indexed families as follows:

Definition A.2. Let $\mathcal{A} = \{A_\alpha : \alpha \in D\}$ be an indexed family of sets. Then:

(i) *Union:* $x \in \bigcup_{\alpha \in D} A_\alpha$ if and only if there exists an index $\alpha_0 \in D$ such that $x \in A_{\alpha_0}$.

(ii) *Intersection:* $x \in \bigcap_{\alpha \in D} A_\alpha$ if and only if $x \in A_\alpha$ for all indices $\alpha \in D$.

If $\mathcal{A} = \{A_\alpha : \alpha \in D\}$ is an indexed family where all A_α belong to the same universal set S and if $B \subset S$, then we have the so called *de Morgan's laws*

$$\left(\bigcup_{\alpha \in D} A_\alpha \right)^c = \bigcap_{\alpha \in D} A_\alpha^c, \qquad \left(\bigcap_{\alpha \in D} A_\alpha \right)^c = \bigcup_{\alpha \in D} A_\alpha^c,$$

and *distributive laws*

$$B \cap \left(\bigcup_{\alpha \in D} A_\alpha \right) = \bigcup_{\alpha \in D} (B \cap A_\alpha), \qquad B \cup \left(\bigcap_{\alpha \in D} A_\alpha \right) = \bigcap_{\alpha \in D} (B \cup A_\alpha).$$

The *Cartesian product* of a finite number of sets is defined as follows: If $A_1, \ldots, A_n$ are sets, then $A_1 \times \cdots \times A_n$ is the set of all n-tuples $(x_1, \ldots, x_n)$ where $x_i \in A_i$ for $i = 1, \ldots, n$. The definition of the Cartesian product of an indexed family requires the Axiom of Choice, which will be discussed in Appendix E.

The father of modern theory of sets is Georg Cantor (1845–1918), who later introduced also the arithmetics of transfinite numbers (See Appendix E). His great disappointment was the appearance of paradoxes in set theory, such as the famous Russell paradox (1902) involving the 'set of all sets'.

A.2   Relations and mappings

If A, B are nonempty sets, any subset $\mathcal{U}$ of $A \times B$ induces a *relation R from A to B* when we define

$$aRb \iff (a, b) \in \mathcal{U}.$$

If $A = B$, we call R a *relation on A*.

A relation R on A is

- *reflexive* if xRx for all $x \in A$,
- *symmetric* if $xRy \implies yRx$,
- *transitive* if xRy and $yRz \implies xRz$.
- *antisymmetric* if xRy and yRx implies $x = y$.

Definition A.3. A relation R on a nonempty set A is called an *equivalence* if it is reflexive, symmetric and transitive. We usually write $x \sim y$ instead of xRy. For any $a \in A$, the set $[a] = \{x \in A : x \sim a\}$ is called the *equivalence class* of a.

We can check that any two equivalence classes $[a], [b]$ in A are either disjoint (if $a \nsim b$) or identical (if $a \sim b$), and that the equivalence classes form a *partition* of A, that is, A is the disjoint union of its equivalence classes.

Example A.4. If X is a vector space and M a subspace of X, we define the relation $x \sim y$ on X by setting $x \sim y \iff x - y \in M$. Then $\sim$ is an equivalence relation on X. The equivalence classes are often written as $[a] = a + M$, and the set of these classes forms a vector space under the operations

$$(a + M) \oplus (b + M) = (a + b) + M, \quad \lambda \odot (a + M) = (\lambda a) + M.$$

(Check that these operations are independent of the choice of representatives for the equivalence class, and verify the vector space axioms.)

Definition A.5. A relation R on a nonempty set A is a *partial order* if it is reflexive, transitive and antisymmetric. We usually write $x \leq y$ in place of xRy. The pair $(A, \leq)$ is called a *partially ordered set*. (More about partially ordered sets can be found in Appendix E.)

Definition A.6. Let A, B be nonempty sets. A *mapping (function)* $f : A \to B$ is a rule that assigns to each element $x \in A$ a unique element $y \in B$, written $y = f(x)$. The set A is the *domain* of f and B is the *codomain* of f. A mapping is determined by the triple (A, B, f). Two mappings (A, B, f) and (C, D, g) are *equal* if $A = C$, $B = D$ and if $f(x) = g(x)$ for all $x \in A = C$.

Let $\mathcal{U} \subset A \times B$ be a set such that

(i) for each $x \in A$ there exists $y \in B$ such that $(x, y) \in \mathcal{U}$,

(ii) if $(x, y_1) \in \mathcal{U}$ and $(x, y_2) \in \mathcal{U}$, then $y_1 = y_2$.

Then the mapping $f\colon A \to B$ can be defined by setting $y = f(x)$ if $(x, y) \in \mathcal{U}$. Thus a mapping is seen as a special case of a relation.

The words 'mapping' and 'function' are interchangeable. We say that a function $f\colon A \to B$ is *injective* if

$$x_1 \neq x_2 \implies f(x_1) \neq f(x_2).$$

Equivalently, $f(x_1) = f(x_2) \implies x_1 = x_2$. A function $f\colon A \to B$ is *surjective* if

$$\text{for each } y \in B \text{ there exists } x \in A \text{ such that } y = f(x).$$

Surjectivity can be described in terms of the range of the function. The *range* of a function $f\colon A \to B$ is the subset $\{y = f(x) : x \in A\}$ of B, written $R(f)$. Then $f\colon A \to B$ is surjective if the range of f coincides with the codomain of f: $R(f) = B$. A function $f\colon A \to B$ is *bijective* if it is both injective and surjective.

Let $f\colon A \to B$ and $g\colon B \to C$ be two functions. The *composition* $g \circ f$ of the two functions is the function $h\colon A \to C$ defined by

$$h(x) = (g \circ f)(x) = g(f(x)), \quad x \in A.$$

The *identity funtion* I_A on A is defined by $I_A(x) = x$ for all $x \in A$. We observe that $f \circ I_A = I_A \circ f = f$. The composition of functions obeys the associative law

$$f \circ (g \circ h) = (f \circ g) \circ h.$$

A function $g\colon B \to A$ is a *left inverse* of a function $f\colon A \to B$ if $g \circ f = I_A$, and $h\colon B \to A$ is a *right inverse* of f if $f \circ h = I_B$. Left and right inverses are not always unique. Suppose a function $f\colon A \to B$ has a left inverse g and a right inverse h. Then, for each $y \in B$,

$$g(y) = (g \circ I_B)(y) = (g \circ (f \circ h))(y) = ((g \circ f) \circ h)(y) = (I_A \circ h)(y) = h(y),$$

which proves that $g = h$. The function $g\colon B \to A$ which is both left and right inverse of f is called the *inverse* of f. The preceding argument shows that an inverse of a function f, if it exists, is unique.

We have the following important characterization of injectivity, surjectivity and bijectivity in terms of one or two sided inverses:

Theorem A.7. *Let $f\colon A \to B$ be a function. Then:*

(i)   *f is injective if and only if f has a left inverse.*

(ii) *f is surjective if and only if f has a right inverse.*

(iii) *f is bijective if and only if f has the inverse.*

Proof. (i) Suppose that f is injective. To construct a left inverse $g \colon B \to A$ of f consider first $y \in R(f)$. The $y = f(x)$ for some $x \in A$; since f is injective, such x is unique, and we set $g(y) = x$. Select an arbitrary point $a \in A$. If $y \in B \setminus R(f)$, set $g(y) = a$. Then g is a function from B to A, and we can verify that $g \circ f = I_A$.

Conversely, let $g \colon B \to A$ be a left inverse of f. Suppose that $f(x_1) = f(x_2)$. Then

$$x_1 = I_A(x_1) = (g \circ f)(x_1) = g(f(x_1)) = g(f(x_2)) = (g \circ f)x_2) = I_A(x_2) = x_2,$$

and f is injective.

(ii) Suppose that f is surjective, that is, that $B = R(f)$. Given any $y \in B$, there exists $x \in A$ such that $y = f(x)$. We make one definite choice of such x for each $y \in B$, call it x_y; this requires the Axiom of Choice which is discussed in Appendix E. Then $h(y) = x_y$ defines a function from B to A, and we can verify that $f \circ h = I_B$.

Conversely, let h be a right inverse of f. If $y \in B$, then $y = I_B(y) = (f \circ h)(y) = f(g(y)) \in R(f)$, which shows that $B = R(f)$, and f is surjective.

(iii) The result follows by combining (i) and (ii). □

A.3 Sequences of sets

In topology and measure theory we often encounter sequences of sets and their 'convergence' is some sense. We call a sequence (A_n) of sets in some universe S *expanding* if $A_n \subset A_{n+1}$ for all n, and a sequence (B_n) *contracting* if $B_n \supset B_{n+1}$ for all n. In this case we write

$$\lim_{n \to \infty} A_n = \bigcup_{n=1}^{\infty} A_n = A \text{ or } A_n \nearrow A, \text{ and } \lim_{n \to \infty} B_n = \bigcap_{n=1}^{\infty} B_n = B \text{ or } B_n \searrow B.$$

For a general sequence (A_n) of subsets of S we define the lower and upper limit of (A_n) by

$$\liminf_{n \to \infty} A_n = \lim_{n \to \infty} \bigcap_{k \geq n} A_k, \qquad \limsup_{n \to \infty} A_n = \lim_{n \to \infty} \bigcup_{k \geq n} A_k.$$

By Problem A.4,

$$\bigcap_{n=1}^{\infty} A_n \subset \liminf_{n \to \infty} A_n \subset \limsup_{n \to \infty} A_n \subset \bigcup_{n=1}^{\infty} A_n; \tag{A.1}$$

if the lower and upper limits are equal, their common value A is called the *limit* of (A_n), written $\lim_{n\to\infty} A_n = A$ or $A_n \to A$. If $A_n \to A$ and $B_n \to B$, then

$$A_n \cap B_n \to A \cap B, \quad A_n \cup B_n \to A \cup B, \quad A_n^c \to A^c \qquad \text{(A.2)}$$

(Problem A.5).

A.4   Inverse images

Let $f, g: A \to B$ be mappings, $C \subset A$ and $U, U_\alpha, V \subset B$. The inverse image of U under f is defined by

$$f^{-1}(U) = \{t \in A : f(t) \in U\}.$$

In this context, f^{-1} is a set function, $f^{-1}: \mathcal{P}(B) \to \mathcal{P}(A)$.

The inverse images play an important role in the global formulation of continuity, and in the definition and properties of measurable functions. The inverse images obey the following rules:

(i) f is injective if and only if $f^{-1}(\{y\})$ is a singleton for each $y \in B$.

(ii) $C \subset f^{-1}(f(C)), \quad f(f^{-1}(U)) \subset U.$

(iii)   If f, g are composable, then $(f \circ g)^{-1} = g^{-1} \circ f^{-1}.$

(iv)   If $U \subset V$, then $f^{-1}(U) \subset f^{-1}(V)$ (the composition of set maps).

(v) $f^{-1}(U \cup V) = f^{-1}(U) \cup f^{-1}(V).$

(vi) $f^{-1}(U \cap V) = f^{-1}(U) \cap f^{-1}(V).$

(vii) $f^{-1}(\bigcup_{\alpha \in D} U_\alpha) = \bigcup_{\alpha \in D} f^{-1}(U_\alpha).$

(viii) $f^{-1}(\bigcap_{\alpha \in D} U_\alpha) = \bigcap_{\alpha \in D} f^{-1}(U_\alpha).$

(ix) $f^{-1}(U^c) = (f^{-1}(U))^c$ (the complements in B and A), $f^{-1}(U \backslash V) = f^{-1}(U) \backslash f^{-1}(V).$

A.5   Problems for Appendix A

1.   If $f: A \to B$ is a function, describe the injectivity and surjectivity of f in terms of a relation induced by the set $\mathcal{U} = \{(x, y) : y = f(x)\} \subset A \times B$.

2.   Let S be a nonempty set and $\mathcal{P}(S)$ the power set of S. Show that the relation $\leq$ defined by

$$A \leq B \iff A \subset B$$

is a partial order on $\mathcal{P}(S)$.

3. Let E be a nonempty set and $\mathcal{F}$ the set of all functions $f \colon E \to \mathbb{R}$. Prove that the relation $\leq$ defined by

$$f \leq g \iff f(t) \leq g(t) \text{ for all } t \in E$$

is a partial order on $\mathcal{F}$.

4. Prove that for any sequence (A_n) of subsets of the universe S,

$$\bigcap_{n=1}^{\infty} A_n \subset \liminf_{n \to \infty} A_n \subset \limsup_{n \to \infty} A_n \subset \bigcup_{n=1}^{\infty} A_n.$$

5. Prove Equation (A.2).

6. Prove the properties of inverse images formulated in Section A.4.

Appendix B

Review of Real Analysis

B.1 Real numbers

The set $\mathbb{R}$ of real numbers forms a *complete ordered field*. We assume that the field axioms are known; they describe the basic properties of addition and multiplication together with the existence of 0 and 1. The axioms for the order relation are as follows:

(i) $x \leq y$ and $y \leq z \implies x \leq z$

(ii) $x \leq y$ and $y \leq x \implies x = y$

(iii) $x, y \in \mathbb{R} \implies$ either $x \leq y$ or $y \leq x$

(iv) $x \leq y \implies x + z \leq y + z$

(v) $0 \leq x$ and $0 \leq y \implies 0 \leq xy$

All other order properties can be derived from these axioms. Now we address the order completeness.

If A is a nonempty set of real numbers, we say that $u \in \mathbb{R}$ is an *upper bound* for A if $x \leq u$ for all $x \in A$. A set which has an upper bound is called *bounded above*. Lower bounds are defined analogously.

Definition B.1. A number $\alpha \in \mathbb{R}$ is a *least upper bound* for A if

(i) α is an upper bound for A, and

(ii) $\alpha \leq u$ for any other upper bound u of A.

A least upper bound of A is also called a *supremum* of A, written $\alpha = \sup A$. Similarly, a number $\beta \in \mathbb{R}$ is a *greatest lower bound* for A if

(i) β is a lower bound for A, and

(ii) $\beta \geq l$ for any other lower bound l of A.

A greatest lower bound of A is also called an *infimum* of A, written $\beta = \inf A$.

It can be proved that a set A has at most one supremum $a = \sup A$. There are two possibilities for the supremum a of A: If $a \in A$, then a is in fact the greatest element of A, also called the *maximum of A*. So, if $\sup A \in A$, then $\sup A = \max A$. If $a \notin A$, then a is a proper supremum of A. For instance, $\sup (0,1) = 1$, but the set has no maximum. On the other hand, $\sup (0,1] = \max (0,1] = 1$.

Similar comments apply also to infimum $\beta = \inf A$. If $\beta \in A$, then β is the least element or *minimum* of A, written $\beta = \min A$.

It is often convenient to use an alternative characterization of the supremum and infimum (see Problem B.2).

A set which is bounded both above and below is called *bounded*. A finite set of real numbers is always bounded and possesses both a maximum and a minimum. The modulus of $x \in \mathbb{R}$ is defined by

$$|x| = \max\{x, -x\}.$$

We review some important properties of the modulus:

(i) $|xy| = |x|\,|y|$

(ii) $\pm x \le |x|$

(iii) $|x + y| \le |x| + |y|$ (the so called *triangle inequality*)

(iv) $|x - y| \ge |\,|x| - |y|\,|$

As a sample we prove the triangle inequality: Adding up the inequalities $x \le |x|$ and $y \le |y|$, and then the inequalities $-x \le |x|$ and $-y \le |y|$ we get

$$|x + y| = \max\{x + y, -(x + y)\} \le |x| + |y|.$$

So far we have not explained what is meant by saying that an ordered field is complete. It means that the field obeys the following axiom:

B.2 (Order completeness axiom). Let A be a nonempty subset of the real numbers which is bounded above. Then A has a least upper bound.

It can be deduced from the order completeness axiom that a nonempty subset A of the real numbers which is bounded below has a greatest lower bound. Further, the least upper bound and the greatest lower bound are unique if they exist.

As a first application of the order completeness of the real numbers we show that the natural numbers are not bounded as a subset of $\mathbb{R}$. For this note that $1 \in \mathbb{R}$, and $n = 1 + \cdots + 1 \in \mathbb{R}$.

Theorem B.3 (Archimedean property of $\mathbb{N}$). *The set $\mathbb{N}$ of natural numbers is not bounded above in $\mathbb{R}$.*

Proof. Suppose that $\mathbb{N}$ is bounded above. By the order completeness axiom, there is a real number $a = \sup \mathbb{N}$. For any natural number n, $n+1$ is also in $\mathbb{N}$, so that $n + 1 \leq a$. Therefore $n \leq a - 1$ for all $n \in \mathbb{N}$, and $a - 1$ is an upper bound for $\mathbb{N}$, which is less than the least upper bound a. This contradiction shows that $\mathbb{N}$ is not bounded above. □

From the Archimedean property of $\mathbb{N}$ we can deduce that the set $\mathbb{Z}$ of all integers is not bounded above or below. In the proof of the following result we use this simple property of integers:

If a, b are integers and $|a - b| < 1$, then $a = b$.

Theorem B.4. *Every nonempty set of integers which is bounded below has the least element. Every nonempty set of integers which is bounded above has the greatest element.*

Proof. Let A be a nonempty set of integers which is bounded below. From the order completeness axiom it follow that A has the greatest lower bound (infimum) a in $\mathbb{R}$. We have to show that $a \in A$. By the definition of infimum, for each $k \in \mathbb{N}$ there is $x_k \in A$ such that $a \leq x_k < a + 1/(2k)$. So $|x_k - a| < 1/(2k)$ for all k. For any $k \in \mathbb{N}$,

$$|x_k - x_1| \leq |x_k - a| + |a - x_1| < 1/(2k) + 1/2 < 1.$$

Since x_k, x_1 are integers, we have $x_k = x_1$ for all k. Consequently, $|a - x_1| < 1/(2k)$ for all k. But then $a = x_1$: If not, we would have $k < 1/(2|a - x_1|)$ for all $k \in \mathbb{N}$, which contradicts the Archimedean property. A similar proof can be given for sets of integers bounded above. □

The order completeness of $\mathbb{R}$ is the deepest of all order properties, and the one that distinguishes $\mathbb{R}$ from $\mathbb{Q}$. It is the axiom that guarantees that the real line is 'connected' with no gaps. The set $\mathbb{Q}$ does not possess the order completeness property as can be seen from the following example.

Example B.5. Let A be the set consisting of all positive rational numbers x such that $x^2 < 2$. Then A is nonempty ($1 \in A$) and bounded above by 2 (verify). We consider A embedded in $\mathbb{R}$. Since A is nonempty and has an upper bound in

$\mathbb{R}$, it has a least upper bound a in $\mathbb{R}$. Exactly one of the relations $a^2 < 2$, $a^2 > 2$ and $a^2 = 2$ must hold. We will exclude the two inequalities by contradiction:

• Suppose that $a^2 < 2$. By the Archimedean property of $\mathbb{N}$ there exists $n \in \mathbb{N}$ such that $n > (2a+1)/(2-a^2)$ (observe that $2 - a^2 > 0$). Then

$$\left(a + \frac{1}{n}\right)^2 = a^2 + \frac{2a}{n} + \frac{1}{n^2} \le a^2 + \frac{2a}{n} + \frac{1}{n} = a^2 + \frac{2a+1}{n} < a^2 + (2 - a^2) = 2.$$

Hence $a + 1/n$ belongs to A, but is greater than the upper bound a, which is impossible.

• Suppose that $a^2 > 2$. Then $a^2 - 2 > 0$, and by the Archimedean property there exists $n \in \mathbb{N}$ such that $n > 2a/(a^2 - 2)$. This implies $a^2 - 2 > 2a/n$, and $a^2 - 2a/n > 2$. The number $a - 1/n$ is not an upper bound for S since a is the least of all upper bounds; so there exists $x \in S$ such that $a \ge x > a - 1/n$. We have

$$x^2 > \left(a - \frac{1}{n}\right)^2 = a^2 - \frac{2a}{n} + \frac{1}{n^2} > a^2 - \frac{2a}{n} > 2,$$

contrary to the definition of A.

• Thus $a^2 < 2$ and $a^2 > 2$ are eliminated. But this leaves $a^2 = 2$, and we know that no such rational number a exists. We have thus shown two things: That $\mathbb{Q}$ is not order complete, and that the equation $a^2 = 2$ has a solution in $\mathbb{R}$.

Theorem B.6 (The density of rationals in $\mathbb{R}$). *For every real number a and every $\varepsilon > 0$ there is a rational number r such that*

$$|a - r| < \varepsilon.$$

Proof. Let $a \in \mathbb{R}$ and let $\varepsilon > 0$. Because $\mathbb{N}$ is unbounded, there is $n \in \mathbb{N}$ with $n > 1/\varepsilon$; so $1/n < \varepsilon$. Keep this n fixed. Let A be the set of integers m for which $m \le na$. The set is nonempty; if not, the integers would have a lower bound na, and this contradicts the Archimedean property. The set A is bounded above by na, and so A has the greatest element p by Theorem B.4. We have

$$p \le na < p + 1$$

and

$$\frac{p}{n} \le a < \frac{p}{n} + \frac{1}{n}.$$

This implies that

$$\left|a - \frac{p}{n}\right| < \frac{1}{n} < \varepsilon.$$

The result then follows on taking $r = p/n$. $\qquad\square$

The order completeness axiom is the key for the existence of irrational numbers. As we have seen in Example B.5, the supremum a of the set $A = \{x \in \mathbb{Q} : x > 0 \text{ and } x^2 < 2\}$ satisfies $a^2 = 2$, while $a \notin \mathbb{Q}$. Similarly we can show that there exists a positive $b \in \mathbb{R}$ such that $b^3 = 2$, etc.

B.1.1 *Complex numbers*

Complex numbers are constructed from real numbers by adjoining a new element i for which $i^2 = -1$, and defining arithmetic operations on the pairs $z = x + iy$, where x and y are real. We do this by requiring the usual laws of arithmetic, that is, the commutative law, the associative law and the distributive law. We define the *complex conjugate*

$$\bar{z} = \overline{x + iy} = x - iy,$$

and the *absolute value*

$$|z| = \sqrt{z\bar{z}} = \sqrt{x^2 + y^2} = |\bar{z}|.$$

Every nonzero complex number z can be written in a polar form $z = |z|e^{i\theta}$, where $e^{i\theta} = \cos\theta + i\sin\theta$, and where θ is an argument of z. The set of all complex numbers is denoted by $\mathbb{C}$. We define the real and imaginary part of $z = x + iy \in \mathbb{C}$ by $\operatorname{Re} z = x$ and $\operatorname{Im} z = y$; observe that

$$\operatorname{Re} z = \tfrac{1}{2}(z + \bar{z}), \quad \operatorname{Im} z = -i\tfrac{1}{2}(z - \bar{z}).$$

We do not define inequalities between complex numbers, but only between their absolute values. We note that in the complex plane, $|z|$ is the distance of z from the origin, and $|z_1 - z_2|$ is the distance between the points z_1 and z_2. The most important inequalities in $\mathbb{C}$ are analogous to those for real numbers:

(i) $|z_1 z_2| = |z_1| |z_2|$: $|z_1 z_2|^2 = z_1 z_2 \overline{z_1 z_2} = z_1 \overline{z_1} z_2 \overline{z_2} = |z_1|^2 |z_2|^2$

(ii) $\pm \operatorname{Re} z \leq |z|$, $\pm \operatorname{Im} z \leq |z|$

(iii) $|z_1 + z_2| \leq |z_1| + |z_2|$ (the *triangle inequality*):

$$|z_1 + z_2|^2 = (z_1 + z_2)\overline{(z_1 + z_2)} = (z_1 + z_2)(\overline{z_1} + \overline{z_2}) = z_1 \overline{z_1} + z_1 \overline{z_2} + \overline{z_1} z_2 + z_2 \overline{z_2}$$
$$= |z_1|^2 + 2\operatorname{Re} z_1 \overline{z_2} + |z_2|^2 \leq |z_1|^2 + 2|z_1||z_2| + |z_2|^2 = (|z_1| + |z_2|)^2$$

(iv) $|z_1 - z_2| \geq |\,|z_1| - |z_2|\,|$

B.2 Real and complex sequences

We say that a sequence (x_n) of real numbers is *increasing* if $x_{n+1} \geq x_n$ for all $n \in \mathbb{N}$. We say that (x_n) is *strictly increasing* if $x_{n+1} > x_n$ for all $n \in$

$\mathbb{N}$. Symmetrically we define *decreasing* and *strictly decreasing* sequences. A sequence is called *monotonic* if it is either increasing or decreasing; it is *strictly monotonic* if it is either strictly increasing or strictly decreasing.

Theorem B.7. *Every sequence (x_n) of real numbers contains a monotonic subsequence.*

Proof. (a) Suppose that the given sequence does not have a greatest term. Then for each term x_n there is another term greater than x_n. In fact, there are infinitely many such terms, otherwise the sequence would have a greatest term. So we can find indices $1 = k_1 < k_2 < k_3 < \cdots$ such that $x_{k_{n+1}} > x_{k_n}$ for all n.

(b) Suppose that the given sequence has a subsequence without a greatest term. By part (a), this subsequence contains an increasing subsubsequence.

(c) Suppose (b) is not true. Then every subsequence of (x_n) contains a greatest term. Define k_n as follows: Let k_1 be the least integer such that x_{k_1} is a greatest term of $x_1, x_2, x_3, \ldots$ Given k_n, define k_{n+1} to be the least integer greater than k_n for which $x_{k_{n+1}}$ is a greatest term of $\{x_j : j > k_n\}$. From this choice it follows that $x_{k_{n+1}} \le x_{k_n}$.   □

A sequence (a_n) of real (or complex) numbers *converges to a* if, for each $\varepsilon > 0$, there is a natural number $N = N(\varepsilon)$ such that

$$|a_n - a| < \varepsilon \text{ for all } n > N.$$

This definition is due to Cauchy. We often write $a = \lim_{n \to \infty} a_n$ or $a_n \to a$. The limits obey a set of rules, the so called *algebra of limits* (Problem B.7).

Theorem B.8 (Monotonic sequence theorem). *Every monotonic bounded sequence of real numbers is convergent.*

Proof. Let (x_n) be an increasing sequence of real numbers which has an upper bound. The set $A = \{x_n : n \in \mathbb{N}\}$ has the least upper bound x in $\mathbb{R}$ by the order completeness axiom. We want to show that (x_n) converges to x. Let $\varepsilon > 0$ be given. By Problem B.2 there exists $N \in \mathbb{N}$ such that $x_N > x - \varepsilon$. So

$$x - \varepsilon < x_n \le x \text{ for all } n > N,$$

that is,

$$|x_n - x| < \varepsilon \text{ for all } n > N.$$

A similar argument can be used for a decreasing sequence.   □

One of the most important methods for determining whether a real (or complex) sequence converges was introduced by Bolzano and later independently by Cauchy, and can be used even when the value of the limit is not known.

Theorem B.9 (Bolzano-Cauchy theorem). *Let (x_n) be a sequence of real numbers satisfying the so called Cauchy condition:*

$$\text{For each } \varepsilon > 0 \text{ there is } N = N(\varepsilon) \in \mathbb{N} \text{ such that}$$
$$|x_m - x_n| < \varepsilon \text{ for all } m, n > N.$$

Then (x_n) converges.

Proof. Let (x_n) be a sequence that satisfies the Cauchy condition. We show that (x_n) is bounded. Take $\varepsilon = 1$. There is N such that $|x_n - x_N| < 1$ for all $n > N$. Therefore $|x_n| \le |x_N| + 1$ for all $n > N$, and

$$|x_n| \le \max\{|x_1|, |x_2|, \ldots, |x_{N-1}|, |x_N| + 1\} \text{ for all } n.$$

By Theorem B.7, (x_n) contains a monotonic subsequence (x_{k_n}). This sequence is bounded, and so it is convergent to some real number x by Theorem B.8. Let $\varepsilon > 0$ be given. By the Cauchy condition there is M such that $|x_m - x_p| < \frac{1}{2}\varepsilon$ if $m, p > M$. If $m > M$, then

$$|x_m - x| \le |x_m - x_{k_n}| + |x_{k_n} - x| < \frac{1}{2}\varepsilon + |x_{k_n} - x|$$

whenever $k_n > M$. Taking the limit as $n \to \infty$, we get $|x_m - x| \le \frac{1}{2}\varepsilon < \varepsilon$ for all $m > M$. $\qquad\square$

An extension of the Bolzano-Cauchy theorem to complex sequences is obtained from these easily proved principles:

(i) $z_n \to z \iff \operatorname{Re} z_n \to \operatorname{Re} z$ and $\operatorname{Im} z_n \to \operatorname{Im} z$;

(ii) (z_n) satisfies the Cauchy condition if and only if $(\operatorname{Re} z_n)$ and $(\operatorname{Im} z_n)$ satisfy the Cauchy condition.

The following theorem gives a very useful characterization of supremum and infimum which is often more convenient than the definition given earlier.

Theorem B.10. *Let A be a bounded nonempty set of real numbers. Then the following is true:*

(i) *A real number a is the supremum of A if and only if a is an upper bound for A and there is a sequence (x_n) in A such that $x_n \to a$.*

(ii) *A real number b is the infimum of A if and only if b is a lower bound for A and there is a sequence (y_n) in A such that $y_n \to b$.*

Proof.   Problem B.5.   □

Definition B.11. Let (a_n) be a real (or complex) sequence. The number a is a *cluster point* of (a_n) if for each $\varepsilon > 0$ there exists an index n such that $|a_n - a| < \varepsilon$.

Clearly, the limit of (a_n), if it exists, is a cluster point of the sequence. We can show that a sequence has a cluster point if and only if it contains a convergent subsequence.

B.3   Real and complex series

A *series* $\sum_{n=1}^{\infty} a_n$ is defined by its terms a_n which are real or complex numbers. With a series $\sum_{n=1}^{\infty} a_n$ we associate the sequence of partial sums defined by

$$s_n = a_1 + \cdots + a_n.$$

We say that the series $\sum_{n=1}^{\infty} a_n$ *converges* if the sequence (s_n) of partial sum converges. The limit of (s_n) is called the *sum* of $\sum_{n=1}^{\infty} a_n$. A series $\sum_{n=1}^{\infty} a_n$ is *absolutely convergent* if the series $\sum_{n=1}^{\infty} |a_n|$ is convergent.

Theorem B.12. *An absolutely convergent real or complex series is itself convergent.*

Proof.   Write $\sigma_n = |a_1| + \cdots + |a_n|$ and $s_n = a_1 + \cdots + a_n$. If $m > n$, then

$$|s_n - s_m| = |a_{n+1} + \cdots + a_m| \leq |a_{n+1}| + \cdots + |a_m| = \sigma_m - \sigma_n.$$

By asumption, (σ_n) is convergent, and so it satisfies the Cauchy condition. In view of the preceding inequality, (s_n) also satisfies the Cauchy condition, and so it is convergent by the Bolzano-Cauchy theorem. It may useful to observe that

$$\left| \sum_{n=1}^{\infty} a_n \right| \leq \sum_{n=1}^{\infty} |a_n|.$$

                                                                                 □

Theorem B.13. *A positive term series $\sum_{n=1}^{\infty} a_n$ converges if and only if the sequence of its partial sums is bounded.*

Proof. The sequence (s_n) is increasing and bounded, and so it converges by the monotonic sequence theorem. □

If $\sum_{n=1}^{\infty} a_n$ is a series and $\pi \colon \mathbb{N} \to \mathbb{N}$ a bijection (a one-to-one correspondence), then the series $\sum_{n=1}^{\infty} a_{\pi(n)}$ is called a *rearrangement* of $\sum_{n=1}^{\infty} a_n$.

Theorem B.14. *If $\sum_{n=1}^{\infty} a_n$ is an absolutely convergent real or complex series, then any rearrangement of $\sum_{n=1}^{\infty} a_n$ is an absolutely convergent series with the same sum.*

Proof. Write $s_n = a_1 + \cdots + a_n$, $\sigma_n = |a_1| + \cdots + |a_n|$, $\tau_n = a_{\pi(1)} + \cdots + a_{\pi(n)}$, and let s be the sum of $\sum_{n=1}^{\infty} a_n$. Given $\varepsilon > 0$, there is an index N such that

$$|\sigma_{N+p} - \sigma_N| = |a_{N+1}| + \cdots + |a_{N+p}| < \varepsilon \tag{B.1}$$

for every integer $p \geq 1$. Choose m so that $1, 2, \ldots, N$ are among $\pi(1), \pi(2), \ldots, \pi(m)$. For any $n > m$, in the expression $\tau_n - s_n$ the terms $a_1, \ldots, a_N$ cancel as they appear in τ_n as well as in s_n. The difference $\tau_n - s_n$ then is the sum of finitely many of the terms $\pm a_{N+1}, \pm a_{N+2}, \ldots$ According to (B.1), $|\tau_n - s_n| < \varepsilon$ whenever $n > m$. So $\tau_n = s_n + (\tau_n - s_n) \to s$. □

Sometimes we have to consider *double series* $\sum_{m,n=1}^{\infty} a_{m,n}$, which is indexed by the ordered pairs $(m, n) \in \mathbb{N} \times \mathbb{N}$. The set $\mathbb{N} \times \mathbb{N}$ is bijective with $\mathbb{N}$; if $\rho \colon \mathbb{N} \to \mathbb{N} \times \mathbb{N}$ is a bijection, the ordinary series $\sum_{j=1}^{\infty} a_{\rho(j)}$ is said to be a *rearrangement* of the double series $\sum_{m,n=1}^{\infty} a_{m,n}$. It is customary to write the rearranged series in a simplified form $\sum_{j=1}^{\infty} b_j$, where $b_j = a_{\rho(j)}$. A double series is said to be *absolutely convergent* if it admits a rearrangement into an ordinary series $\sum_{j=1}^{\infty} b_j$ which is absolutely convergent. The sum of an absolutely convergent double series is then defined by

$$\sum_{m,n=1}^{\infty} a_{m,n} = \sum_{j=1}^{\infty} b_j.$$

In view of the rearrangement theorem for absolutely convergent ordinary series, the sum is independent of any particular rearrangement.

Theorem B.15. *Let $\sum_{m,n=1}^{\infty} a_{m,n}$ be an absolutely convergent real or complex double series. Then*

$$\sum_{m=1}^{\infty} \left(\sum_{n=1}^{\infty} a_{m,n} \right) = \sum_{m,n=1}^{\infty} a_{m,n} = \sum_{n=1}^{\infty} \left(\sum_{m=1}^{\infty} a_{m,n} \right).$$

Proof.   The theorem says that in the scheme

$$a_{1,1} + a_{1,2} + a_{1,3} + \cdots$$
$$+ a_{2,1} + a_{2,2} + a_{2,3} + \cdots$$
$$+ a_{3,1} + a_{3,2} + a_{3,3} + \cdots$$
$$+ \cdots$$
$$+ a_{m,1} + a_{m,2} + a_{m,3} + \cdots$$

we can sum first by the rows and then add up the row sums or we can first sum by the columns and then add up the column sums.

Each row series $\sum_{n=1}^{\infty} a_{m,n}$ is absolutely convergent; let A_m be its sum. Arrange the double series into an ordinary series $\sum_{j=1}^{\infty} b_j$. Given $\varepsilon > 0$ there is an index N such that $\sum_{j=N+1}^{\infty} |b_j| < \varepsilon$. For any $j \in \{1, \ldots, N\}$, the term b_j appears in one of the series $\sum_{n=1}^{\infty} a_{m_j,n}$; let $M = \max(m_1, \ldots, m_N, N)$. For any $m > M$, $A_1 + \cdots + A_m - (b_1 + \cdots + b_N)$ is the sum of a series whose terms are some of the b_j for $j > N$ as $b_1, \ldots, b_N$ cancel out in this sum. So, for $m > M$, $\sum_{j=N+1}^{\infty} |b_j| < \varepsilon$, and

$$\left| A_1 + \cdots + A_m - \sum_{j=1}^{\infty} b_j \right| \leq \left| A_1 + \cdots + A_m - (b_1 + \cdots + b_N) \right| + \left| \sum_{j=N+1}^{\infty} b_j \right|$$

$$\leq \sum_{j=N+1}^{\infty} |b_j| + \sum_{j=N+1}^{\infty} |b_j| < 2\varepsilon.$$

This proves that

$$\sum_{m=1}^{\infty} \left(\sum_{n=1}^{\infty} a_{m,n} \right) = \sum_{m=1}^{\infty} A_m = \sum_{j=1}^{\infty} b_j = \sum_{m,n=1}^{\infty} a_{m,n}.$$

The other part of the theorem is proved similarly.   □

B.4   Closed bounded intervals

Let a, b be real numbers, $a < b$. The closed bounded interval $[a, b]$ in $\mathbb{R}$ is defined by

$$[a, b] = \{x \in \mathbb{R} : a \leq x \leq b\}.$$

B.4.1 *Compactness property*

Closed bounded intervals are important in real analysis. In particular, continuous functions 'behave well' on closed bounded intervals, while they may not on other types of intervals. (The precise meaning of 'well behaved' is explained below.) The key to this is the following *compactness* property of bounded intervals.

Theorem B.16. *Every sequence contained in a closed bounded interval* $[a, b]$ *has a subsequence convergent to a point contained in* $[a, b]$.

Proof. Let (x_n) be a sequence contained in a closed bounded interval $[a, b]$. Then the sequence is bounded, and contains a monotonic subsequence (x_{k_n}) by Theorem B.7. This subsequence is also bounded, and therefore convergent by Theorem B.8. Let x be the limit of (x_{k_n}). Since $a \leq x_{k_n} \leq b$ for all $n \in \mathbb{N}$, we have $a \leq x \leq b$ in view of Problem B.7 (iv). □

Let J be an interval in $\mathbb{R}$ (not necessarily closed or bounded). Recall that a function $f \colon J \to \mathbb{C}$ is continuous at a point $c \in J$ if, for each $\varepsilon > 0$, there is $\delta > 0$ such that

$$x \in J \text{ and } |x - c| < \delta \implies |f(x) - f(c)| < \varepsilon.$$

It can be shown that this definition, due to Cauchy, is equivalent to the following criterion using sequences, which was introduced by Heine:

Theorem B.17. *A real or complex valued function* f *is continuous at a point* c *if and only if the following condition is satisfied:*

$$x_n \in J \text{ and } x_n \to c \implies f(x_n) \to f(c).$$

We say that a function f is continuous on the interval J if it is continuous at each point of J. A function $f \colon J \to \mathbb{R}$ is *bounded* on J if there is a constant $K > 0$ such that $|f(x)| \leq K$ for all $x \in J$. This is equivalent to saying that the set $\{f(x) : x \in J\}$ is bounded in $\mathbb{C}$.

Theorem B.18. *Every real or complex valued function* f *continuous on a closed bounded interval* $[a, b]$ *is bounded on* $[a, b]$.

Proof. For a proof by contradiction assume that there is a function f which is continuous, but not bounded, on the interval $[a, b]$. For each $n \in \mathbb{N}$ there is a point $x_n \in [a, b]$ such that $|f(x_n)| > n$. (This is obtained by negation of the definition of boundedness.) By Theorem B.16, (x_n) has a subsequence (x_{k_n}) convergent to a point $x \in [a, b]$. By the Heine criterion of continuity (Theorem B.17), $f(x_{k_n}) \to f(x)$. This, however, contradicts the inequalities $|f(x_{k_n})| > k_n$, as $k_n \to \infty$. □

Theorem B.19. *Every real valued function continuous on a closed bounded interval $[a, b]$ attains its infimum and supremum on $[a, b]$.*

Proof. Let $f \colon [a, b] \to \mathbb{R}$ be a continuous function, and let $A = \{f(x) : x \in [a, b]\}$. Then A is nonempty (as $f(a) \in A$), and bounded in view of the foregoing theorem. By the order completeness axiom, the numbers $m = \inf A$ and $M = \sup A$ exist. By Theorem B.10, there are sequences $(u_n) \in A$ and $(v_n) \in A$ such that $u_n \to m$ and $v_n \to M$. By the definition of A, $u_n = f(x_n)$ for some $x_n \in [a, b]$ and $v_n = f(y_n)$ for some $y_n \in [a, b]$. By Theorem B.16, the sequences (x_n) and (y_n) contain subsequences (x_{k_n}) and (y_{p_n}) which converge to points s and t in $[a, b]$, respectively.

By the sequential criterion of continuity, $f(x_{k_n}) \to f(s)$ and $f(y_{p_n}) \to f(t)$. From the uniqueness of limits we get $f(s) = m$ and $f(t) = M$. This means that f attains its minimum on $[a, b]$ at the point s, and its maximum at the point t. □

Theorem B.20 (Intermediate value property). *Let $f \colon [a, b] \to \mathbb{R}$ be a continuous function on a closed bounded interval $[a, b]$. If w is between $f(a)$ and $f(b)$, then there is $c \in [a, b]$ such that $w = f(c)$.*

Proof. Without loss of generality we may assume that $f(a) < w < f(b)$. Define $A = \{x \in [a, b] : f(x) < w\}$. Then A is nonempty since $f(a) < w$, and bounded above as $A \subset [a, b]$. Then $c = \sup A$ exists and $c \le b$. Choose (x_n) in A with $x_n \to c$ (Theorem B.10). We have $f(x_n) < w$ by the definition of A. By the continuity of f, $f(c) = \lim_{n \to \infty} f(x_n) \le w$. Hence $c < b$ as $f(b) > w$. Choose any sequence (y_n) with $c < y_n < b$ and $y_n \to c$. Then $f(y_n) \ge w$ as $y_n \notin A$. By Heine's criterion of continuity,

$$f(c) = \lim_{n \to \infty} f(y_n) \ge w.$$

Therefore $f(c) = w$. □

Theorem B.21. *A continuous real valued function maps an arbitrary interval onto an interval, a closed bounded interval onto a closed bounded interval.*

Proof. Follows from Theorems B.20 and B.19. □

B.4.2   *Full covers and Cousin's lemma*

Many theorems of real analysis can be given simple proofs using properties of covers consisting of closed bounded intervals.

Definition B.22. Let $[a, b]$ be a closed bounded interval and $\delta \colon [a, b] \to \mathbb{R}$ a strictly positive function. A family $\mathcal{C}$ of closed subintervals of $[a, b]$ is a *full δ-cover of $[a, b]$* if, for any subinterval $[u, v]$ of $[a, b]$,

$$x \in [u, v] \text{ and } v - u < \delta(x) \implies [u, v] \in \mathcal{C}.$$

A *full cover* of $[a, b]$ is a full δ-cover of $[a, b]$ for some positive function $\delta \colon [a, b] \to \mathbb{R}$.

The following lemma is a version of the result obtained in 1895 by Pierre Cousin. This very useful result is indispensable in the theory of the Kurzweil–Henstock integral (see [6]). The method used in the following proof is called *successive interval bisection*. At each step one half of a given interval is selected according to a preset condition. Successive bisection is a very useful technique often used in analysis.

Lemma B.23 (Cousin's lemma). *Let $\mathcal{C}$ be a full cover for $[a, b]$. Then each closed subinterval of $[a, b]$ has a partition whose subintervals lie in $\mathcal{C}$.*

Proof. Let $\mathcal{C}$ be a full δ-cover of $[a, b]$ for some positive function $\delta \colon [a, b] \to \mathbb{R}$. For a proof by contradiction assume that $\mathcal{C}$ contains no finite partition of $[u, v] \subset [a, b]$. Let w be the midpoint of $[u, v]$. Then $\mathcal{C}$ contains no partition of at least one of the intervals $[u, w]$, $[w, v]$; call this interval $[u_1, v_1]$. If the intervals $[u_1, v_1], \ldots, [u_n, v_n]$ have been selected so that $\mathcal{C}$ contains no partition of any of them, let w_n be the midpoint of $[u_n, v_n]$; then $\mathcal{C}$ contains no partition of at least one of the intervals $[u_n, w_n]$, $[w_n, v_n]$; denote this interval by $[u_{n+1}, v_{n+1}]$. Then

$$[u_n, v_n] \supset [u_{n+1}, v_{n+1}], \quad v_n - u_n = (\tfrac{1}{2})^n (v - u), \quad n \in \mathbb{N}.$$

By the monotonic sequence theorem, both (u_n) and (v_n) converge to the same limit: $u_n \nearrow x$ and $v_n \searrow x$. Hence there exists a point x common to all intervals $[u_n, v_n]$. Since $v_n - u_n \to 0$, there exists $k \in \mathbb{N}$ such that $v_k - u_k < \delta(x)$. By the definition of a full δ-cover, $[u_k, v_k] \in \mathcal{C}$, which is the desired contradiction. $\qquad \square$

We use Cousin's lemma to deduce the following deep theorem of real analysis.

Theorem B.24 (Heine–Borel theorem). *Every family $\{J_\alpha : \alpha \in \Delta\}$ of open intervals whose union contains the closed bounded interval $[a, b]$ has a finite subfamily whose union contains $[a, b]$.*

Proof.   For each $x \in [a, b]$ there exists $\alpha \in \Delta$ and $\delta(x) > 0$ such that $(x - \delta(x), x + \delta(x)) \subset J_\alpha$. Let $\mathcal{C}$ be the family of intervals $[u, v] \subset [a, b]$ defined as follows:

$$[u, v] \in \mathcal{C} \iff \text{ there exists } \alpha \in \Delta \text{ such that } [u, v] \subset J_\alpha.$$

Suppose that $x \in [a, b]$, and that $[u, v] \subset [a, b]$ satisfies $x \in [u, v]$ and $v - u < \delta(x)$. Then $[u, v] \subset (x - \delta(x), x + \delta(x)) \subset J_\alpha$ for some $\alpha \in \Delta$. Hence $\mathcal{C}$ is a full δ-cover of $[a, b]$, and by Cousin's lemma there exists a partition $\{x_0, x_1, \ldots, x_n\}$ of $[a, b]$ such that each subinterval $[x_{k-1}, x_k]$ belongs to $\mathcal{C}$. There exists indices $\alpha_1, \ldots, \alpha_n$ such that $[x_{k-1}, x_k] \subset J_{\alpha_k}$, $k = 1, \ldots, n$, and $\{J_{\alpha_1}, \ldots, J_{\alpha_n}\}$ is a finite cover of $[a, b]$. $\square$

As another application of Cousin's lemma we give a proof of the following result.

Theorem B.25. *Let $F\colon (a, b) \to \mathbb{R}$ be continuous on (a, b), and let $F'(x) \geq 0$ for all $x \in (a, b) \setminus D$, where D is a countable subset of (a, b). Then F is increasing in (a, b).*

Proof.   Let $\{s_n\}$ be an enumeration of D, and let $\varepsilon > 0$ be given. We define families $\mathcal{C}_m$, $m = 0, 1, 2, \ldots$, as follows. An interval $[u, v] \subset [a, b]$ belongs to $\mathcal{C}_0$ if

$$F(v) - F(u) \geq -\varepsilon(v - u),$$

and to $\mathcal{C}_m$, $m = 1, 2, \ldots$, if

$$s_m \in [u, v] \text{ and } |F(v) - F(u)| < (\tfrac{1}{2})^m \varepsilon.$$

From the definition of the derivative $F'(x)$ for $x \in (a, b) \setminus D$, and from the continuity of F at each s_m we deduce that $\mathcal{C} = \bigcup_{m=0}^{\infty} \mathcal{C}_m$ is a full δ-cover for some positive function δ. By Cousin's lemma, every subinterval $[c, d]$ of $[a, b]$ has a partition $\{x_0, x_1, \ldots, x_m\}$ such that each subinterval $J_k = [x_{k-1}, x_k]$ belongs to $\mathcal{C}$. Then

$$F(d) - F(c) \geq \sum_{J_k \in \mathcal{C}_0} (F(x_k) - F(x_{k-1})) - \sum_{m=1}^{\infty} \sum_{J_k \in \mathcal{C}_m} |F(x_k) - F(x_{k-1})|$$

$$\geq -\varepsilon \sum_{J_k \in \mathcal{C}_0} (x_k - x_{k-1}) - \varepsilon \sum_{m=1}^{\infty} 2(\tfrac{1}{2})^m$$

$$\geq -\varepsilon(d - c + 2)$$

(as each $\mathcal{C}_m$, $m \geq 1$, contains at most two subintervals of the partition). Since $\varepsilon > 0$ was arbitrary, we have $F(d) - F(c) \geq 0$. Hence F is increasing on (a, b). $\square$

Corollary B.26. *Let* $F: (a,b) \to \mathbb{C}$ *be continuous on* (a,b), *and let* $F'(x) = 0$ *for all* $x \in (a,b) \setminus D$, *where* D *is a countable subset of* (a,b). *Then* F *is constant on* (a,b).

Proof. Write $F = F_1 + iF_2$, where F_1, F_2 are real valued. Then $F'_k = 0$ ($k = 1, 2$), and by the preceding theorem, F_k and $-F_k$ are increasing for $k = 1, 2$. Thus F_k is constant for $k = 1, 2$, and so is F. $\qquad \square$

The preceding theorem and corollary have important generalizations, crucial for Lebesgue integration of derivatives. We say that a function F is *absolutely continuous* on an interval I if, for each $\varepsilon > 0$, there is $\delta > 0$ such that for any finite pairwise disjoint family of subintervals $I_k = (a_k, b_k)$ of I for which $\sum_{k=1}^{n} (b_k - a_k) = \sum_{k=1}^{n} |I_k| < \delta$ we have $\sum_{k=1}^{n} |F(b_k) - F(a_k)| < \varepsilon$. A set $A \subset \mathbb{R}$ is called a *null set* if for every $\varepsilon > 0$ there exists a sequence of open intervals I_n such that $A \subset \bigcup_{n=1}^{\infty} I_n$ and $\sum_{n=1}^{\infty} |I_n| < \varepsilon$. (This is equivalent to A being a set of Lebesgue measure zero.) The proof of the following theorem is adapted from Botsko [9].

Theorem B.27. *Let* $F: (a,b) \to \mathbb{R}$ *be an absolutely continuous function on* (a,b) *such that* $F'(x) \geq 0$ *for all* $x \in (a,b) \setminus D$, *where* D *is a null set in* (a,b). *Then* F *is increasing on* (a,b).

Proof. Let $[a_0, b_0]$ be a closed bounded subinterval of (a,b), and let $\varepsilon > 0$ be given. Since F is absolutely continuous, there exists $\delta > 0$ such that for any finite family of nonoverlapping subintervals $[a_k, b_k]$ of $[a_0, b_0]$ with the total length less that δ we have $\sum_k |F(b_k) - F(a_k)| < \varepsilon$. Since D is a null set, there exists a sequence of open intervals J_n such that $D \subset \bigcup_{n=1}^{\infty} J_n$, and $\sum_{n=1}^{\infty} |J_n| < \delta$. Let $\mathcal{B}$ be the family of all intervals $[u, v] \subset [a_0, b_0]$ such that $F(v) - F(u) > -\varepsilon(v - u)$, and $\mathcal{E}$ the family of all intervals $[u, v] \subset [a_0, b_0]$ such that $[u, v] \subset J_n$ for some n. We can prove that $\mathcal{C} = \mathcal{B} \cup \mathcal{E}$ is a full cover of $[a_0, b_0]$ for some positive function η. By Cousin's lemma there exists a partition $\{x_0, x_1, \ldots, x_m\}$ of $[a_0, b_0]$ such that each $I_k = [x_{k-1}, x_k]$ belongs to $\mathcal{C}$. Then

$$F(b_0) - F(a_0) \geq \sum_{J_k \in \mathcal{B}} (F(x_k) - F(x_{k-1})) - \sum_{I_s \subset J_n} |F(x_s) - F(x_{s-1})|$$

$$> -\varepsilon \sum_{\mathcal{B}} (x_k - x_{k-1}) - \varepsilon \geq -\varepsilon(b_0 - a_0 + 1)$$

(as $\sum_{\mathcal{E}} |I_s| \leq \sum_n |J_n| < \delta$). Since $\varepsilon > 0$ was arbitrary, $F(b_0) \geq F(a_0)$. The preceding argument applies to any closed interval $[a_0, x] \subset [a_0, b_0]$, which implies $F(x) \geq F(a_0)$ for all $x \in [a_0, b_0]$, that is, F is increasing on $[a_0, b_0]$.

Choosing a sequence of closed bounded intervals such that $[a_n, b_n] \nearrow (a, b)$, we conclude that F is increasing on (a, b). $\qquad\qquad\qquad\qquad\qquad\qquad\square$

Corollary B.28. *Let $F \colon (a, b) \to \mathbb{C}$ be an absolutely continuous function on (a, b) such that $F'(x) = 0$ for all $x \in (a, b) \setminus D$, where D is a null set in (a, b). Then F is constant on (a, b).*

B.5 ⋆ Covers and Lebesgue's differentiation theorem ⋆

In this section we assume basic concepts of topology on the real line, continuity of functions, the definition and properties of Borel subsets of $\mathbb{R}$, and the basic properties of the Lebesgue measure and Lebesgue measurable functions on $\mathbb{R}$.

The following result is one of the most celebrated theorems in theory of functions of a real variable. It was proved by Lebesgue in the last chapter of his book on integration [33] in 1904 for continuous functions, and was later extended to arbitrary monotonic functions. It is a stunning and unexpected result.

Theorem B.29. *A real valued monotonic function f on $[a, b]$ possesses a derivative f' almost everywhere in $[a, b]$.*

The proof given here is adapted from Hagood [21], and is based on a special type of interval covers.

Definition B.30. *Let E be a subset of $\mathbb{R}$. A family $\mathcal{A}$ of closed intervals is a right adapted interval cover for E if for each $x \in E$ there exists an interval $[L(x), R(x)]$ in $\mathcal{A}$ such that $x \in (L(x), R(x))$ and*

$$s \in [L(x), x] \implies [s, R(x)] \in \mathcal{A}.$$

A *left adapted interval cover* is defined similarly; an *adapted interval cover* is either right or left adapted interval cover.

An adapted interval cover satifies an analogue of Cousin's lemma, in which intervals with nonintersecting interiors are called *nonoverlapping*.

Lemma B.31. *Let $\mathcal{A}$ be an adapted interval cover of a compact set $K \subset \mathbb{R}$. Then there exists a finite cover of K consisting of nonoverlapping intervals in $\mathcal{A}$.*

Proof. Suppose that $\mathcal{A}$ is right adapted. Let $a = \min K$, $b = \max K$ and let A be the set of all $t \in [a, b]$ such that $[a, t] \cap K$ has a finite cover consisting of nonoverlapping intervals in $\mathcal{A}$. Then A is nonempty as $a \in A$. Let $\beta = \sup A$. We show that $\beta \in K$. For a proof by contradiction assume that $\beta \in [a, b] \setminus K$. Let (c, d) be the largest interval satisfying $\beta \in (c, d) \subset [a, b] \setminus K$; then $c, d \in K$. Further, $c \in A$ and there exists a finite cover $\mathcal{B}$ of $[a, c]$ consisting of nonoverlapping intervals in $\mathcal{A}$. Let $\mathcal{B}_1$ be the family obtained from $\mathcal{B}$ by deleting all intervals lying to the right of c, and adding the interval $[d, R(d)]$, which belongs to $\mathcal{A}$ by definition. Then $\mathcal{B}_1$ is a finite cover of $[a,d] \cap K$ consisting of intervals in $\mathcal{A}$ contrary to the definition of β. Hence $\beta \in K$.

Let $t \in (L(\beta), \beta] \cap A$ and choose any finite cover $\mathcal{C}$ of $[a,t] \cap K$ by nonoverlapping intervals in $\mathcal{A}$. If $[r,s]$ is the right-most interval of $\mathcal{C}$ that contains t, then either $s \geq b$ in which case $b \in A$ as required, or $s \leq \beta$ and $\mathcal{C}$ can be modified to include $[s, R(\beta)]$. Then $\min\{R(\beta), b\} \in A$, which is impossible unless $b = \beta \in A$.

Let $f: [a,b] \rightarrow \mathbb{R}$. We define four *Dini derivates* $D^+ f$, $D_+ f$, $D^- f$, $D_- f$, called upper right, lower right, upper left, lower left Dini derivates, respectively, by

$$D^+ f(x) = \inf_{\varepsilon > 0} \ \sup_{0 < h < \varepsilon} \frac{f(x+h) - f(x)}{h},$$

$$D_+ f(x) = \sup_{\varepsilon > 0} \ \inf_{0 < h < \varepsilon} \frac{f(x+h) - f(x)}{h},$$

$$D^- f(x) = \inf_{\eta < 0} \ \sup_{\eta < h < 0} \frac{f(x+h) - f(x)}{h},$$

$$D_- f(x) = \sup_{\eta < 0} \ \inf_{\eta < h < 0} \frac{f(x+h) - f(x)}{h},$$

where we admit infinite values of sup and inf. The right derivative of f at $x \in [a,b)$ exists if and only if $D^+ f(x)$ and $D_+ f(x)$ are both finite and equal. A similar statement holds for the left derivative at $x \in (a,b]$.

We now prove the following growth lemma for the Dini derivates of a monotonic function.

Lemma B.32. *Let $f: [a,b] \rightarrow \mathbb{R}$ be strictly increasing, and let C be the set of all points of continuity of f. Let E be a Borel subset of C, p, q real numbers, let Df stand for any of the four Dini derivates of f, and m for the Lebesgue measure on $\mathbb{R}$. Then:*

(i) *If $Df(x) > q$ for all $x \in E$, then $m(f(E)) \geq q \, m(E)$.*

(ii) *If $Df(x) < p$ for all $x \in E$, then $m(f(E)) \leq p \, m(E)$.*

Proof. The points of discontinuity of a monotonic function form a countable set; hence C is a Borel set. Let $g: C \rightarrow f(C)$ be the domain and codomain restriction of f. Since f is strictly increasing, g is bijective, and both g and the inverse g^{-1} are continuous relative to the subspace topologies of C and $f(C)$, and therefore Borel measurable.

First we give the proof under the assumption that $D = D^+$.

(i) Let E be a Borel subset of C on which $D^+ f > q$. Then $f(E) = (g^{-1})^{-1}(E)$ is a Borel set (the Borel measurability of g^{-1}). Let $\varepsilon > 0$. Then there exists a compact set $K \subset E$ and an open set $U \supset f(E)$ such that

$$m(E \setminus K) < \varepsilon, \qquad m(U \setminus f(E)) < \varepsilon.$$

We construct a right adapted interval cover $\mathcal{A}$ for K as follows. For each $x \in K$ there exists an open interval $I \subset (a, b)$ containing x such that $f(I) \subset U$ (the continuity of f at x). Since $D^+ f(x) > q$, there exists a real number $R(x) \in I$ such that $x < R(x)$ and

$$f(R(x)) - f(x) > q(R(x) - x).$$

By the continuity of f at x there exists $L(x) \in I$ such that $L(x) < x$ and

$$f(R(x)) - f(s) > q(R(x) - s)$$

whenever $L(x) \le s \le x$. Let

$$\mathcal{A} = \{[s, R(x)] : x \in K, \ L(x) \le s \le x\}.$$

Then $\mathcal{A}$ is a right adapted interval cover of K, and by Lemma B.31 there exists a finite cover of K consisting of nonoverlapping intervals $[c_i, d_i]$ $(i = 1, \ldots, n)$ in $\mathcal{A}$. For each $i \in \{1, \ldots, n\}$ there is an associated point $x_i \in K$ such that $L(x_i) \le c_i \le x_i < d_i = R(x_i)$. The intervals $[f(c_i), f(d_i)]$ are also nonoverlapping and lie in U. Then

$$m(f(E)) > m(U) - \varepsilon \ge m\left(\bigcup_{i=1}^{n} [f(c_i), f(d_i)] \right) - \varepsilon = \sum_{i=1}^{n} (f(d_i) - f(c_i)) - \varepsilon$$

$$\ge \sum_{i=1}^{n} q(d_i - c_i) - \varepsilon \ge q \, m(K) - \varepsilon > q \, m(E) - \varepsilon(1 + q).$$

Since ε was arbitrary, $m(f(E)) \ge q \, m(E)$.

(ii) Let $\varepsilon > 0$ and choose a compact set $K \subset f(E)$ and an open set $U \supset E$ such that

$$m(f(E) \setminus K) < \varepsilon, \qquad m(U \setminus f(E)) < \varepsilon.$$

For some finite or countable set of points $\alpha_i, \beta_i \in E$,

$$K = [f(\alpha_0), f(\beta_0)] \setminus \bigcup_i (f(\alpha_i), f(\beta_i));$$

hence

$$f^{-1}(K) = [\alpha_0, \beta_0] \setminus \bigcup_i (\alpha_i, \beta_i),$$

which is a closed subset of E. This enables us to apply the technique from the proof of part (i) to $f^{-1}(K)$ and U to show that $m(f(E)) \le p \, m(E)$.

The proof for $D = D^-$ is similar. The implications for $D_+ f$ and $D_- f$ can be then deduced from the preceding results. $\qquad\square$

Proof of Theorem B.29. If f is strictly increasing, then $-\infty < Df \le \infty$ for any Dini derivate D. Let $A = \{x \in C : Df(x) = \infty\}$. Then A is a Borel set, and for any $q > 0$ we have $Df > q$ on A, so that

$$f(b) - f(a) \geq m(f(A)) \geq q \, m(A).$$

Thus $m(A) = 0$, so that the Dini derivates of f are finite a.e. The sets of the form

$$B = \{x \in C : D_+ f(x) < p < q < D^- f(x)\}$$

satisfy $q \, m(B) \leq m(f(B)) \leq p \, m(B)$, so that $m(B) = 0$. That is, $D^- f \leq D_+ f$ a.e. Similarly, $D^+ f \leq D_- f$ a.e. Hence the derivative of f exists a.e. in $[a, b]$ provided f is strictly increasing.

If f is merely increasing, we set $g(x) = f(x) + \alpha x$ for some $\alpha > 0$. Then g is strictly increasing and $Dg(x) = Df(x) + \alpha$ for all $x \in C$ and any Dini derivate D. Then apply the preceding result. If f is decreasing, consider $-f$. $\qquad\square$

The statement of Lebesgue's differentiation theorem given in terms of null sets does not involve any measure theory: *A real valued monotonic function f on $[a, b]$ possesses a derivative f' except on a null subset of $[a, b]$.* (Recall that $A \subset \mathbb{R}$ is a null set if for every $\varepsilon > 0$ there exists a sequence of open intervals I_n such that $A \subset \bigcup_{n=1}^{\infty} I_n$ and $\sum_{n=1}^{\infty} |I_n| < \varepsilon$.)

A proof that does not use any measure theory was given by F. Riesz and is based on his 'rising sun lemma'; see Riesz and Sz.-Nagy [40]. An elementary (but rather long) proof was given recently by Botsko [10]. A standard measure theoretical proof of the theorem uses the so called Vitali covering theorem.

B.6 Problems for Appendix B

1. Use the order axioms to prove the following:

(i) $0 \leq x$ and $0 \leq y \implies 0 \leq x + y$

(ii) $0 < x \leq y \implies 1/y \leq 1/x$

2. Prove the following characterization of supremum. A number $\alpha \in \mathbb{R}$ is the supremum of a nonempty set $A \subset \mathbb{R}$ if and only if the following two conditions are satisfied:

(i) $x \leq \alpha$ for all $x \in A$;

(ii) if $\varepsilon > 0$, then there exists $x \in A$ such that $\alpha - \varepsilon < x \leq \alpha$.

State and prove an analogous characterization of infimum.

3. Assume that A and B are nonempty bounded sets of real numbers. Define the sets $A + B = \{a + b : a \in A, b \in B\}$, $AB = \{ab : a \in A, b \in B\}$, $-A = \{-1\}A$, $A^{-1} = \{a^{-1} : a \in A\}$ provided $0 \notin A$. Prove the following properties of supremum and infimum:

(i) $A \subset B \implies \sup A \leq \sup B$, $\quad \inf A \geq \inf B$.

(ii) $\sup(A \cup B) = \max\{\sup A, \sup B\}$.

(iii) $\inf(A \cup B) = \min\{\inf A, \inf B\}$.

(iv) $\sup(A + B) = \sup A + \sup B$, $\inf(A + B) = \inf A + \inf B$.

(v) If $\inf A \geq 0$ and $\inf B \geq 0$, then $\sup AB = \sup A \cdot \sup B$ and $\inf AB = \inf A \cdot \inf B$.

(vi) $\sup(-A) = -\inf A$, $\inf(-A) = -\sup A$.

(vii) If $\inf A > 0$, then $\sup A^{-1} = (\inf A)^{-1}$, $\inf A^{-1} = (\sup A)^{-1}$.

4. Find $\sup A$, if it exists, and decide if it is $\max A$:

(i) $A = \{x \in \mathbb{R} : 0 \leq x < 1\}$ (ii) $A = \{x \in \mathbb{R} : 0 \leq x \leq 1\}$

(iii) $A = \{x \in \mathbb{R} : x^2 < 5\}$ (iv) $A = \{x \in \mathbb{Q} : x^2 < 5\}$

5. Prove Theorem B.10 using Problem B.2.

6. Write $x \vee y = \max(x, y)$ and $x \wedge y = \min(x, y)$. Show that

$$x \vee y = \tfrac{1}{2}(x + y + |x - y|), \quad x \wedge y = \tfrac{1}{2}(x + y - |x - y|).$$

7. (The algebra of limits.) Let (x_n) and (y_n) be convergent sequences of real numbers. Prove that:

(i) $\lim\limits_{n \to \infty} (x_n + y_n) = \lim\limits_{n \to \infty} x_n + \lim\limits_{n \to \infty} y_n$.

(ii) $\lim\limits_{n \to \infty} (x_n y_n) = \left(\lim\limits_{n \to \infty} x_n \right) \left(\lim\limits_{n \to \infty} y_n \right)$.

(iii) If $\lim\limits_{n \to \infty} x_n \neq 0$, then $x_n \neq 0$ from a certain index on, and
$\lim\limits_{n \to \infty} x_n^{-1} = \left(\lim\limits_{n \to \infty} x_n \right)^{-1}$.

(iv) If $\lim\limits_{n \to \infty} x_n$ exists and, from a certain index on, $x_n \neq 0$ and x_n^{-1} are bounded, then $\lim\limits_{n \to \infty} x_n \neq 0$ and $\lim\limits_{n \to \infty} x_n^{-1} = \left(\lim\limits_{n \to \infty} x_n \right)^{-1}$.

(v) If $x_n \leq y_n$ for all $n \in \mathbb{N}$, then $\lim\limits_{n \to \infty} x_n \leq \lim\limits_{n \to \infty} y_n$.

8. Let (x_n) be a real sequence. Show that the limit

$$\lim\limits_{n \to \infty} \sup\limits_{k \geq n} x_k \qquad (\text{respectively } \lim\limits_{n \to \infty} \inf\limits_{k \geq n} x_k)$$

exists if (x_n) is bounded above (respectively bounded below). These limits are usually denoted by $\lim \sup_{n \to \infty} x_n$ (upper limit) and $\lim \inf_{n \to \infty} x_n$ (lower limit), respectively.
Show that $\lim \sup$ is *subadditive* and $\lim \inf$ is *superadditive*, that is

$$\lim\sup\limits_{n \to \infty}(x_n + y_n) \leq \lim\sup\limits_{n \to \infty} x_n + \lim\sup\limits_{n \to \infty} y_n,$$

$$\lim\inf\limits_{n \to \infty}(x_n + y_n) \geq \lim\inf\limits_{n \to \infty} x_n + \lim\inf\limits_{n \to \infty} y_n.$$

If (x_n), (y_n) are nonnegative sequences, show that lim sup is *submultiplicative* and lim inf *supermultiplicative*. (Give definitions!)

9. Let (x_n) be a bounded real sequence. Prove the following facts:

(i) (x_n) has at least one cluster point.

(ii) $\inf_{n \in \mathbb{N}} x_n \leq \liminf_{n \to \infty} x_n \leq \limsup_{n \to \infty} x_n \leq \sup_{n \in \mathbb{N}} x_n$.

(iii) $\limsup_{n \to \infty} x_n$ is the greatest cluster point of (x_n).

(iv) $\liminf_{n \to \infty} x_n$ is the least cluster point of (x_n).

10. Show that a real sequence (x_n) is convergent if and only if it is bounded and $\limsup_{n \to \infty} x_n \leq \liminf_{n \to \infty} x_n$.

11. Let $f : [a, b] \to \mathbb{R}$ be continuous on $[a, b]$ and let $f(c) > 0$ for some point $c \in (a, b)$. Prove that there exists an interval $(u, v) \subset [a, b]$ containing c such that $f(t) > 0$ for all $t \in (u, v)$.

12. Use the intermediate value property to prove that every real valued function continuous and injective on an interval is strictly monotonic.

13. (Uniform continuity.) Let $f : [a, b] \to \mathbb{R}$ be continuous on $[a, b]$ and let $\varepsilon > 0$. Define $\mathcal{C}$ to be the family of all intervals $[c, d] \subset [a, b]$ such that

$$[u, v] \subset [c, d] \implies |f(v) - f(u)| < \varepsilon.$$

Prove that $\mathcal{C}$ is a full cover of $[a, b]$ for some positive function η on $[a, b]$. Use this to conclude that for each $\varepsilon > 0$ there exists $\delta > 0$ such that for each $x \in [a, b]$,

$$t \in [a, b] \text{ and } |t - x| < \delta \implies |f(t) - f(x)| < \varepsilon.$$

Observe that δ depends only on ε, not on x (the so called uniform δ).

14. *The mean value theorem.* Prove the following three forms of the mean value theorem:

Let $-\infty \leq a < b \leq \infty$, $F : (a, b) \to \mathbb{C}$ and let F satisfy any one of the following conditions with a constant $M \geq 0$:

(A) $|F'| \leq M$ in (a, b).

(B) F is continuous in (a, b) and $|F'| \leq M$ in (a, b) except for a countable subset of (a, b).

(C) F is absolutely continuous in (a, b) and $|F'| \leq M$ in (a, b) except for a null subset of (a, b).

Then $|F(y) - F(x)| \leq M(y - x)$ whenever $a < x < y < b$.

15. Let $F : (a, b) \to \mathbb{C}$ satisfy any one of the conditions (A), (B) or (C) of Problem B.14 with $M = 0$ (that is, $F' = 0$ on an appropriate subset of (a, b)). Prove that F is constant in (a, b).

Appendix C

Inequalities

C.1 Introduction

Inequalities can be defined only between real numbers; there is no sensible definition of order for complex numbers. Recall the axioms for inequalities for real numbers and the properties of the absolute value from Section B.1.

When we deal with complex numbers, we can only consider inequalities between their absolute values, which are nonnegative real numbers. The following two inequalities are particularly useful: Let $a_i \in \mathbb{C}$, $i = 1, \ldots, n$. Then

$$\left| \sum_{i=1}^{n} a_i \right| \leq \sum_{i=1}^{n} |a_i|, \qquad \left| \sum_{i=1}^{n} a_i \right| \geq |a_k| - \sum_{i \neq k} |a_i|. \tag{C.1}$$

In the second inequality we try to choose a_k with the largest absolute value. We also note that

$$|\operatorname{Re} z| \leq |z|, \qquad |\operatorname{Im} z| \leq |z|.$$

C.2 Finite sums

Jensen's inequality

Let X be a real vector space and D a convex subset of X; this means that if x and y belong to D, then the line segment $\{tx + (1 - t)y : 0 \leq t \leq 1\}$ lies in D. A function $f \colon D \to \mathbb{R}$ is *convex* if

$$f\big(tx + (1 - t)y\big) \leq tf(x) + (1 - t)f(y)$$

for all $x, y \in D$ and all $t \in [0, 1]$.

Theorem C.1. *Let $f: D \to \mathbb{R}$ be a convex function defined on a convex subset D of a real vector space X. Then*

$$f\left(\sum_{i=1}^{n} t_i x_i\right) \leq \sum_{i=1}^{n} t_i f(x_i), \quad x_i \in D, \ t_i \in [0,1], \ \sum_{i=1}^{n} t_i = 1. \qquad (C.2)$$

Proof.   Problem C.1.   □

From Jensen's inequality we can deduce the so called arithmetic and geometric mean (AM-GM) inequality: *Let $a_1, \ldots, a_n \geq 0$, $p_1, \ldots, p_n > 0$ and $p_1 + \cdots + p_n = 1$. Then*

$$\prod_{i=1}^{n} a_i^{p_i} \leq \sum_{i=1}^{n} p_i a_i. \qquad (C.3)$$

The classical case: $\sqrt{a_1 a_2} \leq \frac{1}{2}(a_1 + a_2)$.

Proof.   Problem C.2.   □

Auxiliary inequality involving conjugate indices

Theorem C.2. *Let $p > 1$ be a real number, and let q be the* conjugate index *of p, that is, $1/p + 1/q = 1$. Then, for any real numbers $\alpha \geq 0$, $\beta \geq 0$,*

$$\alpha\beta \leq \frac{\alpha^p}{p} + \frac{\beta^q}{q}. \qquad (C.4)$$

Proof.   Problem C.3.   □

Another auxiliary inequality

Theorem C.3. *Let $0 < p < 1$. Then for any $\alpha, \beta \geq 0$,*

$$(\alpha + \beta)^p \leq \alpha^p + \beta^p. \qquad (C.5)$$

Proof.   Write $f(x) = x^p + \beta^p - (x + \beta)^p$, and show that $f'(x) \geq 0$ and $f(0) = 0$.   □

Hölder's inequality

Theorem C.4. *Let p be a real number, $p > 1$, and let q be the conjugate index of p, that is, $1/p + 1/q = 1$. Then, for any real or complex numbers $\xi_1, \ldots, \xi_N, \ \eta_1, \ldots, \eta_N$,*

$$\sum_{k=1}^{N} |\xi_k \eta_k| \leq \left(\sum_{k=1}^{N} |\xi_k|^p\right)^{1/p} \left(\sum_{k=1}^{N} |\eta_k|^q\right)^{1/q}. \qquad (C.6)$$

Proof. Set

$$A = \left(\sum_{k=1}^{N} |\xi_k|^p\right)^{1/p}, \qquad B = \left(\sum_{k=1}^{N} |\eta_k|^q\right)^{1/q}.$$

The inequality is obviously true if $A = 0$ or $B = 0$. Hence we assume that $A > 0$, $B > 0$, and set $\alpha_k = \xi_k/A$, $\beta_k = \eta_k/B$ for $k = 1, \ldots, N$. Note that

$$\sum_{k=1}^{N} |\alpha_k|^p = \sum_{k=1}^{N} |\beta_k|^q = 1.$$

Using inequality (C.4), we have

$$\sum_{k=1}^{N} |\xi_k \eta_k| = AB \sum_{k=1}^{N} |\alpha_k||\beta_k| \le AB\left(\frac{1}{p}\sum_{k=1}^{N} |\alpha_k|^p + \frac{1}{q}\sum_{k=1}^{N} |\beta_k|^q\right)$$

$$= AB\left(\frac{1}{p} + \frac{1}{q}\right) = AB.$$

$\square$

Minkowski's inequality for finite sums

Theorem C.5. *Let $p \in \mathbb{R}$ with $p \ge 1$. Then, for any real or complex numbers $\xi_1, \ldots, \xi_N$, $\eta_1, \ldots, \eta_N$,*

$$\left(\sum_{k=1}^{N} |\xi_k + \eta_k|^p\right)^{1/p} \le \left(\sum_{k=1}^{N} |\xi_k|^p\right)^{1/p} + \left(\sum_{k=1}^{N} |\eta_k|^p\right)^{1/p}. \qquad (C.7)$$

Proof. Suppose that $p > 1$, and that $q = p/(p-1)$ is the conjugate index of p. Define

$$A = \left(\sum_{k=1}^{N} |\xi_k|^p\right)^{1/p}, \quad B = \left(\sum_{k=1}^{N} |\eta_k|^p\right)^{1/p}, \quad C = \left(\sum_{k=1}^{N} |\xi_k + \eta_k|^p\right)^{1/p}.$$

The inequality is obviously true if $C = 0$. Assume therefore that $C \ne 0$. In the following calculation we use (C.6):

$$C^p = \sum_{k=1}^{N} |\xi_k + \eta_k|^p = \sum_{k=1}^{N} |\xi_k + \eta_k||\xi_k + \eta_k|^{p-1}$$

$$\le \sum_{k=1}^{N} (|\xi_k| + |\eta_k|)|\xi_k + \eta_k|^{p-1}$$

$$= \sum_{k=1}^{N} |\xi_k||\xi_k + \eta_k|^{p-1} + \sum_{k=1}^{N} |\eta_k||\xi_k + \eta_k|^{p-1}$$

$$\leq \left(\left(\sum_{k=1}^{N} |\xi_k|^p \right)^{1/p} + \left(\sum_{k=1}^{N} |\eta_k|^p \right)^{1/p} \right) \left(\sum_{k=1}^{N} |\xi_k + \eta_k|^{q(p-1)} \right)^{1/q}$$

$$= (A+B)\left(\sum_{k=1}^{N} |\xi_k + \eta_k|^p \right)^{1/q} = (A+B)C^{p/q}.$$

Dividing by $C^{p/q}$, we get $C^{p-p/q} = C \leq A + B$.

The case $p = 1$ is left as an exercise. □

Another useful inequality

Theorem C.6. *For any real or complex numbers x, y, we have*

$$\frac{|x+y|}{1+|x+y|} \leq \frac{|x|}{1+|x|} + \frac{|y|}{1+|y|}. \tag{C.8}$$

C.3　Infinite sums and integrals

Hölder's inequality for infinite sums

Theorem C.7. *Let $p > 1$ be a real number, and let $q = p/(p-1)$ be the conjugate index of p, so that $1/p + 1/q = 1$. If $\xi_1, \xi_2, \xi_3, \ldots, \eta_1, \eta_2, \eta_3, \ldots$ are real or complex sequences such that $\sum_{k=1}^{\infty} |\xi_k|^p$, $\sum_{k=1}^{\infty} |\eta_k|^q$ are convergent, then also the series $\sum_{k=1}^{\infty} |\xi_k \eta_k|$ is convergent, and*

$$\sum_{k=1}^{\infty} |\xi_k \eta_k| \leq \left(\sum_{k=1}^{\infty} |\xi_k|^p \right)^{1/p} \left(\sum_{k=1}^{\infty} |\eta_k|^q \right)^{1/q}. \tag{C.9}$$

Proof. By (C.6), for any positive integer N we have

$$\sum_{k=1}^{N} |\xi_k \eta_k| \leq \left(\sum_{k=1}^{N} |\xi_k|^p \right)^{1/p} \left(\sum_{k=1}^{N} |\eta_k|^q \right)^{1/q} \leq \left(\sum_{k=1}^{\infty} |\xi_k|^p \right)^{1/p} \left(\sum_{k=1}^{\infty} |\eta_k|^q \right)^{1/q}.$$

This means that the positive term series on the far left has bounded partial sums, and therefore is convergent. The inequality (C.9) follows on taking the limit as $N \to \infty$ in

$$\sum_{k=1}^{N} |\xi_k \eta_k| \leq \left(\sum_{k=1}^{\infty} |\xi_k|^p \right)^{1/p} \left(\sum_{k=1}^{\infty} |\eta_k|^q \right)^{1/q}.$$

□

Minkowski's inequality for infinite sums

Theorem C.8. *Let* p *be a real number with* $p \geq 1$. *If* $\xi_1, \xi_2, \xi_3, \ldots,$ $\eta_1, \eta_2, \eta_3, \ldots$ *are real or complex sequences such that* $\sum_{k=1}^{\infty} |\xi_k|^p$, $\sum_{k=1}^{\infty} |\eta_k|^p$ *are convergent, then also the series* $\sum_{k=1}^{\infty} |\xi_k + \eta_k|^p$ *is convergent, and*

$$\left(\sum_{k=1}^{\infty} |\xi_k + \eta_k|^p \right)^{1/p} \leq \left(\sum_{k=1}^{\infty} |\xi_k|^p \right)^{1/p} + \left(\sum_{k=1}^{\infty} |\eta_k|^p \right)^{1/p}. \qquad (C.10)$$

Proof. For any positive integer N we have

$$\left(\sum_{k=1}^{N} |\xi_k + \eta_k|^p \right)^{1/p} \leq \left(\sum_{k=1}^{N} |\xi_k|^p \right)^{1/p} + \left(\sum_{k=1}^{N} |\eta_k|^p \right)^{1/p}$$

$$\leq \left(\sum_{k=1}^{\infty} |\xi_k|^p \right)^{1/p} + \left(\sum_{k=1}^{\infty} |\eta_k|^p \right)^{1/p}.$$

The result then follows as in the preceding proof. □

Hölder's inequality for integrals

Theorem C.9. *Let* $p > 1$ *be a real number, and let* $q = p/(p-1)$ *be the conjugate index of* p. *If* $f \in L^p$ *and* $h \in L^q$, *then the product function* fh *is in* L^1, *and*

$$\int |fh| \leq \left(\int |f|^p \right)^{1/p} \left(\int |h|^q \right)^{1/q}. \qquad (C.11)$$

Minkowski's inequality for integrals

Theorem C.10. *Let* $p \geq 1$ *be a real number, and let* $f, g \in L^p$. *Then* $f + g \in L^p$, *and*

$$\left(\int |f + g|^p \right)^{1/p} \leq \left(\int |f|^p \right)^{1/p} + \left(\int |g|^p \right)^{1/p}. \qquad (C.12)$$

A complete proof of the preceding two inequalities for functions defined on a general measure space is given in Chapter 17, Section 17.1.

C.4   Problems for Appendix C

1. Prove Jensen's inequality. (Induction on n.)

2. Prove the AM-GM inequality C.3.

Suggestion. The function $f(t) = -\log t$ is convex on $(0, \infty)$.

3. Prove the inequality (C.4).

Suggestion. Apply the AM-GM inequality with $a_1 = \alpha^p$, $a_2 = \beta^q$, $p_1 = 1/p$, $p_2 = 1/q$. Alternatively, use calculus to show that the function $f(t) = t^p/p + \beta^q/q - \beta t$ attains its minimum at $t_0 = \beta^{q-1}$ and that $f(t_0) = 0$.

4. Let $n \in \mathbb{N}$ and let $a_1, \dots, a_n$ be complex numbers. Prove the following inequalities:

(i) If $0 < p < 1$, then

$$\left| \sum_{k=1}^{n} a_k \right|^p \leq \left(\sum_{k=1}^{n} |a_k| \right)^p \leq \sum_{k=1}^{n} |a_k|^p.$$

(ii) If $p \geq 1$, then

$$\left(\sum_{k=1}^{n} |a_k| \right)^p \geq \sum_{k=1}^{n} |a_k|^p.$$

Also

$$\sum_{k=1}^{n} |a_k| \leq \sqrt[p]{n} \left(\sum_{k=1}^{n} |a_k|^p \right)^{1/p}.$$

5. Prove the inequality (C.8).

Suggestion. Show first that the function $f(t) = t/(1+t)$ on $[0, \infty)$ is increasing and subadditive, that is, $f(t_1 + t_2) \leq f(t_1) + f(t_2)$.

6. A miniproject: *Integral form of Jensen's inequality.* Let $g \colon [a, b] \to \mathbb{R}$ be continuous and $f \colon (c, d) \to \mathbb{R}$ convex, where $(c, d) \supset g([a, b])$. Prove that

$$f \left(\frac{1}{b-a} \int_a^b g(t)\, dt \right) \leq \frac{1}{b-a} \int_a^b f(g(t))\, dt.$$

Can the discrete Jensen's inequality be derived from its integral counterpart?

Suggestion: (i) Any convex function on an open interval is continuous.

(ii) If $w \in (c, d)$, show that there is $\alpha \in \mathbb{R}$ such that

$$f(u) \geq \alpha(u - w) + f(w) \quad \text{for all } u \in (c, d)$$

(the line through $(w, f(w))$ with slope α is always below or on the graph of f).

(iii) If $w := (b-a)^{-1} \int_a^b g(t)\, dt$, show that $w \in (c, d)$. According to (ii),

$$f(g(t)) \geq \alpha(g(t) - w) + f(w) \quad \text{for all } t \in [a, b].$$

Integrate both sides of the inequality.

Appendix D

Survey of Metric Spaces

In this survey we concentrate on standard metrics, with pseudometrics mentioned occasionally. Also discussed are some typical topological spaces.

D.1 Euclidean spaces

They are the familiar spaces of linear algebra and solid geometry. Some of the terminology used in the metric space theory is motivated by the geometry of these spaces. However, we should be on our guard against attributing properties of Euclidean spaces to general metric spaces, and even less so to pseudometric spaces.

Euclidean spaces $\mathbb{R}^n$ and $\mathbb{C}^n$

The points of Euclidean spaces are n-dimensional vectors $x = (\xi_1, \dots, \xi_n)$ with real or complex coordinates. The Euclidean metric is defined by

$$d(x, y) = \left(\sum_{k=1}^{n} |\xi_k - \eta_k|^2 \right)^{1/2}.$$

Proof of the triangle inequality **M4** follows from Minkowski's inequality (C.7) proved in Appendix C when we set $p = 2$. These spaces are complete.

D.2 Sequence spaces

In this section we look at subspaces of the space $\mathbb{R}^{\mathbb{N}}$ (or $\mathbb{C}^{\mathbb{N}}$) which consist of infinite dimensional vectors $x = (\xi_1, \xi_2, \xi_3, \dots)$ with $\xi_k \in \mathbb{R}$ (or $\xi_k \in \mathbb{C}$).

The space ℓ^p, $p \geq 1$

Let p be a real number, $p \geq 1$. The elements of ℓ^p are $\mathbb{R}^{\mathbb{N}}$ or $\mathbb{C}^{\mathbb{N}}$ vectors $x = (\xi_1, \xi_2, \xi_3, \dots)$ with the property that the infinite series $\sum_{k=1}^{\infty} |\xi_k|^p$ converges. The metric is defined by

$$d_p(x, y) = \left(\sum_{k=1}^{\infty} |\xi_k - \eta_k|^p \right)^{1/p}.$$

In this case we have to prove first that the series on the right converges. This fact and the triangle inequality **M4** follow from Minkowski's inequality (C.10) for infinite sums established in Appendix C. The ℓ^p spaces are complete.

Two special cases deserve a mention: The space ℓ^2 can be regarded as an infinite dimensional version of a Euclidean space, in which the metric is given by

$$d_2(x, y) = \left(\sum_{k=1}^{\infty} |\xi_k - \eta_k|^2 \right)^{1/2}.$$

This space is often called the *Hilbert sequence space*.

The space ℓ^1 is useful as a source of nontrivial examples of metric space properties. The metric in ℓ^1 is given by

$$d_1(x, y) = \sum_{k=1}^{\infty} |\xi_k - \eta_k|.$$

Let us observe that spaces $\mathbb{R}^n$ and $\mathbb{C}^n$ can be equipped with finite versions of the ℓ^p metrics.

The space ℓ^p, $0 < p < 1$

Let p be a real number, $0 < p < 1$. The elements of ℓ^p are $\mathbb{R}^{\mathbb{N}}$ or $\mathbb{C}^{\mathbb{N}}$ vectors $x = (\xi_1, \xi_2, \xi_3, \dots)$ with the property that the infinite series $\sum_{k=1}^{\infty} |\xi_k|^p$ converges. The metric is defined by

$$d_p(x, y) = \sum_{k=1}^{\infty} |\xi_k - \eta_k|^p.$$

The triangle inequality follows from (C.5). The space is complete.

The space ℓ^∞

An $\mathbb{R}^{\mathbb{N}}$ (or $\mathbb{C}^{\mathbb{N}}$) vector $x = (\xi_1, \xi_2, \xi_3 \ldots)$ belongs to ℓ^∞ if there is a constant $c > 0$ such that $|\xi_n| \leq c$ for all $n \in \mathbb{N}$; that is, the coordinates of x form a bounded set in $\mathbb{R}$ (or $\mathbb{C}$). The metric is given by

$$d_\infty(x, y) = \sup_{n \in \mathbb{N}} |\xi_n - \eta_n|.$$

The space ℓ^∞ is complete.

The Hilbert cube

A metric subspace of ℓ^2 consisting of all $\mathbb{R}^{\mathbb{N}}$ (or $\mathbb{C}^{\mathbb{N}}$) vectors $x = (\xi_1, \xi_2, \xi_3, \ldots)$ with $|\xi_k| \leq 1/k$ for all k; the metric is inherited from ℓ^2. The space is complete; it is also compact.

The space c

This is a subspace of ℓ^∞ which again comes in two varieties, real and complex. The space **c** consists of all these $x = (\xi_1, \xi_2, \xi_3, \ldots)$ for which $\lim_{n \to \infty} \xi_n$ exists and is finite. The metric in **c** is inherited from ℓ^∞. The space **c** is complete being a closed subspace of ℓ^∞.

The space c_0

The elements of c_0 are those $x = (\xi_1, \xi_2, \xi_3, \ldots)$ for which $\lim_{n \to \infty} \xi_n = 0$. The metric is again inherited from ℓ^∞. The space c_0 is complete being a closed subspace of ℓ^∞.

The space E_0

The space of all $\mathbb{R}^{\mathbb{N}}$ (or $\mathbb{C}^{\mathbb{N}}$) vectors $x = (\xi_1, \xi_2, \xi_3, \ldots)$ which have only finitely many nonzero coordinates. E_0 can be equipped with an ℓ^p metric d_p for $p \geq 1$, or with the ℓ^∞ metric d_∞. The space (E_0, d_∞) is incomplete.

The space $\mathbf{bv}_0$

The space of all $\mathbb{R}^{\mathbb{N}}$ (or $\mathbb{C}^{\mathbb{N}}$) vectors $x = (\xi_1, \xi_2, \xi_3, \ldots)$ such that $\sum_{k=1}^{\infty} |\xi_{k+1} - \xi_k| < \infty$, and $\lim_{n \to \infty} \xi_n = 0$. The metric is defined by

$$d(x,y) = \sum_{k=1}^{\infty} |(\xi_{k+1} - \xi_k) - (\eta_{k+1} - \eta_k)|.$$

The space $\mathbf{bv}_0$ is isometrically isomorphic to ℓ^1, and so it is complete.

The space $\mathbf{bv}$

This is the space of all $x = (\xi_1, \xi_2, \xi_3, \ldots) \in \mathbb{R}^{\mathbb{N}}$ (resp. $\mathbb{C}^{\mathbb{N}}$) for which the series $\sum_{k=1}^{\infty} |\xi_{k+1} - \xi_k|$ converges. The metric is defined by

$$d(x,y) = |\xi_1 - \eta_1| + \sum_{k=1}^{\infty} |(\xi_{k+1} - \xi_k) - (\eta_{k+1} - \eta_k)|.$$

The space $\mathbf{bv}$ is isometrically isomorphic to the direct sum $\mathbb{K} \oplus \mathbf{bv}_0$ ($\mathbb{K}$ is the scalar field) equipped with the sum metric, and so it is complete.

D.3   Function spaces

By function spaces we understand metric spaces whose elements are real (or complex) valued functions defined on the same domain E.

The space $B(E)$ with the uniform metric

Suppose E is a nonempty set and by $B(E)$ denote the space of all bounded real (or complex) valued functions on E. This means $f \in B(E)$ if there is a constant $c > 0$ such that $|f(t)| \leq c$ for all $t \in E$. Define

$$d(f,g) = \sup_{t \in E} |f(t) - g(t)|.$$

This metric is known as the *uniform metric* on $B(E)$. The space $B(E)$ is complete.

   If X is a complete normed space, we can generalize $B(E)$ to the space $(B(E), X)$ of all bounded functions $f : E \to X$ with the metric $d(f,g) = \sup_{t \in E} \|f(t) - g(t)\|$. This is also a complete space.

The space $C[a, b]$ with the uniform metric

By $C[a, b]$ we denote the set of all continuous real (or complex) valued functions on the closed bounded interval $[a, b]$. It is known from classical analysis that these function are bounded. The metric introduced on $C[a, b]$ is the uniform metric of the preceding example. In fact, every real valued continuous function attains its maximum on a closed bounded interval, and so we have

$$d(f, g) = \max_{t \in [a,b]} |f(t) - g(t)|.$$

The space $C[a, b]$ is complete being a closed subspace of $B[a, b]$.

If X is a complete normed space, we can generalize $C[a, b]$ to the space $(C[a, b], X)$ of all continuous functions $f : [a, b] \to X$ with the metric $d(f, g) = \sup_{t \in E} \|f(t) - g(t)\|$. This is also a complete space being a closed subspace of $(B[a, b], X)$.

The space $QC[a, b]$ with the uniform metric

This is the space of all quasicontinuous real (or complex) valued functions on $[a, b]$. A function f is *quasicontinuous* if the one sided limits $f(x+)$ and $f(x-)$ exist at each point $x \in [a, b]$. (At the end points we set $f(a-) = f(a)$ and $f(b+) = f(b)$.) Equipped with the uniform metric, $QC[a, b]$ is complete.

If X is a complete normed space, we can generalize $QC[a, b]$ to the space $(QC[a, b], X)$ of all quasicontinuous functions $f : [a, b] \to X$ with $d(f, g) = \sup_{t \in E} \|f(t) - g(t)\|$. This is also a complete space being a closed subspace of $(B[a, b], X)$.

The space $CB(\mathbb{R})$ with the uniform metric

This is the space of all continuous and bounded real (or complex) valued functions on $\mathbb{R}$ equipped with the uniform metric. This space is complete being a closed subspace of $B(\mathbb{R})$.

The space $UCB(\mathbb{R})$ with the uniform metric

The space consists of all uniformly continuous and bounded real (or complex) valued functions on $\mathbb{R}$ equipped with the uniform metric. It plays an important role in theory of operator semigroups. Observe: $f(t) = \sin t$

belongs to $UCB(\mathbb{R})$, but $g(t) = \sin t^2$ belongs to $CB(\mathbb{R})$ but not to $UCB(\mathbb{R})$. This space is complete.

The space $C[a, b]$ with the L^1 metric

The metric defined on $C[a, b]$ is given by

$$\rho(f, g) = \int_a^b |f(t) - g(t)|\, dt.$$

A verification of the Hausdorff axioms is left as an exercise. Let us remark that **M2** requires the following consequence of continuity: *If a continuous function is nonzero at some point t_0, then it is nonzero at some interval containing t_0.* (See Problem B.11.) This space is incomplete.

The space L^1

The space $L^1 = L^1(\mathbb{R}^k)$ of all Lebesgue integrable functions on $\mathbb{R}^k$ is defined in Section 2.2.2 of Chapter 2. A function f on $\mathbb{R}^k$ is *Lebesgue integrable* if there is a Cauchy sequence (f_n) of step functions on $\mathbb{R}^k$ such that $\lim_{n \to \infty} f_n = f$ almost everywhere in $\mathbb{R}^k$; such a sequence is called an *approximating sequence* for f. The integral of f is defined by

$$\int f = \lim_{n \to \infty} \int f_n.$$

Two Lebesgue integrable function which are equal almost everywhere are regarded as equal. With this convention, L^1 is a metric space with the metric defined by

$$d(f, g) = \int |f - g|.$$

We often consider spaces $L^1(a, b)$, where (a, b) is a real interval. They arise by a completion of the space of step functions on (a, b) under the integral metric. The space L^1 is complete.

The general theory of the space $L^1(\mu)$ on measure spaces is given in Chapter 13.

The L^p spaces

Let $p \geq 1$ be a real number. The space $L^p = L^p(\mathbb{R}^k)$ consists of all those functions f on $\mathbb{R}^k$ which are pointwise limits of step functions and for which $|f|^p$ is Lebesgue integrable. The equality in L^p is the equality almost everywhere. The L^p metric is defined by

$$d_p(f,g) = \left(\int |f - g|^p \right)^{1/p}.$$

The triangle inequality follows from the integral form of Minkowski's inequality given in the preceding Appendix. For $p = 1$ we obtain the space L^1 discussed above. In applications we often consider spaces $L^p(a, b)$. The spaces L^p are complete.

The more general Lebesgue spaces $L^p(\mu) = L^p(S, \Sigma, \mu)$ are introduced and discussed in detail in Chapter 17.

The space $C_{2\pi}$

This space consists of all (complex or real valued) functions f on $[0, 2\pi]$ satisfying $f(0) = f(2\pi)$, equipped with the uniform norm. This space is important in the theory of Fourier series. The trigonometric polynomials $\sum_{k=-n}^{n} \alpha_k e^{ikt}$, $0 \le t \le 2\pi$, form a total subset of $C_{2\pi}$ (they are dense in $C_{2\pi}$). (See Theorem 5.37.) The space is complete.

D.4 Other useful metric spaces

These spaces, especially the discrete space, are very useful for producing counterexamples.

The discrete space

If S is an arbitrary nonempty set, define

$$d(x, y) = \begin{cases} 1 & \text{if } x \neq y, \\ 0 & \text{if } x = y. \end{cases}$$

This is the so called *discrete metric* on S. The discrete space is complete.

The space $(\mathbb{R}, \rho)$

The metric on $\mathbb{R}$ is defined by

$$\rho(x, y) = |f(x) - f(y)|, \quad \text{where } f(x) = x/(1 + |x|). \tag{D.1}$$

The space $(\mathbb{R}, \rho)$ is incomplete.

This example points to a general method for introducing a metric (or pseudometric) on an arbitrary set S. If $f\colon S \to \mathbb{R}$ is a function on S, the formula $\rho(x, y) = |f(x) - f(y)|$ defines a pseudometric on S; if f is injective, ρ is a metric. More generally, if (T, d) is a metric space and $f\colon S \to T$ a function on S, $\rho(x, y) = d(f(x), f(y))$ defines a pseudometric on S; ρ is a metric if f is injective.

The space $(\mathbb{R}, \sigma)$

The metric is defined by

$$\sigma(x, y) = \frac{|x - y|}{1 + |x - y|}.$$

M4 follows from the inequality (C.8) of Appendix C. The metric σ is equivalent to the Euclidean metric $d(x, y) = |x - y|$ on $\mathbb{R}$. The space $(\mathbb{R}, \sigma)$ is complete.

This is an example of a general method for a remetrization of a metric space (S, d) by introducing an equivalent bounded metric on S:

$$\rho(x, y) = \frac{d(x, y)}{1 + d(x, y)}$$

The Fréchet space s

The space consists of all $\mathbb{R}^{\mathbb{N}}$ (or $\mathbb{C}^{\mathbb{N}}$) vectors $x = (\xi_1, \xi_2, \xi_3, \dots)$ with the metric defined by

$$d(x, y) = \sum_{k=1}^{\infty} (\tfrac{1}{2})^k \frac{|\xi_k - \eta_k|}{1 + |\xi_k - \eta_k|};$$

the factors $(\tfrac{1}{2})^k$ are introduced to make the series convergent. **M4** follows from the inequality (C.8) of Appendix C and consideration of convergence. The Fréchet space **s** is complete, and the convergence in the metric is equivalent to the coordinatewise convergence.

D.5   Normed spaces

A *norm* on X is a real valued function on a vector space X satisfying the following axioms:

N1. $\|x\| \geq 0$ for all $x \in X$.

N2. $\|x\| = 0$ implies $x = 0$.

N3. $\|\alpha x\| = |\alpha| \, \|x\|$ for all $x \in X$ and all scalars α.

N4. $\|x + y\| \leq \|x\| + \|y\|$ for all $x, y \in X$.

A normed space can be turned into a metric space by defining the *metric induced by the norm* $\|\cdot\|$:

$$d(x, y) = \|x - y\|.$$

If (X, d) is a metric space and if X is not just a set but a vector space, we may ask whether the metric d is induced by a norm. If such a norm existed, it would have to satisfy

$$\|x\| = d(x, 0).$$

It is now enough to check whether so defined $\|\cdot\|$ satisfies the norm axioms.

We now look at the metric spaces defined earlier, and check which metrics are induced by a norm. It turns out that these spaces are in majority. The spaces in Sections D.1, D.2 and D.3 are vector spaces and, in each case with the exception of ℓ^p, $0 < p < 1$, the metric is induced by a norm. As an example, we observe that, for any $p \geq 1$, the ℓ^p metric d_p is induced by the norm

$$\|x\|_p = \left(\sum_{k=1}^{\infty} |\xi_k|^p \right)^{1/p}.$$

The ℓ^∞ metric on ℓ^∞ is induced by the norm $\|x\|_\infty = \sup_{n \in \mathbb{N}} |\xi_n|$. The metric on **bv** is induced by the norm $\|x\| = |\xi_1| + \sum_{k=1}^{\infty} |\xi_{k+1} - \xi_k|$. A verification of the norm axioms is left as an exercise.

The spaces of Section D.4 are vector spaces, with the exception of the discrete space, but the metrics are not induced by any norm. To check this for the space $(\mathbb{R}, \rho)$ defined in (D.1) consider $\|x\| = \rho(x, 0) = |x|/(1 + |x|)$; we see immediately that axiom **N3** does not hold.

D.6 Topological spaces

Given a set X, a *topology* on X is a family $\mathcal{T}$ of subsets of X containing $\emptyset$ and X, which is closed under arbitrary unions and finite intersections; the sets belonging to $\mathcal{T}$ are called *open*. Every metric or pseudometric space (X, d) gives rise to a topological space if $\mathcal{T}$ is taken to be the family of all d-open sets. A topology $\mathcal{T}$ is called *(pseudo)metrizable* if there is a (pseudo)metric d on X such that a set belongs to $\mathcal{T}$ if and only it is d-open.

Discrete topology

On any set X the discrete topology has all subsets of X open. It is the strongest topology on X. The discrete topology is metrizable by the discrete metric.

Indiscrete topology

This is weakest topology on X consisting of $\emptyset$ and X. The indiscrete topology is pseudometrizable, but not metrizable.

Particular point topology

Let X be a set and p a particular point of X. The open sets are $\emptyset$ and all subsets of X containing p. The *Sierpinski space* is the set $[0,1]$ with the particular point 0.

Excluded point topology

The open sets are X and all subsets of X which do not contain a given point $p \in X$.

Finite complement topology

Let X be infinite. The open sets are $\emptyset$, X and all subsets of X with finite complements. This is the weakest topology on X in which the singletons are closed.

Countable complement topology

Let X be uncountable. The open sets are $\emptyset$, X and all subsets of X with countable complements.

Appendix E

Cardinal Numbers, Axiom of Choice

E.1 Cardinal numbers

In Analysis as well as in other areas of mathematics we need to deal with infinite sets, and compare their sizes. Problems associated with infinite sets were considered by many mathematicians, such as Carl Friedrich Gauss (1777–1855) and Bernhard Bolzano (1781–1848). At the end of 19th century Georg Cantor (1845–1918) founded a theory of the infinite that is now an important part of mathematical heritage.

The key idea is that two sets A, B have the same 'size' if and only if they are bijective, that is, if there exists a bijection $f \colon A \to B$ (and therefore a bijection $g \colon B \to A$). The relation of bijectivity of sets is an equivalence relation. A cardinal number is a label attached to all sets of the same size. Relevant properties of cardinal numbers are summarized as follows.

C1 With each set A we associate a *cardinal number* denoted by card A. For each cardinal number a there is a set A such that card $A = a$.

C2 card $A = 0$ if and only if $A = \emptyset$.

C3 If A is a nonempty finite set bijective to the set $\{1, 2, \ldots, n\}$ for some $n \in \mathbb{N}$, then card $A = n$.

C4 For any two sets A, B, card $A =$ card B if and only if A is bijective to B.

We use the following notation:

$$\text{card } \mathbb{N} = \aleph_0, \qquad \text{card } \mathbb{R} = \mathfrak{c}.$$

The letter $\aleph$ is 'aleph', the first letter of the Hebrew alphabet.

Inequalities for cardinal numbers. Let a, b be two cardinal numbers. We say that $a \leq b$ if there exist sets A, B such that $a =$ card A and $b =$ card B and that A is bijective to a subset of B. We say that $a < b$ if

$a \leq b$ and $a \neq b$. (This definition is independent of the choice of A and B as long as $a = \text{card } A$ and $b = \text{card } B$.)

A set A is called *countable* if $\text{card } A \leq \aleph_0$. This means that a set is countable if and only if it is finite or bijective to $\mathbb{N}$. A typical example of an uncountable set is $\mathbb{R}$ (Example E.2).

Theorem E.1. *Let a, b, d be cardinal numbers. Then:*

(i) $a \leq a$.

(ii) $a \leq b$ *and* $b \leq d \implies a \leq d$.

(iii) *The Schröder-Bernstein theorem:* $a \leq b$ *and* $b \leq a \implies a = b$.

(iv) *For any cardinal numbers a, b either $a \leq b$ or $b \leq a$.*

(Proofs of these properties as well as of all the following results in this Appendix can be found in the book [35] by You-Feng Lin and Schwu-Yeng T. Lin. The proof of the Schöder-Bernstein theorem is fairly technical.)

Example E.2. We have $\aleph_0 < \mathfrak{c}$, that is, $\text{card } \mathbb{N} < \text{card } \mathbb{R}$. This means that $\mathbb{R}$ is not countable.

Clearly, $\aleph_0 \leq \mathfrak{c}$ since $\mathbb{N}$ is bijective to a subset of $\mathbb{R}$. The following procedure known as the *Cantor diagonal argument* shows that $\aleph_0 \neq \mathfrak{c}$. For a proof by contradiction we assume that $\mathbb{R}$ is bijective to $\mathbb{N}$. Then also the interval $(0, 1)$ is bijective to $\mathbb{N}$. We can express each number $x \in (0, 1)$ as a decimal expansion

$$x = .x_1 x_2 x_3 \ldots$$

where $x_i \in \{0, 1, \ldots, 9\}$ for all $i \in \mathbb{N}$. To ensure the uniqueness of such an expansion, we write all terminating expansions in the form with the digit 9 repeating from a certain index on, for instance .25 will have the expansion .24999.... According to our assumption there is a bijection $f : \mathbb{N} \to (0, 1)$, and we can list all the elements of $(0, 1)$ in this form:

$$f(1) = .a_{11} a_{12} a_{13} \ldots$$
$$f(2) = .a_{21} a_{22} a_{23} \ldots$$
$$\ldots\ldots \quad \ldots$$
$$f(n) = .a_{n1} a_{n2} a_{n3} \ldots$$
$$\ldots\ldots \quad \ldots$$

with $a_{ij} \in \{0, 1, 2, \ldots, 9\}$. Let us write

$$z = .z_1 z_2 z_3 \ldots,$$

where $z_i = 5$ if $a_{ii} \neq 5$, and $z_i = 1$ if $a_{ii} = 5$. Then $z \in (0, 1)$, but $f(k) \neq z$ for all $k \in \mathbb{N}$; this is a contradiction.

Example E.3 (Cantor's theorem). *For any set A, card $A <$ card $\mathcal{P}(A)$, where $\mathcal{P}(A)$ is the power set of A.*

Before the next definition we introduce the following notation for sets of functions: If A, B are sets, then A^B denotes the family of all functions $f \colon B \to A$.

Algebraic operation with cardinal numbers. Let a, b be cardinal numbers and A, B be sets such that $a = $ card A and $b = $ card B. We define:

(i) $a + b := $ card $(A \cup B)$ provided A, B are *disjoint*.

(ii) $ab := $ card $A \times B$.

(iii) $a^b := $ card A^B.

(In connection with the definition of the sum we note that for any cardinal numbers a, b there do exist disjoint sets of cardinality a and b: If necessary take $A \times \{\alpha\}$ and $B \times \{\beta\}$ with $\alpha \neq \beta$ in place of A and B.) The definitions are independent of the choice of A, B.

Example E.4. Let a, b be finite cardinals (that is, natural numbers). Then the algebraic operations for a, b as cardinals agree with the usual operations on natural numbers. (Prove the exponentiation result as an exercise.)

Theorem E.5. *Let a, b, d be cardinals. Then the following hold:*

$$a + (b + d) = (a + b) + d, \qquad a + b = b + a$$
$$a(b + d) = ab + ad, \qquad a(bd) = (ab)d$$
$$ab = ba, \qquad a^b a^d = a^{b+d}$$
$$(a^b)^d = a^{bd}, \qquad a^d b^d = (ab)^d$$
$$a \leq b \implies a + d \leq b + d, \qquad a \leq b \implies ad \leq bd$$
$$a \leq b \implies a^d \leq b^d, \qquad a \leq b \implies d^a \leq d^b.$$

Theorem E.6. *Let a, b be infinite cardinals. Then $a + b = ab = \max\{a, b\}$.*

Theorem E.7. *A countable union of countable sets is countable.*

Proof. Follows from the equation $\aleph_0 \aleph_0 = \aleph_0$. $\square$

Theorem E.8. *For any set A, card $\mathcal{P}(A) = 2^{\text{card } A}$.*

Proof.   To prove this we set $B = \{0, 1\}$, and assign to each set $D \in \mathcal{P}(A)$ the characteristic function $\chi_D \colon A \to B$. Then show that $\mathcal{P}(A)$ is bijective to B^A under $D \mapsto \chi_D$. $\square$

Example E.9. $2^{\aleph_0} = \mathfrak{c}$ and $\aleph_0 < \mathfrak{c}$. (Exercise.)

Example E.10. $\aleph_0^{\aleph_0} = \mathfrak{c} = \mathfrak{c}^{\aleph_0}$. (Exercise.)

Example E.11. $\aleph_0 \mathfrak{c} = \mathfrak{c}$. (Exercise.)

The following result is essential in the demonstration that there are more Lebesgue measurable sets than Borel sets. We state it without proof.

Example E.12. Let $\mathcal{F}$ be a family of subsets of S with $\mathrm{card}\,(\mathcal{F}) = a$, and let $\mathcal{A}$ be the least σ-algebra of subsets of S containing $\mathcal{F}$. Then $\mathrm{card}\,(\mathcal{A}) \leq a^{\aleph_0}$.

E.2   Axiom of Choice and Zorn's Lemma

First we consider an axiom which is usually assumed in Analysis. It has a nonconstructive nature, and sometimes its application leads to counter-intuitive results. However, without it we could not prove that there exist subsets of $\mathbb{R}$ which are not Lebesgue measurable, and we could not define the Cartesian product of uncountably many sets.

E.13 (Axiom of choice). *Let $\mathcal{F}$ be a nonempty family of nonempty pairwise disjoint sets. Then there exists a set C which contains exactly one element from each member of $\mathcal{F}$.*

If the family is countable, we do not need the full force of the Axiom of choice to construct the set C.

Definition E.14. Let $\{S_\alpha : \alpha \in \Delta\}$ be an indexed family of sets. The *Cartesian product* of $\{S_\alpha : \alpha \in \Delta\}$, denoted $\prod_{\alpha \in \Delta} S_\alpha$, is the set of all functions $s \colon \Delta \to \bigcup_{\alpha \in \Delta} S_\alpha$ such that $s(\alpha) \in S_\alpha$ for each $\alpha \in \Delta$.

If each of the sets S_α is nonempty, then the Axiom of choice guarantees that the product space $\prod_{\alpha \in \Delta} S_\alpha$ is nonempty.

Example E.15. Let A, B be nonempty sets. Then A^B can be expressed in terms of the Cartesian product as $A^B = \prod_{\alpha \in B} A_\alpha$, where $A_\alpha = A$ for all α.

In this section we also need some basic facts about order. Recall that a relation $\leq$ on a set A is a *partial order* (Definition A.5) if it is reflexive, transitive and antisymmetric: (i) $a \leq a$ for all $a \in A$, (ii) $a \leq b$ and $b \leq d$ implies $a \leq d$, (iii) $a \leq b$ and $b \leq a$ implies $a = b$.

• An *upper bound* for a subset B of A is an element $b \in A$ such that $x \leq b$ for all $x \in B$.

• A *least upper bound* for B is an upper bound $s \in A$ for B such that $s \leq b$ for any other upper bound b of B. (A *greatest lower bound* is defined similarly.)

• An element $d \in A$ is a *maximal element* if, for each $x \in A$, $d \leq x$ implies $x = d$. (A *minimal element* is defined analogously.)

• A partially ordered set $(A, \leq)$ is *well ordered* if every nonempty subset B of A contains a (unique) minimal element (called the *least element* of B).

• We also define $a < b$ to mean $a \leq b$ and $a \neq b$.

An example of a partial order is the inclusion $\subset$ on the power set $\mathcal{P}(S)$ of some set S. A *linear order* is a partial order such that for any two elements $a, b \in A$, either $a \leq b$ or $b \leq a$. The following result is traditionally called Zorn's lemma, but was originally given by Hausdorff.

Theorem E.16 (Zorn's lemma). *Let $(A, \leq)$ be a partially ordered set in which every linearly ordered subset has an upper bound. Then A has a maximal element.*

Zorn's lemma is equivalent to the Axiom of choice and also to the Hausdorff maximality principle. A family $\mathcal{C}$ of subsets of a universal set S is a *chain* if it is ordered linearly relative to the set inclusion.

Theorem E.17 (Hausdorff maximality principle). *Let $\mathcal{F}$ be a nonempty family of subsets of S in which every chain has an upper bound. Then $\mathcal{F}$ has a maximal element.*

We give an application of the Hausdorff maximality principle. For the following result recall that a *Hamel basis* of a vector space X is a linearly independent subset of X whose span is X. For example, the so called *standard Hamel basis* in $\mathbb{R}^n$ is formed by the vectors

$$e_1 = (1, 0, \ldots, 0), \ e_2 = (0, 1, \ldots, 0), \ \ldots, \ e_n = (0, 0, \ldots, 1).$$

Hamel bases can be finite, countable or uncountable. If B is a Hamel basis of X, then for each nonzero vector $x \in X$ there is a unique finite subset

B_x of B and a unique set of nonzero scalars $\{\lambda(u) : u \in B_x\}$ such that $x = \sum_{u \in B_x} \lambda(u)u$.

Theorem E.18. *Every nonzero vector space possesses a Hamel basis.*

Proof. Let $\mathcal{H}$ be the family of all linearly independent subsets of the vector space X. It is nonempty as it contains some $\{a\}$, where $a \neq 0$. To apply the Hausdorff maximality principle, assume that $\mathcal{C}$ is a chain in $\mathcal{H}$, and define $B = \bigcup_{A \in \mathcal{C}} A$. Let $a_1, \ldots, a_n$ be elements of B. Then there exist sets $A_1, \ldots, A_n \in \mathcal{C}$ such that $a_i \in A_i$ for $i = 1, \ldots, n$. Since $\mathcal{C}$ is a chain, one of the sets, say A_n, contains the remaining sets. Then $a_i \in A_n$ for $i = 1, \ldots, n$, and $\sum_{i=1}^{n} \lambda_i a_i = 0$ implies $\lambda_i = 0$ for $i = 1, \ldots, n$ since A_n is a linearly independent set. Hence B is linearly independent, and therefore a bound for $\mathcal{C}$ in $\mathcal{H}$. By the Hausdorff maximality principle $\mathcal{H}$ has a maximal element, say W. We want to show that $\mathsf{sp}(W) = X$. For a proof by contradiction suppose that $\mathsf{sp}(W) \neq X$. Then there exists $u \in X \setminus \mathsf{sp}(W)$. But the set $W \cup \{u\}$ is linearly independent (otherwise u would be contained in $\mathsf{sp}(W)$), and $W \subset W \cup \{u\}$, while $W \neq W \cup \{u\}$. This contradicts the maximality of W, and proves that W is a Hamel basis for X. $\qquad\square$

Theorem E.19. *Any two Hamel bases of a vector space X have the same cardinal number.*

Proof. We assume that X does not possess a finite basis, as this case is covered in Linear Algebra. Let B_1 and B_2 be two Hamel bases of X. For each $x \in B_1$ let $B_2(x)$ be the set of all elements of B_2 needed to expand x as a finite linear combination of elements of B_2. If $y \in B_2$, show that there exists $x \in B_1$ such that $y \in B_2(x)$. Hence

$$B_2 = \bigcup_{x \in B_1} B_2(x).$$

This implies that $\mathsf{card}\, B_2 \leq \aleph_0\, \mathsf{card}\, B_1 = \mathsf{card}\, B_1$. A symmetrical argument shows that $\mathsf{card}\, B_1 \leq \mathsf{card}\, B_2$. By the Schröder–Bernstein theorem, $\mathsf{card}\, B_1 = \mathsf{card}\, B_2$. $\qquad\square$

A further example of the Hausdorff maximality principle is the proof of the theorem that every nonzero Hilbert space H possesses an orthonormal basis (see Chapter 8, Theorem 8.19). It is proved in Theorem 8.20 that any two orthonormal bases of the same Hilbert space have the same cordinality.

Yet another equivalent of the Axiom of choice, the well ordering principle, was introduced by Zermelo in 1904. This result was regarded as controversial at the time, and led to the examination of those results in mathematics that depended on the Axiom of choice.

Theorem E.20 (Well ordering principle). *Every nonempty set can be well ordered.*

The principle of transfinite induction is also equivalent to the Axiom of choice.

Theorem E.21 (Principle of transfinite induction). *Let $(A, \leq)$ be a well ordered set and let B be a subset of A with the following properties:*

(i) *The least element of A is contained in B.*

(ii) *If $x \in A$ and $A_x = \{u \in A : u \leq x\} \subset B$, then the least element of $A \setminus A_x$ (the successor of x) belongs to B.*

Then $B = A$.

E.3   Problems for Appendix E

1. Let $(A, \leq)$ be a partially ordered set and let $a, b \in A$. Prove that a least upper bound of $\{a, b\}$ is unique if it exists. Repeat for a greatest lower bound. We use the notation $\sup\{a, b\} = a \vee b$ and $\inf\{a, b\} = a \wedge b$ for the least upper bound and the greatest lower bound of $\{a, b\}$, respectively.

2. Let $A = \{1, 2, 3, 4, 5, 6, 7\}$ and let $x \leq y$ means 'x divides y'. Show that $(A, \leq)$ is a partially ordered set and find the maximal elements of A.

3. Let E be a nonempty set and $\mathcal{F} = \mathbb{R}^E$ the family of all functions $f : E \to \mathbb{R}$. For $f, g \in \mathcal{F}$ define $f \leq g$ if $f(t) \leq g(t)$ for all $t \in E$.

(i) Show that $\leq$ is a partial order in $\mathcal{F}$.

(ii) If $f, g \in \mathcal{F}$, show that $f \vee g$ and $f \wedge g$ always exist.

4. Let X be vector space and M a subspace of X. Show that there is a subspace N of X such that $X = M \oplus N$. (This means that every $x \in X$ has a unique decomposition $x = u + v$, where $u \in M$ and $v \in N$.)

Suggestion. If A is a Hamel basis of M, prove that there exists a Hamel basis B of X containing A. Set $N = \mathsf{sp}(B \setminus A)$.

5. Let X be a vector space and let $X = M \oplus N$, where M, N are subspaces. Prove that N is isomorphic to the quotient space X/M. The dimension of X/M

is called the *codimension* of the subspace M.

Suggestion. If $X = M \oplus N$ is a direct sum of subspaces, let Px be the projection of x onto its N-component; then P is linear operator on X to X. Set $T(x+M) = Px$ for all $x \in X$. Check that T is a well defined linear operator on X/M to N. Verify that T is bijective.

Appendix F

Answers to Selected Odd Numbered Problems

Chapter 1

1.1 M5. From $d(x,z) \leq d(x,y) + d(y,z)$ and $d(y,z) \leq d(y,x) + d(x,z)$ it follows that

$$d(x,y) \geq d(x,z) - d(y,z), \quad d(x,y) \geq d(y,z) - d(x,z);$$

hence $d(x,y) \geq \max\{d(x,z) - d(y,z), d(y,z) - d(x,z)\} = |d(x,z) - d(y,z)|$.
M6 is proved similarly.

1.3 In $\mathbb{R}$: (i), (ii) and (vi) are metrics, (iii), (iv) are pseudometrics but not metrics, (v) is not a pseudometric as $\psi(2,0) > \psi(2,1) + \psi(1,0)$.
In $\mathbb{C}$: (i) is a metric, (ii), (iii), (iv) and (vi) are pseudometrics but not metrics (the functions in (ii) and (vi) are no longer injective), (v) is not a pseudometric.

1.5 Verification of the Hausdorff axioms is fairly routine except **M2**. The function $h(t) := |f(t) - g(t)|$ is continuous. Suppose that $d(f,g) = 0$, that is, $\int_a^b h(t)\,dt = 0$. Suppose that for some $t_0 \in [a,b]$, $h(t_0) \neq 0$, say $h(t_0) > 0$. Let $\varepsilon = \frac{1}{2}h(t_0)$. By the continuity of h there exists $\delta > 0$ such that $|h(t) - h(t_0)| < \varepsilon$ and

$$h(t) = h(t_0) + (h(t) - h(t_0)) \geq h(t_0) - |h(t) - h(t_0)| > h(t_0) - \varepsilon = \varepsilon > 0$$

for all $t \in J := [a,b] \cap (t_0 - \delta, t_0 + \delta)$. Then

$$d(f,g) = \int_a^b |h(t)|\,dt \geq \int_J |h(t)|\,dt \geq |J|\varepsilon > 0,$$

where $|J|$ is the length of J. This contradicts $d(f,g) = 0$.

1.7 As a sample we show how to proceed in part (iii) for ρ: a point in the ball satisfies $\rho(x,1) < \frac{1}{4}$, that is, $|f(x) - f(1)| = |f(x) - \frac{1}{2}| < \frac{1}{4}$. This inequality is equivalent to $\frac{1}{4} < f(x) < \frac{3}{4}$, which in turn is equivalent to $\frac{1}{3} < x < 3$.
For ρ: (i) $(-1,1)$ (ii) $\mathbb{R}$ (iii) $(\frac{1}{3}, 3)$ (iv) $(0, \infty)$ (v) $(-1, \infty)$

For σ: (i) $\left(-\frac{1}{\sqrt{2}}, \frac{1}{\sqrt{2}}\right)$ (ii) $(-1, 1)$ (iii) $\left(-\frac{\sqrt{5}}{2}, -\frac{\sqrt{3}}{2}\right) \cup \left(\frac{\sqrt{3}}{2}, \frac{\sqrt{5}}{2}\right)$
(iv) $\left(-\frac{\sqrt{3}}{\sqrt{2}}, -\frac{1}{\sqrt{2}}\right) \cup \left(\frac{1}{\sqrt{2}}, \frac{\sqrt{3}}{\sqrt{2}}\right)$ (v) $(-\sqrt{2}, 1) \cup (1, \sqrt{2})$

1.9 Let A_i $(i = 1, \ldots, n)$ be bounded subsets of a pseudometric space (S, d). Then there exist balls $B(a_i; r_i)$ such that $A_i \subset B(a_i; r_i)$ for each $i \in \{1, \ldots, n\}$. Let a be an arbitrary point of S. We show that there exists a ball centred at a which contains the set $A = \bigcup_{i=1}^{n} A_i$. Let $x \in A$. Then $x \in A_k$ for some k, and $d(x, a) \leq d(x, a_k) + d(a_k, a) < r_k + d(a_k, a)$. Let

$$r = \max \{r_k + d(a_k, a) : 1 \leq k \leq n\};$$

then $d(x, a) < r$. Hence $A \subset B(a; r)$.

1.11 Let $A = \{u_1, \ldots, u_n\}$. Consider a point $u_k \in A$. If d is a metric, $d(u_k, u_i) \neq 0$ if $i \neq k$. Then
$$r = \min \{d(u_i, u_k) : 1 \leq i \leq n, \ i \neq k\}$$
is positive. Hence $B(u_k; r) \cap A = \{u_k\}$, that is, u_k is isolated in A.

Example 1.49 exhibits a finite set A in a pseudometric space in which one point is a limit point of A.

1.15 All sets are open. In (i)–(iv) the boundary is obtained when we replace the strict inequality by equality, and the closure is obtained when the strict inequality is replaced by nonstrict inequality of the same direction. In (v), the closure is $[\frac{1}{2}, 1] \times [\frac{1}{2}, 1]$. The boundary consists of two straight line segments; the first connects the points $(\frac{1}{2}, \frac{1}{2})$ and $(1, \frac{1}{2})$, the second connects $(\frac{1}{2}, \frac{1}{2})$ and $(\frac{1}{2}, 1)$. Observe that the points of the form $(x, 1)$, $\frac{1}{2} < x \leq 1$, and $(1, y)$, $\frac{1}{2} < y \leq 1$, are not boundary points of the given set.

1.17 In S: Each A_k is open: If $x \in (2k, 2k+1]$, then $B_S(x; 1) = (2k, 2k+1] = A_k$. Each A_k is closed since $B_k := \bigcup_{i \neq k} A_i$ is open and $A_k = B_k^c = S \setminus B_k$.
In $\mathbb{R}$: A_k is not open in $\mathbb{R}$ since $2k + 1$ is not an interior point of A_k, and it is not closed since $2k$ is in $\overline{A}_k$ but not in A_k.

1.23 Let $a \in S$. Then $a \in \overline{A} \iff (\forall r > 0) \ B(a; r) \cap A \neq \emptyset$. Hence

$$a \in \overline{A}^c \iff (\exists r > 0) \ B(a; r) \cap A = \emptyset \iff (\exists r > 0) \ B(a; r) \subset A^c \iff a \in (A^c)^\circ$$

Similarly, $a \in A^\circ \iff (\exists r > 0) \ B(a; r) \cap A^c = \emptyset$. Then

$$a \in (A^\circ)^c \iff (\forall r > 0) \ B(a; r) \cap A^c \neq \emptyset \iff a \in \overline{A^c}.$$

Finally, $(A^c)^\circ \subset \overline{A^c}$.

1.27 $A^\circ = \emptyset$, $\partial A = A \cup \{0\} = \overline{A}$.

1.31 Let $(a, b) \in A \times B$. There exist positive numbers r_1, r_2 such that $d(s, a) < r_1$ implies $s \in A$ and $\rho(t, b) < r_2$ implies $t \in B$. Let $r = \min\{r_1, r_2\}$. Then the ball $B((a, b); r)$ in $S \times T$ is contained in $A \times B$: Let $\sigma((s, t), (a, b)) = d(s, a) + \rho(t, b) < r$. Then $d(s, a) < r \leq r_1$ gives $s \in A$, and $\rho(t, b) < r \leq r_2$ gives $t \in B$, that is, $(s, t) \in A \times B$.

1.35 Boundedness: $|f_a(x)| = |d(x, a) - d(x, x_0)| \leq d(a, x_0) = \text{const}$.
Isometry: $|f_a(x) - f_b(x)| = |d(a, x) - d(b, x)| \leq d(a, b)$, and $|f_a(b) - f_b(b)| = d(a, b)$.

Chapter 2

2.1 If $a, b \in \mathbb{Z}$ and $|a - b| < 1$, then $a = b$. Thus if $a_n \to a$ in $\mathbb{Z}$, there exists $N \in \mathbb{N}$ such that $a_n = a$ for all $n \geq N$.
Let (a_n) be a Cauchy sequence in $\mathbb{Z}$. Then there exists $K \in \mathbb{N}$ such that $|a_m - a_n| < \frac{1}{2}$ if $m, n \geq K$. Hence $a_n = a_K$ for all $n \geq K$. The space is complete.

2.3 Let x_n ($n \in \mathbb{N}$) and y be elements of Q such that for each $k \in \mathbb{N}$, $\xi_{nk} \to \eta_k$ as $n \to \infty$. Given $\varepsilon > 0$, there exists $N \in \mathbb{N}$ such that $\sum_{k=N+1}^{\infty} 1/k^2 < \frac{1}{4}\varepsilon^2$. Then

$$d(x_n, y) = \left(\sum_{k=1}^{\infty} |\xi_{nk} - \eta_k|^2 \right)^{1/2}$$

$$\leq \left(\sum_{k=1}^{N} |\xi_{nk} - \eta_k|^2 \right)^{1/2} + \left(\sum_{k=N+1}^{\infty} |\xi_{nk}|^2 \right)^{1/2} + \left(\sum_{k=N+1}^{\infty} |\eta_k|^2 \right)^{1/2}$$

$$\leq \left(\sum_{k=1}^{N} |\xi_{nk} - \eta_k|^2 \right)^{1/2} + 2\left(\sum_{k=N+1}^{\infty} \frac{1}{k^2} \right)^{1/2} \leq \left(\sum_{k=1}^{N} |\xi_{nk} - \eta_k|^2 \right)^{1/2} + \varepsilon,$$

and $\limsup_{n\to\infty} d(x_n, y) \leq \varepsilon$. Since $\varepsilon > 0$ was arbitrary, $d(x_n, y) \to 0$. The converse implication is clear.

2.5 (i) $f(x) \equiv 0$; $d(f_n, f) = 1/n$ on $[0, 1]$, uniform.
(ii) $f(x) = 1$ for $x > 0$, $f(0) = 0$; $d(f_n, f) = 1$ on $[0, \infty)$, not uniform; $d(f_n, f) = 1/(1 + na)$ on (a, ∞), uniform.
(iii) $f(x) = x$; $d(f_n, f) = 1$ on $(0, 1)$, not uniform; $d(f_n, f) = (1 - a)/(1 + na)$ on $(a, 1)$, uniform.
(iv) $f(x) \equiv 0$; $d(f_n, f) = n^{n+2}/(n+1)^{n+1} \to \infty$ on $(0, 1)$, not uniform; $d(f_n, f) = f_n(a) \to 0$ on $(a, 1)$, uniform.

2.7 Given $\varepsilon > 0$ there is $N(\varepsilon) := N$ such that $\|f_n - f_m\| \leq \varepsilon$ if $n, m \geq N$. From $|f_n(t) - f_m(t)| \leq \|f_n - f_m\|$ we conclude that $(f_n(t))$ is Cauchy in $\mathbb{R}$ for any $t \in E$. Define $f(t) := \lim_{n\to\infty} f(t)$ for all $t \in E$. Then for any $n \geq N$,

$$|f_n(t) - f(t)| = \lim_{m\to\infty} |f_n(t) - f_m(t)| \leq \varepsilon;$$

since ε was arbitrary, $\|f_n - f\| = \sup_{t \in E} |f_n(t) - f(t)| \leq \varepsilon$ if $n \geq N$. Then f is bounded as $|f(t)| \leq |f_N(t)| + |f_N(t) - f(t)| \leq \|f_N\| + \varepsilon$.

2.11 Every real valued function f continuous on $[a, b]$ is bounded on $[a, b]$; thus $C[a, b] \subset B[a, b]$. Let f be contained in the closure of $C[a, b]$ in the space $B[a, b]$ equipped with the uniform norm $\|\cdot\|$. We show that f is continuous on $[a, b]$. Let $\varepsilon > 0$. Then there exists $g \in C[a, b]$ with $\|f - g\| < \frac{1}{3}\varepsilon$. Let $s \in [a, b]$. Since g is continuous at s, there exists $\delta > 0$ such that $|g(t) - g(s)| < \frac{1}{3}\varepsilon$ if $|t - s| < \delta$ in $[a, b]$. Let $|t - s| < \delta$ in $[a, b]$. Then

$$|f(t) - f(s)| \leq |f(t) - g(t)| + |g(t) - g(s)| + |g(s) - f(s)|$$
$$\leq \|f - g\| + |g(t) - g(s)| + \|g - f\| < \tfrac{1}{3}\varepsilon + \tfrac{1}{3}\varepsilon + \tfrac{1}{3}\varepsilon = \varepsilon.$$

2.15 Let $x_n = (1, \frac{1}{2}, \frac{1}{3}, \dots, \frac{1}{n}, 0, 0, \dots)$ in E_0. Then (x_n) is a Cauchy sequence. If $x \in E_0$, there exists N such that $\xi_k = 0$ for all $k \geq N$. For any $n \geq N$, $d(x_n, x) = 1/N = \text{const}$, and $d(x_n, x) \not\to 0$.

2.17 Write $S = [0, 2]$ and $T = \{0, 4\} \cup [1, 2) \cup (2, 3)$. Then f is a bijection from S to T, and the metric spaces (S, ρ) and (T, d), where d is the Euclidean metric $d(u, v) = |u - v|$, are isometric. The completion of (T, d) is the closure of T in the complete space $(\mathbb{R}, d)$: $\overline{T} = \{0, 4\} \cup [1, 3]$. Hence we have to add two ideal points α and β to S, corresponding to 2 and 3 in T, respectively: $S^\sharp = S \cup \{\alpha, \beta\}$. To extend the metric ρ to $S^\sharp$, extend first f to $S^\sharp$ by setting $f^\sharp(\alpha) = 2$ and $f^\sharp(\beta) = 3$, and define $\rho^\sharp(x, y) = |f^\sharp(x) - f^\sharp(y)|$ for all $x, y \in S^\sharp$. Note that 0 and 2 are isolated points of S and $S^\sharp$, and that

$$\tfrac{1}{n} \xrightarrow{\rho} 1, \quad 1 - \tfrac{1}{n} \xrightarrow{\rho} \alpha, \quad 2 - \tfrac{1}{n} \xrightarrow{\rho} \alpha, \quad 1 + \tfrac{1}{n} \xrightarrow{\rho} \beta.$$

2.19 According to Equation (2.5) with $x = x_m$ and $y = x^*$, $d(x_m, x^*) \leq (1 - \alpha)^{-1}\alpha^m d(x_1, x_0)$. Write $\kappa := (1 - \alpha)^{-1} d(x_1, x_0)$. To obtain $d(x_m, x^*) \leq \kappa \alpha^m < \varepsilon$, we need $m > \log(\varepsilon/\kappa)/\log \alpha$ iterations.

2.21 Use the first derivative $f'(x) = (2\sqrt{x} - 1)/(2\sqrt{x})$: $\alpha := \sup\{|f'(x)| : x \in J\}$.

J	$f(J)$	selfmap	α	contr.	J compl.	fix'd point
$[0, 1]$	$[\frac{7}{4}, 2]$	no	∞	no	yes	no
$[\frac{1}{4}, 5]$	$[\frac{7}{4}, 7 - \sqrt{5}]$	yes	$1 - \sqrt{5}/10$	yes	yes	yes
$(\frac{1}{16}, 9]$	$[\frac{7}{4}, 8]$	yes	1	no	no	yes
$[1, 3]$	$[2, 5 - \sqrt{3}]$	no	$1 - \sqrt{3}/6$	yes	yes	no
$(1, 4]$	$[2, 4]$	yes	$\frac{3}{4}$	yes	no	yes

2.23 A condition equivalent to contraction is $\sup d(f(x), f(y))/d(x, y) = \alpha < 1$ $(x \neq y)$. For $x, y \geq 1$,

$$\sup_{x \neq y} \frac{d(f(x), f(y))}{d(x, y)} = \sup_{x \neq y} |1 - (xy)^{-1}| = 1.$$

Hence f is not a contraction.

2.25 $f'(x) = 1/(4f(x)\sqrt{x}) > 0$. Thus f is strictly increasing, and $\sqrt{3} < f(\sqrt{3}) < f(2) < 2$. So f is a selfmap of $[\sqrt{3}, 2]$. The function f' is decreasing, so $f'(\sqrt{3})$ is the maximum value of f' on $[\sqrt{3}, 2]$: $\alpha = f(\sqrt{3}) < f(1) = \sqrt{3}/12 < 1$. The functions satisfies the hypotheses of the contraction mapping theorem on $[\sqrt{3}, 2]$, and there is a unique fixed point x^* in that interval. For the rest of the problem we observe that $x = f(x)$ in $[\sqrt{3}, 2]$ if and only if $x^4 - 4x^2 - x + 4 = 0$.

2.27 (i) For this solution we use the full covers of closed bounded intervals discussed in Appendix B. Let $\varepsilon > 0$. We define the family $\mathcal{C}$ of subintervals $[u, v]$ of $[a, b]$ as follows:

$$[u, v] \in \mathcal{C} \iff \left(x \in [u, v] \implies \begin{cases} \|f(t) - f(x-)\| < \varepsilon & \text{if } u < t < x, \\ \|f(t) - f(x+)\| < \varepsilon & \text{if } x < t < v. \end{cases} \right)$$

From the definition of one sided limits we conclude that $\mathcal{C}$ is a full δ-cover of $[a, b]$ for some positive function $\delta \colon [a, b] \to \mathbb{R}$. By Cousin's lemma there is a partition $a = u_0 < u_1 < \cdots < u_n = b$ of $[a, b]$ with $[u_{k-1}, u_k] \in \mathcal{C}$ for all $k = 1, \ldots, n$. Select a point x_k in each subinterval (u_{k-1}, u_k), and for each $k \in \{1, \ldots, n\}$ define

$$g(t) = \begin{cases} f(x_k-) & \text{if } t \in (u_{k-1}, x_k), \\ f(x_k+) & \text{if } t \in (x_k, u_k), \\ f(x_k) & \text{if } t = x_k. \end{cases}$$

Then g is a step function, and $\|f(t) - g(t)\| < \varepsilon$ for all $t \in [a, b]$. Choosing ε to be $1, 1/2, 1/3, \ldots$ in succession, we construct a sequence (g_n) of step functions uniformly convergent to f on $[a, b]$.

Conversely assume that $\lim_{n \to \infty} f_t(t) = f(t)$ uniformly for $t \in [a, b]$, where f_n are X-valued step functions. To prove the existence of $f(t_0+)$ for $t_0 \in [a, b)$ we use the interchange of limits principle for the uniform convergence (Problem 2.10),

$$\lim_{n \to \infty} \left(\lim_{t \to t_0+} f_n(t) \right) = \lim_{t \to t_0+} \left(\lim_{n \to \infty} f_n(t) \right) = \lim_{t \to t_0+} f(t);$$

the existence of the left limit $f(t_0-)$ for $t_0 \in (a, b]$ is proved analogously.

(ii) Let $f \colon [a, b] \to X$ be quasicontinuous. Then f is the uniform limit on $[a, b]$ of a sequence (f_n) of X-valued step functions. The set D_n of discontinuities of f_n is finite, and the set $D = \bigcup_{n=1}^{\infty} D_n$ is countable. At every point $t_0 \in [a, b] \setminus D$ the interchange of limits principle applies, which means that f is continuous at every such t_0. For a counterexample consider the function $f \colon [0, 1] \to \mathbb{R}$ defined by $f(t) = 0$ ($f(t) = 1$) if t irrational (rational).

(iii) If $f \colon [a, b] \to \mathbb{R}$ is increasing, then $f(x-) = \sup_{t < x} f(t)$ for every $x \in (a, b]$, and $f(x+) = \inf_{t > x} f(t)$ for every $x \in [a, b)$. For f decreasing apply the preceding result to $-f$.s

Chapter 3

3.3 (i) We use the closed set characterization of continuity: A mapping $f: S \to T$ is continuous if and only if $f^{-1}(V)$ is closed in S for all closed $V \subset T$.

Suppose first that $f: S \to T$ is continuous. Let $A \subset S$. Then

$A \subset f^{-1}(f(A))$	true for any mapping, continuous or not
$A \subset f^{-1}(\overline{f(A)})$	since $f(A) \subset \overline{f(A)}$
$\overline{A} \subset f^{-1}(\overline{f(A)})$	since $f^{-1}(\overline{f(A)})$ is closed by the criterion of continuity
$f(\overline{A}) \subset \overline{f(A)}$	applying f to both sides

Conversely assume that $f: S \to T$ is such that $f(\overline{A}) \subset \overline{f(A)}$ for all subsets A of S. Let V be a closed subset of T; we show that $A := f^{-1}(V)$ is closed in S:

$f(A) \subset V$	definition of $f^{-1}(V)$
$f(\overline{A}) \subset \overline{f(A)} \subset \overline{V} = V$	hypothesis and the fact that V is closed
$\overline{A} \subset f^{-1}(V) = A$	definition of $f^{-1}(V)$

Hence A is closed. (ii) is proved similarly.

3.5 A point $a \in S$ is isolated in S if and only if there exists $\delta > 0$ such that $B(a; \delta) = \{a\}$. For any open ball $B(f(a); \varepsilon)$ in T,

$$f(B(a; \delta)) = f(\{a\}) = \{f(a)\} \subset B(f(a); \varepsilon).$$

3.9 The functions are continuous on the following sets, discontinuous everywhere else:

(i) $\mathbb{R}^2 \setminus \{(\xi_1, \xi_2) : \xi_2 = 0\}$.

(ii) $\mathbb{R}^2 \setminus \{(\xi_1, \xi_2) : \xi_1 = 1\}$.

(iii) $\mathbb{R}^2$: At the origin we have $f(x) = \|x\|_2^2 / \log(1 + \|x\|_1) \leq \|x\|_2^2 / \log(1 + \|x\|_2) \to 0$ as $\|x\|_2 \to 0$. (Here $\|\cdot\|_p$ refers to the ℓ^p norm on $\mathbb{R}^3$, $p = 1, 2$.)

(iv) $\mathbb{R}^2 \setminus \{(\xi_1, \xi_2) : 2\xi_1^2 + \xi_2 = 0\}$. For the discontinuity at the origin set $\xi_2 = -2\xi_1^2 + \xi_1^4$.

(v) $\mathbb{R}^2 \setminus \{(\xi_1.\xi_2) : \xi_2 = |\xi_1|\}$. (Continuous on $\xi_2 = -|\xi_1| \neq 0$.)

3.11 For the unboundedness set $\xi_2 = \xi_1^4$.

3.15 Let G be open in T. Then $f^{-1}(G)$ is open in the discrete space S, as all subsets of that space are open (and closed).

3.17 Let $t_0 \in S$, and let $\varepsilon > 0$ be given. There exists $\delta_n(\varepsilon) > 0$ such that $\rho(f_n(t), f_n(t_0)) < \varepsilon$ whenever $d(t, t_0) < \delta_n(\varepsilon)$. Since $f_n \to f$ uniformly on S, there exists $N(\varepsilon) \in \mathbb{N}$ such that $\rho(f_n(t), f(t)) < \varepsilon$ for all $t \in S$ whenever $n \geq N(\varepsilon)$. Pick $k \geq N(\frac{1}{3}\varepsilon)$ and keep it fixed. Now set $\delta(\varepsilon) := \delta_k(\frac{1}{3}\varepsilon)$. If $d(t, t_0) < \delta(\varepsilon)$, then

$$\rho(f(t), f(t_0)) \leq \rho(f(t), f_k(t)) + \rho(f_k(t), f_k(t_0)) + \rho(f_k(t_0), f(t_0))$$
$$< \tfrac{1}{3}\varepsilon + \tfrac{1}{3}\varepsilon + \tfrac{1}{3}\varepsilon = \varepsilon.$$

Chapter 4

4.3 (ii) Let A_α be a family of sets in $\mathcal{T}$, and let $A = \bigcup_\alpha A_\alpha$. By de Morgan's law, $A^c = \bigcap_\alpha A_\alpha^c$, where each set A_α^c is countable. Then A^c is also countable, and $A \in \mathcal{T}$.

Let $A_1, \ldots, A_n$ be sets in $\mathcal{T}$, and let $B = \bigcap_{i=1}^n A_i$. Then $B^c = \bigcup_{i=1}^n A_i^c$, where each A_i^c is countable. Hence B^c is countable, and $B \in \mathcal{T}$. $X \in \mathcal{T}$ as $X^c = \emptyset$ is countable (cardinality zero).

(i) is proved similarly.

4.9 The indiscrete topology on X is $\mathcal{T} = \{\emptyset, X\}$. Let $a \neq b$ be two points in X. If d is any metric on X, then $B(a; \frac{1}{2}d(a,b))$ is an open set containing a but not b. In $(X, \mathcal{T})$, X is the only open set containing a, and it also contains b.

4.11 4.1: If $X = \{a\}$, then vacuously Hausdorff. If $\mathsf{card}\,(X) > 1$, then not Hausdorff as any point other than a is contained in one only open set X.

4.2: Not Hausdorff: the only open set containing d is X.

4.3: Not Hausdorff: If $a \in A$ and $b \in B$, where A, B are two disjoint sets, then $B \subset A^c$. This shows that A, B cannot be both open.

4.15 First show that $f_n \to f$ in $(X, \mathcal{T})$ if and only if $f_n(x) \to f(x)$ for all $x \in \mathbb{R}$. Let A be the subset of X consisting of those functions $f \colon \mathbb{R} \to \mathbb{R}$ with $f(x) = 0$ for all x in a finite subset of $\mathbb{R}$, and $f(x) = 1$ elsewhere. Show that the function g which is 0 everywhere in $\mathbb{R}$ is in the closure $\overline{A}$ but no sequence in A converges to it.

4.17 Let (S_i, d_i) $(i = 1, \ldots, n)$ be metric spaces, and let d be the product metric on $S := S_1 \times \cdots \times S_n$ defined by $d(x,y) = \sum_{i=1}^n d_i(x_i, y_i)$. Further, let $\mathcal{T}_i$ be the topology induced on S_i by d_i $(i = 1, \ldots, n)$ and $\mathcal{U}$ the topology on S induced by d, and $\mathcal{T}$ the product topology on S.

   Let $V \in \mathcal{T}$ and $a \in V$. Then there exist $U_i \in \mathcal{T}_i$ such that $a \in U_1 \times \cdots \times U_n \subset V$. Since $\mathcal{T}_i$ is induced by the metric d_i, for each i there exists $\varepsilon_i > 0$ such that $B_{d_i}(a_i; \varepsilon_i) \subset U_i$ $(i = 1, \ldots, n)$. Let $\varepsilon = \min_i \varepsilon_i$. Then $B_d(a; \varepsilon) \subset B_{d_1}(a_1; \varepsilon_1) \times \cdots \times B_{d_n}(a_n; \varepsilon_n) \subset U_1 \times \cdots \times U_n \subset V$. Thus V is open in the topology $\mathcal{U}$ induced by d. This proves that $\mathcal{T} \subset \mathcal{U}$.

   Conversely, let $U \in \mathcal{U}$, and let $a \in U$. Then there exists $\varepsilon > 0$ such that $B_d(a; \varepsilon) \subset U$. The set $V := B_{d_1}(a_1; \varepsilon/n) \times \cdots \times B_{d_n}(a_n; \varepsilon/n)$ is in $\mathcal{T}$, and $a \in V \subset B(a; \varepsilon) \subset U$. Hence $U \in \mathcal{T}$, and $\mathcal{U} \subset \mathcal{T}$.

4.19 We need the fact that the projections $\pi_i \colon X \to X_i$ are continuous and open (Problem 4.20).

(i) Let the X_i be Hausdorff for $i = 1, \ldots, n$. Let a, b be distinct points in X. Then for some k, $a_k \neq b_k$. Since X_k is Hausdorff, there exist disjoint open neighbourhoods U_k, V_k in X_k of a_k, b_k, respectively. From the continuity of π_k, the sets $U = \pi_k^{-1}(U_k)$ and $V = \pi_k^{-1}(V_k)$ are open, disjoint and $a \in U$, $b \in V$. Hence X is Hausdorff.

Conversely, let X be Hausdorff. If a_k, b_k are distinct points in X_k, choose points $a \in \pi_i^{-1}(\{a_i\})$ and $b \in \pi_i^{-1}(\{b_i\})$. Then a and b are distinct, and there exist disjoint open neighbourhoods U, V of a, b, respectively. Since the projection π_k is open, the sets $U_k = \pi_k(U)$ and $V_k = \pi_k(V)$ are open and disjoint in X_k, and $a_k \in U_k$, $b_k \in V_k$.

(ii) Suppose that for each i the space X_i is separable with a countable dense subset Y_i. Then $Y = Y_1 \times \cdots \times Y_n$ is countable. If $a \in X$ and U is an open neighbourhood of a in X, there exist open sets $U_i \subset X_i$ such that $a \in U_1 \times \cdots \times U_n \subset U$. For each i there exists $y_i \in U_i \cap Y_i$; so $y \in U \cap Y$, and X is separable with a countable dense subset Y.

Conversely, let X be separable with a countable dense subset Y. For a given $k \in \{1, \ldots, n\}$, $\pi_k(Y)$ is a countable dense subset of X_k.

4.21 Since $(\overline{B})^c$ is an open set, for each $a \in A$ there exists $r(a) > 0$ such that $B(a; r(a)) \subset (\overline{B})^c$. Similarly, for each $b \in B$ there exists $s(b) > 0$ such that $B(b; s(b)) \subset (\overline{A})^c$. If $a \in A$ and $b \in B$, we can show that

$$B(a; \tfrac{1}{2}r(a)) \cap B(b; \tfrac{1}{2}s(b)) = \emptyset. \tag{F.1}$$

Set $U = \bigcup_{a \in A} B(a; \tfrac{1}{2}r(a))$ and $V = \bigcup_{b \in B} B(b; \tfrac{1}{2}s(b))$. Then $A \subset U$, $B \subset V$, and U, V are disjoint in view of (F.1).

Chapter 5

5.7 Suppose that X is compact. For a proof by contradiction assume that there exists a family $\{F_\alpha\}$ of closed sets with the finite intersection property whose intersection is empty. Then $\{F_\alpha^c\}$ is an open cover for X. Any finite subfamily of $\{F_\alpha\}$ has nonempty intersection by assumption, which means that the open cover $\{F_\alpha^c\}$ has no open subcover.

Conversely, assume that every family of closed sets with the finite intersection property has nonempty intersection. Suppose there is an open cover $\{U_\alpha\}$ of X which has no finite subcover. Then $\{U_\alpha^c\}$ is a family of closed sets with the finite intersection property but empty intersection.

5.9 By the definition of infimum there exists a sequence (u_n) is A such that $d(s, u_n) \to d(s, A)$ (a so called *minimizing sequence*). Since A is compact, there exists a subsequence (u_{k_n}) convergent to a point $a \in A$. Then $d(s, u_{k_n}) \to d(s, a) = d(s, A)$ by the continuity of the function $x \mapsto d(s, x)$. If S is the Euclidean plane, A the unit circle and s the origin, every point a in A satisfies $d(s, a) = d(s, A) = 1$. If S is the Euclidean line, $A = (1, 2]$ and $s = 0$, then $d(s, A) = 1$, but $d(s, a) > 1$ for every $a \in A$.

5.13 (i) Let $\{U_\alpha\}$ be an open cover for a set $A \subset X$. Pick a particular U_{α_0}. Then the set $A \cap U_{\alpha_0}^c \subset U_{\alpha_0}^c$ is finite, say $\{a_1, \ldots, a_n\}$. For each i there exists U_{α_i} which contains a_i; then $\{U_{\alpha_0}, U_{\alpha_1}, \ldots, U_{\alpha_n}\}$ is a finite subcover for A.

(ii) If A is a proper infinite subset of X, it is compact by part (i), but A is not closed (only finite subsets and X are closed).

5.15 (i) Contraposition: Let X be a compact topological space. Suppose that A is a subset of X with no limit point. Then for each $x \in X$ there exists an open neighbourhood $U(x)$ such that $U(x) \cap A = \{x\} \cap A$. Since $\{U(x) : x \in X\}$ is an open cover for X, there exists a finite subcover $\{U(x_i) : i = 1, \ldots, n\}$ of X. But each $U(x_i)$ contains at most one point of A, and A is finite.

(ii) Contraposition: Suppose S is a metric space which is not compact. Then S contains an injective sequence (x_n) which has no convergent subsequence. Thus $\{x_n : n \in \mathbb{N}\}$ is an infinite set with no limit point.

5.17 Using the intermediate value theorem, we can prove that a continuous injective function f on an interval I is strictly monotonic. Suppose that f is strictly increasing. Let $y_0 \in f(I)$ be given, and let $x_0 = f(y_0)$. If $[u, v]$ is any closed bounded interval such that $x_0 \in (u, v)$, then $y_0 \in f([u, v]) = [f(u), f(v)]$. The restriction $f_0 : [u, v] \to [f(u), f(v)]$ of f is a continuous bijection between compact spaces, and hence g is continuous at y_0.

5.21 Show first that $x_n \to y$ in Q if and only if $\xi_{nk} \to \eta_k$ for each index k. Then prove that Q is totally bounded and complete.

5.23 Let $S_0 = S$ and $S_{n+1} = f(S_n)$, $n \in \mathbb{N}$. If $K = \bigcap_{n=1}^{\infty} S_n$, show that $f(K) = K$ using the continuity of f; K is nonempty by the finite intersection property of compact spaces. Then conclude that K is a singleton using the assumption on the metric.

5.25 We prove by induction that for all $t \in [0, 1]$, $0 \le p_n(t) \le \sqrt{t}$, $n = 1, 2, \ldots$ The case $n = 1$ is clear. Assume that the inequality is true for some n. Then

$$\sqrt{t} - p_{n+1}(t) = \sqrt{t} - p_n(t) - \tfrac{1}{2}(t - p_n^2(t)) = (\sqrt{t} - p_n(t))(1 - \tfrac{1}{2}(\sqrt{t} + p_n(t))) \ge 0$$

since $\tfrac{1}{2}(\sqrt{t} + p_n(t)) \le \sqrt{t} \le 1$. Consequently, $p_{n+1}(t) \ge p_n(t)$ for all $t \in [0, 1]$ and all n. Since (p_n) is increasing and bounded, it converges pointwise to some limit $f(t)$; from the inductive definition of p_n we get $t - [f(t)]^2 = 0$, and so $f(t) = \sqrt{t}$ as $f(t) \ge 0$. By Dini's theorem (p_n) converges uniformly on $[0, 1]$.

5.27 The coordinates of a multiindex $\alpha = (\alpha_1, \ldots, \alpha_k)$ in $\mathbb{R}^k$ are nonnegative integers, and $|\alpha| = \alpha_1 + \cdots + \alpha_k$. A polynomial in k variables can be written as $p(x) = \sum_{|\alpha|=0}^{n} c_\alpha x^\alpha$, where $x^\alpha = x_1^{\alpha_1} \cdots x_k^{\alpha_k}$, and c_α are real coefficients. The degree of $p(x)$ is the maximum $|\alpha|$ for which $c_\alpha \ne 0$. Adding summands with zero coefficients if necessary we can write any two polynomials as sums from $|\alpha| = 0$ to $|\alpha| = n$ for n the maximum degree of the two polynomials. Then

$$p(x) + q(x) = \sum_{|\alpha|=0}^{n} (c_\alpha + d_\alpha) x^\alpha, \quad p(x)q(x) = \sum_{|\gamma|=0}^{2n} \left(\sum_{\gamma = \alpha + \beta} c_\alpha d_\beta \right) x^\gamma.$$

From these expressions we deduce that $\mathcal{P}$ is an algebra. Further, $\mathcal{P}$ contains $p(x) = x^{(0,\ldots,0)} \equiv 1$. Let a, b be two distinct points of K, say $a_1 \neq b_1$. Then $p(a) = a_1 \neq b_1 = p(b)$ for $p(x) = x^{(1,0,\ldots,0)}$.

Chapter 6

6.3 Suppose first that X is disconnected. Then there exists a proper subset A of X which is both open and closed. Set $B = A^c$. Then $X = A \cup B$, and

$$\overline{A} \cap B = A \cap B = \emptyset, \quad A \cap \overline{B} = A \cap B = \emptyset.$$

Conversely assume that $X = A \cup B$ with nonempty sets A, B satisfying $\overline{A} \cap B = \emptyset = A \cap \overline{B}$. From $\overline{A} \cap B = \emptyset$ we obtain $\overline{A} \subset B^c = A$ as $X = A \cup B$. Thus A is closed. Similarly, B is closed, which makes A open. Hence A is a proper subset of X which is both open and closed, and so X is disconnected.

6.5 Contrapositive: Let X be disconnected. There is a proper subset A of X which is both open and closed. Set $B = A^c$, and let $f = \chi_A$ be the characteristic function of A in X. Then f is continuous as $f^{-1}(U)$ is open for every open subset of $Y = \{0, 1\}$: $f^{-1}(Y) = X$, $f^{-1}(\emptyset) = \emptyset$, $f^{-1}(\{1\}) = A$, $f^{-1}(\{0\}) = B$.
Conversely, let $f : X \to \{0, 1\}$ be a nonconstant continuous function. Then the sets $A = f^{-1}(\{1\})$ and $B = f^{-1}(\{0\})$ are nonempty and open, and $X = A \cup B$. So X is disconnected.

6.7 Let $f : B \to \{0, 1\}$ be a continuous function from B to the discrete space $\{0, 1\}$. Since A is connected and the restriction of f to A is continuous, f is constant on A, say $f(A) = \{0\}$. From the continuity of f on B,

$$f(B) = f(\overline{A} \cap B) \subset f(\overline{A}) \subset \overline{f(A)} = \overline{\{0\}} = \{0\}.$$

Thus every continuous function $f : B \to \{0, 1\}$ is constant, and B is connected by Theorem 6.10.

6.9 Contrapositive: Let A be a finite subset of X, $A = \{a_1, \ldots, a_n\}$. Since X is Hausdorff, there are disjoint open sets U_i such that $a_i \in U_i$ ($i = 1, \ldots, n$). Then each singleton $\{a_k\} = A \cap U_k$ is open in the subspace topology of A, so that $B_k = \bigcup_{i \neq k} \{a_i\}$ is open in A. Then $\{a_k\} = A \setminus B_k$ is closed, and A is disconnected.
For a counterexample consider the plane $\mathbb{R}^2$ with the pseudometric $p(x, y) = |\xi_1 - \eta_1|$. The ball $A = \{(\xi_1, \xi_2) : -1 < \xi_1 < 1\}$ in $(\mathbb{R}^2, p)$ is connected (Problem 6.5), yet it contains infinitely many points.

6.15 Let $\varphi : [0, 1] \to S$ be a path in S. Then the set $\varphi^{-1}(B)$ is closed as B is closed. It is also open: let $t \in \varphi^{-1}(B)$, and let U be a neighbourhood of $\varphi(t)$ in S small enough to ensure that U is not a connected subset of S. From the continuity of φ it follows that there is an open interval $I \subset [0,1]$ containing t such that $f(I) \subset U$. But $\varphi(I)$ is connected, so $\varphi(I) \subset B$ as $\varphi(t) \in \varphi(I) \cap B$. Hence $I \subset \varphi^{-1}(B)$ and $\varphi^{-1}(B)$ is open.

If φ were a path from a point in A to a point in B, then $\varphi^{-1}(B)$ would be a proper subset of $[0,1]$, both open and closed, which is impossible as $[0,1]$ is connected. This contradiction shows that there is no such path, and S is not path connected.

Chapter 7

7.1 (i) Triangle inequality: $\|x\| = \|y + (x-y)\| \le \|y\| + \|x-y\|$, and $\|y\| = \|x + (y-x)\| \le \|x\| + \|y-x\|$. Hence

$$\|x - y\| \ge \max\{\|x\| - \|y\|, \|y\| - \|x\|\} = |\,\|x\| - \|y\|\,|.$$

7.3 Only (i) is induced by a norm; (ii)–(iv) do not satisfy **N3**.

7.5 The first inequality follows from $\left(\sum_{i=1}^{n} |\xi_i|\right)^2 = \sum_{i,j=1}^{n} |\xi_i \xi_j| \ge \sum_{i=1}^{n} |\xi_i|^2$ by taking the square root. For the second inequality we need the Cauchy-Schwarz inequality:

$$\sum_{i=1}^{n} |\xi_i| = \sum_{i=1}^{n} 1 \cdot |\xi_i| \le \left(\sum_{i=1}^{n} 1^2\right)^{1/2} \left(\sum_{i=1}^{n} |\xi_i|^2\right)^{1/2} = \sqrt{n}\,\|x\|_2.$$

7.7 Write $\alpha = \|x\|_p$. Since $\alpha^{-1}|\xi_i| \le 1$ for each i, we have $(\alpha^{-1}|\xi_i|)^p \le \alpha^{-1}|\xi_i|$, and $\alpha^{-p} \sum_{i=1}^{n} |\xi_i|^p \le \alpha^{-1} \sum_{i=1}^{n} |\xi_i|$, that is, $\alpha^{-p}\alpha^p \le \alpha^{-1}\|x\|_1$, or $\|x\|_p = \alpha \le \|x\|_1$. Next we need Hölder's inequality with $q = p/(p-1)$ (see (C.6)):

$$\|x\|_1 = \sum_{i=1}^{n} 1 \cdot |\xi_i| \le \left(\sum_{i=1}^{n} 1^q\right)^{1/q} \left(\sum_{i=1}^{n} |\xi_i|^p\right)^{1/p} = n^{1/q}\|x\|_p.$$

So far we have proved that $n^{(1-p)/p}\|x\|_1 \le \|x\|_p \le \|x\|_1$ (as $-1/q = (1-p)/p$). To prove that the constants $\lambda_n = n^{(1-p)/p}$ and $\mu_n = 1$ are the best possible: For $x = (1,1,\ldots,1)$, $\|x\| = n$ and $\|x\|_p = n^{1/p}$; then $n^{(1-p)/p}\|x\|_1 = \|x\|_p$. For $y = (1,0,\ldots,0)$, $\|y\|_1 = 1 = \|y\|_p$.

7.9 Let $x, y \in \overline{M}$ and let $\lambda \ne 0$. Given $\varepsilon > 0$, there exist $u, v, w \in M$ such that $\|x - u\| < \frac{1}{2}\varepsilon$, $\|y - v\| < \frac{1}{2}\varepsilon$ and $\|x - w\| < |\lambda|^{-1}\varepsilon$. Then

$$\|(x+y) - (u+v)\| \le \|x-u\| + \|y-v\| < \tfrac{1}{2}\varepsilon + \tfrac{1}{2}\varepsilon = \varepsilon,$$

and $\|\lambda x - \lambda w\| = |\lambda|\|x - w\| < \varepsilon$. Since $u + v \in M$ and $\lambda w \in M$, $x + y \in \overline{M}$ and $\lambda x \in \overline{M}$.

7.11 All spaces are infinite dimensional with the exception of (v) when E is a finite set. All spaces are complete with the exception of (vii): Let f_n be the polygonal (piecewise linear) function with the vertices $(a, 0)$, $(\frac{1}{2}(a+b) - (\frac{1}{2})^n, 0)$, $(\frac{1}{2}(a+b), 1)$, $(b, 1)$ (for n sufficiently large). Show that $(f_n) \subset C[a, b]$ is Cauchy, has a limit in $L^1[a, b]$, but no limit in $C[a, b]$.

7.15 If $x \in \ell^p \setminus \{0\}$ for some $p \geq 1$ and $s > p$, set $\alpha_i = \xi_i / \|x\|_p$ ($i \in \mathbb{N}$). Then $|\alpha_i|^s \leq |\alpha_i|^p$ as $|\alpha_i| \leq 1$ for all i. Then

$$\frac{|\xi_i|^s}{\|x\|_p^s} \leq \frac{|\xi_i|^p}{\|x\|_p^p} \implies \|x\|_p^p \sum_{i=1}^{\infty} |\xi|^s \leq \|x\|_p^s \sum_{i=1}^{\infty} |\xi_i|^p,$$

$$\|x\|_p^{p/s} \|x\|_s \leq \|x\|_p \|x\|_p^{p/s} \implies \|x\|_s \leq \|x\|_p.$$

This shows that $\ell^s \subset \ell^p$ for $s > p$, and that the norm decreases.

By the preceding argument, $\ell^p \subset \ell^{p_0} \subset \ell^{\infty}$ if $p > p_0$. Let $x \in \ell^{p_0}$. Since $\varphi(p) = \|x\|_p$ is decreasing, $\lim_{p \to \infty} \varphi(p)$ exists. Let $\varepsilon > 0$. Then there exists an index $N \in \mathbb{N}$ such that for all $n \geq N$, $\sum_{n+1}^{\infty} |\xi_i|^{p_0} < \varepsilon^{p_0}$. Let $p \geq p_0$. By the preceding problem,

$$\left(\sum_{n+1}^{\infty} |\xi_i|^p \right)^{1/p} \leq \left(\sum_{n+1}^{\infty} |\xi_i|^{p_0} \right)^{1/p_0} < \varepsilon \quad \text{if } n \geq N.$$

We observe that

$$\left(\sum_{i=1}^{n} |\xi_i|^p \right)^{1/p} \leq n^{1/p} \max_{1 \leq i \leq n} |\xi_i| \leq n^{1/p} \|x\|_{\infty}.$$

If $n \geq N$, then

$$\|x\|_p \leq n^{1/p} \|x\|_{\infty} + \left(\sum_{k=n+1}^{\infty} |\xi_k|^p \right)^{1/p} < n^{1/p} \|x\|_{\infty} + \varepsilon;$$

therefore $\lim_{p \to \infty} \|x\|_p \leq \|x\|_{\infty} + \varepsilon$. Since $\varepsilon > 0$ was arbitrary, $\lim_{p \to \infty} \|x\|_p \leq \|x\|_{\infty}$.

On the other hand, for each i and each $p \geq p_0$, $|\xi_i| \leq \|x\|_p$, and $\|x\|_{\infty} \leq \|x\|_p$. The result then follows on taking the limit as $p \to \infty$.

7.17 Both ℓ^{∞} and $\mathbf{c}$ contain the element $e = (1, 1, 1, \dots)$, for which

$$\left\| e - \sum_{i=1}^{n} e_i \right\| = \|(0, \dots, 0, 1, 1, \dots)\| = 1 \not\to 0.$$

Let $x = (\xi_1, \xi_2, \dots) \in \mathbf{c}_0$; then $\lim_{i \to \infty} \xi_i = 0$. For any $n \in \mathbb{N}$,

$$\lim_{n \to \infty} \left\| x - \sum_{i=1}^{n-1} \xi_i e_i \right\| = \lim_{n \to \infty} \|(0, \dots, 0, \xi_n, \xi_{n+1}, \dots)\|$$

$$= \lim_{n \to \infty} \sup_{i \geq n} |\xi_i| = \limsup_{n \to \infty} |\xi_n| = \lim_{n \to \infty} |\xi_n| = 0.$$

7.21 It is enough to prove that an n-dimensional Hausdorff TVS $(X, \mathcal{T})$ is linearly homeomorphic to $(\mathbb{K}^n, \|\cdot\|_2)$, where $\|\cdot\|_2$ is the Euclidean norm. Let $a_1, \ldots, a_n$ be a basis of X and $T(\xi_1, \ldots, \xi_n) = \xi_1 a_1 + \cdots + \xi_n a_n \in X$. Show that $T \colon \mathbb{K}^n \to X$ is continuous. Let $\varepsilon > 0$ be given and let $S = \{(\xi_1, \ldots, \xi_n) \in \mathbb{K}^n : \|(\xi_1, \ldots, \xi_n)\|_2 = \varepsilon\}$. Show that $T(S)$ is compact and closed. Show that it is possible to choose a balanced neighbourhood U of 0 in X such that $U \cap T(S) = \emptyset$. Prove that for every $x \in U$, $(\eta_1, \ldots, \eta_n) = T^{-1}(x)$ satisfies $\|(\eta_1, \ldots, \eta_n)\|_2 < \varepsilon$.

Chapter 8

8.1 By **I4**, $\langle 0, x \rangle = \langle 0 \cdot x, x \rangle = 0 \cdot \langle x, x \rangle = 0$. Then $\langle x, 0 \rangle = \overline{\langle 0, x \rangle} = 0$. Further by **I3**

$$\langle x, \lambda y \rangle = \overline{\langle \lambda y, x \rangle} = \overline{\lambda \langle y, x \rangle} = \overline{\lambda} \, \overline{\langle y, x \rangle} = \overline{\lambda} \langle x, y \rangle.$$

8.3 We may assume that $y \neq 0$ and find $\alpha \in \mathbb{C}$ such that $x - \alpha y \perp y$: from $\langle x - \alpha y, y \rangle = \langle x, y \rangle - \alpha \langle y, y \rangle = 0$ we find $\alpha = \langle x, y \rangle / \|y\|^2$. Applying Pythagoras's theorem we get

$$\|x\|^2 = \|(x - \alpha y) + \alpha y\|^2 = \|x - \alpha y\|^2 + |\alpha|^2 \|y\|^2$$

$$\geq |\alpha|^2 \|y\|^2 = \frac{|\langle x, y \rangle|^2}{\|y\|^4} \|y\|^2 = \frac{|\langle x, y \rangle|^2}{\|y\|^2}$$

which implies $|\langle x, y \rangle|^2 \leq \|x\|^2 \|y\|^2$, and $|\langle x, y \rangle| \leq \|x\| \|y\|$. Suppose that $|\langle x, y \rangle| = \|x\| \|y\|$ and $y \neq 0$, and let $\alpha = \langle x, y \rangle / \|y\|^2$. By the preceding calculation,

$$\|x\|^2 = \|x - \alpha y\|^2 + |\alpha|^2 \|y\|^2 = \|x - \alpha y\|^2 + \|x\|^2,$$

that is, $x - \alpha y = 0$.

8.5 The verification of the inner product axioms is routine except in (ix). For this case we recall that a hermitian matrix A can be unitarily diagonalized, that is, $A = U^H D U$, where U is unitary ($U^H U = I = U U^H$) and D diagonal with the eigenvalues λ_i of A on the diagonal. Let $x \in \mathbb{C}^n$ and $y = U x = (\eta_1, \ldots, \eta_n)$. Then

$$\langle x, x \rangle = x^H U^H D U x = (U x)^H D U x = \lambda_1 |\eta_1|^2 + \cdots + \lambda_n |\eta_n|^2 \geq 0.$$

Suppose that $\langle x, x \rangle = 0$. Then $\eta_i = 0$ for all i, that is, $y = U x = 0$. Hence $x = U^H U x = 0$. All spaces are complete with the exception of (v) and (viii).

8.9 (i) $n = 1$, $p \geq 1$: the norm $\|\xi\| = |\xi|$ is induced by inner product $\langle \xi, \eta \rangle = \xi \overline{\eta}$. $n > 1$ and $p = 2$: The norm $\|x\| = (|\xi_1|^2 + \cdots + |\xi_n|^2)^{1/2}$ is induced by the inner product $\langle x, y \rangle = \xi_1 \overline{\eta}_1 + \cdots + \xi_n \overline{\eta}_n$. $n > 1$ and $p \neq 2$: The vectors $x = (1, 1, 0, \ldots, 0)$ and $y = (1, -1, 0, \ldots, 0)$ do not satisfy the parallelogram identity (check), which means that the norm is not induced by any inner product.

(ii) $E = \{a\}$ is a singleton. The norm $\|f\| = |f(a)|$ is induced by the inner product $\langle f, g \rangle = f(a)\overline{g(a)}$.

E contains points $a \neq b$. Let $u(a) = 1$ and $u(t) = 0$ if $t \neq a$, and $v(b) = 1$ and $v(t) = 0$ if $t \neq b$. The vectors $f = u + v$ and $g = u - v$ do not satisfy the parallelogram identity (check), and therefore the norm is not induced by any inner product.

(iii) Let f be the linear function through the points $(a, -1)$ and $(b, 1)$, and let $g = |f|$. Check that these two vectors do not satisfy the parallelogram identity; therefore the norm is not induced by any inner product.

8.11 Let $a, b \in X$. The continuity of the inner product at $(a, b) \in X \times X$ follows from

$$|\langle x, y \rangle - \langle a, b \rangle| = |\langle x - a, y \rangle + \langle a, y - b \rangle| \leq \|x - a\|\|y\| + \|y - b\|\|a\|.$$

8.13 The central idea: Pick $k \in \{1, \ldots, n\}$ and write $\xi_i = \langle x, a_i \rangle$ for any i. Then

$$\langle Px, a_k \rangle = \left\langle \sum_{i=1}^{n} \xi_i a_i, a_k \right\rangle = \sum_{i=1}^{n} \xi_i \langle a_i, a_k \rangle = \xi_k = \langle x, a_k \rangle.$$

8.15 Recall that for any $k \in \mathbb{Z}$, $e^{i2k\pi} = 1$. If m, n are distinct integers, then

$$\langle e_m, e_n \rangle = \frac{1}{2\pi} \int_0^{2\pi} e^{imt}\overline{e^{int}}\, dt = \frac{1}{2\pi} \int_0^{2\pi} e^{i(m-n)t}\, dt = \frac{1}{2\pi} \frac{1}{i(m-n)} \left[e^{i(m-n)t} \right]_0^{2\pi},$$

which is equal to 0.

8.17 Observe that $\cos t = \mathsf{Re}\, e^{it}$ and use the procedure from the solution to Problem 8.15.

8.19 Let $x \in X$ and let $a_1, \ldots, a_k \in A$ be such that $|\langle x, a \rangle| > 1/m$ for a positive integer m. By Bessel's inequality,

$$\frac{k}{m^2} \leq \left\| \sum_{i=1}^{k} |\langle x, a_i \rangle|^2 \right\| \leq \|x\|^2,$$

and $k \leq m^2\|x\|^2$. Hence the set A_m of the elements of A satisfying $|\langle x, a \rangle| > 1/m$ contains at most $m^2\|x\|^2$ vectors. The set of all $a \in A$ satisfying $\langle x, a \rangle \neq 0$ is equal to $\bigcup_{m=1}^{\infty} A_m$, which is a countable set (Theorem E.7).

8.23 For each $x \in H$ order the set $S_x = \{u \in S : \langle x, u \rangle \neq 0\}$ in a sequence. In (iii) consider $S_x \cup S_y$.

8.25 Use Problem 8.23: From $S \subset \text{sp } S \subset \overline{\text{sp } S}$ we get $\{0\} = S^{\perp} \supset \overline{\text{sp } S}^{\perp}$. Since $\overline{\text{sp } S}$ is a closed subspace, $\overline{\text{sp } S}^{\perp\perp} = \overline{\text{sp } S}$. Therefore

$$H = \{0\}^{\perp} \subset \overline{\text{sp } S}^{\perp\perp} = \overline{\text{sp } S}.$$

8.27 If $d = \text{dist}\,(x, C) = \inf\{\|x - a\| : a \in C\}$, then there exists a so called minimizing sequence (a_n) in C such that $\|x - a_n\| \to d$ (Theorem B.10). We show that (a_n) is Cauchy by a clever trick. First, $\|a_n - a_m\| = \|(a_n - x) - (a_m - x)\|$. By the parallelogram law,

$$\|(a_n - x) + (a_m - x)\|^2 + \|a_n - a_m\|^2 = 2\|a_n - x\|^2 + 2\|a_m - x\|^2.$$

Since C is convex, $a_{mn} = \frac{1}{2}(a_n + a_m)$ is in C and $\|(a_n - x) + (a_m - x)\|^2 = 4\|a_{mn} - x\|^2 \geq 4d^2$. Therefore

$$0 \leq \|a_n - a_m\|^2 \leq -4d^2 + 2\|a_n - x\|^2 + 2\|a_m - x\|^2.$$

Taking the limit as $m \wedge n = \min\{m, n\} \to \infty$, we get $\lim_{m \wedge n \to \infty} \|a_n - a_m\|^2 = 0$, that is, (a_n) is Cauchy. Since C is complete (a closed subset of a complete space), (a_n) converges to a point $x_0 \in C$, and $d = \lim_n \|a_n - x\| = \|x_0 - x\|$. The parallelogram law can be used to prove the uniqueness of x_0.

Chapter 9

9.1 Let $T, S \in \mathcal{B}(X, Y)$ and let $\|x\| = 1$. Then

$$\|(T + S)x\| = \|Tx + Sx\| \leq \|Tx\| + \|Sx\| \leq \|T\|\|x\| + \|S\|\|x\| = \|T\| + \|S\|,$$

which shows that $T + S \in \mathcal{B}(X, Y)$ and that $\|T + S\| \leq \|T\| + \|S\|$. The equation $\|\lambda T\| = |\lambda|\|T\|$ is proved similarly.

Let Y be a Banach space and let (T_n) be a Cauchy sequence in $\mathcal{B}(X, Y)$. We can show that for any $x \in X$, $(T_n x)$ is Cauchy, and therefore convergent in Y. Define $Tx = \lim_{n \to \infty} T_n x$. From the properties of limits we can verify that T is linear. To show that T is bounded, we recall that Cauchy sequences are bounded, and proceed from there.

Finally we show that (T_n) converges to T in the operator norm. Let $\varepsilon > 0$. Then there exists $N(\varepsilon)$ such that

$$\|T_n - T_m\| \leq \varepsilon \quad \text{if } m, n \geq N(\varepsilon).$$

Let $x \in X$. Then

$$\|T_n x - T_m x\| \leq \|T_n - T_m\|\,\varepsilon\,\|x\|.$$

Choose $n \geq N(\varepsilon)$ and keep it fixed, and let m vary. Then

$$\|T_n x - T x\| = \lim_{m \to \infty} \|T_n x - T_m x\| \leq \varepsilon\,\|x\|$$

for each $x \in X$. Thus $\|(T_n - T)x\|/\|x\| \le \varepsilon$ if $n \ge N(\varepsilon)$, $x \ne 0$; so $\|T_n - T\| \le \varepsilon$ if $n \ge N(\varepsilon)$.

9.3 The operator $\widehat{T}$ is well defined: If $x_1 + N = x_2 + N$, then $x_2 - x_1 \in N$, and $Tx_2 = Tx_1 + T(x_2 - x_1) = Tx_1$. The quotient space is normed with $\|x + N\| = \inf\{\|x + u\| : u \in N\}$ since N is closed.

$\widehat{T}$ is linear: $\widehat{T}((x_1 + N) + (x_2 + N)) = \widehat{T}((x_1 + x_2) + N) = T(x_1 + x_2) = Tx_1 + Tx_2 = \widehat{T}(x_1 + N) + \widehat{T}(x_2 + N)$; similarly $\widehat{T}(\lambda(x + N)) = \lambda\widehat{T}(x + N)$.

The norm of $\widehat{T}$: $\|\widehat{T}(x + N)\| = \|Tx\| = \|T(x + u)\| \le \|T\|\|x + u\|$ for each $u \in N$. Taking infimum as $u \in N$, we get $\|\widehat{T}(x + N)\| \le \|T\|\|x + N\|$, that is $\|\widehat{T}\| \le \|T\|$.

To prove the reverse inequality, let $\varepsilon > 0$ be arbitrary. There exists $x \in X\backslash\{0\}$ such that $\|Tx\|/\|x\| > \|T\| - \varepsilon$. Also $\|x + N\| \le \|x\|$ (definition of infimum). Hence

$$\frac{\|\widehat{T}(x + N)\|}{\|x + N\|} = \frac{\|Tx\|}{\|x + N\|} \ge \frac{\|Tx\|}{\|x\|} > \|T\| - \varepsilon,$$

and $\|\widehat{T}\| \ge \|T\|$.

9.7 (i) The linearity holds as $(f + g)' = f' + g'$ and $(\lambda f)' = \lambda f'$ for any $f, g \in X$ and any scalar λ. If $f \in X$, then

$$\|Df\| = \|f'\| \le \|f\| + \|f'\| = \|f\|_1,$$

and D is bounded with $\|D\| \le 1$. If $f_n(t) = t^n$, then $\|f_n\|_1 = \|t^n\| + \|nt^{n-1}\| = 1 + n$, and $\|Df_n\|/\|f_n\|_1 = n/(n+1)$. Taking the limit as $n \to \infty$ we get $\|D\| = 1$.

(ii) If both spaces are equipped with the same uniform norm $\|\cdot\|$, then for $f_n(t) = t^n$ we have $\|Df_n\|/\|f_n\| = n$. This proves that D is not bounded.

9.9 For a general diagonal operator T the norm of T is given by $\|T\| = \sup_n |\alpha_n|$. To avoid square roots, we consider $|\alpha_n|^2$; we also consider continuous variable x in place of the discrete variable n in order to apply calculus:

$$|\alpha_n|^2 = \frac{(n + 1)^2 + 1}{(n - 1)^2 + 1}; \quad \text{set } f(x) = \frac{(x + 1)^2 + 1}{(x - 1)^2 + 1}, \ x \ge 1.$$

From

$$f'(x) = \frac{-4x^2 + 8}{((x - 1)^2 + 1)^2} = \frac{4(\sqrt{2} + x)(\sqrt{2} - x)}{((x - 1)^2 + 1)^2}$$

we see that $f(x)$ is increasing on $[1, \sqrt{2}]$ and decreasing on $[\sqrt{2}, \infty)$; the maximum of $f(x)$ occurs at $x = \sqrt{2}$. For the discrete variable n the maximum occurs at $n = 1$ or $n = 2$: $|\alpha_1|^2 = 5 = |\alpha_2|^2$ (by a coincidence, the values are the same). Hence $\|T\| = \sqrt{5}$.

9.11 (i) Vectors in $\mathbb{C}^n$ and $\mathbb{C}^m$ are written as columns; $(Tx)_j$ denotes the jth coordinate of Tx. Let $x \in \mathbb{C}^n$. Then

$$\|Tx\|_\infty = \max_{1 \le j \le m} |(Tx)_j| = \max_{1 \le j \le m} \left| \sum_{k=1}^n \alpha_{jk}\xi_k \right|$$

$$\le \max_{1 \le j \le m} \sum_{k=1}^n |\alpha_{jk}||\xi_k| \le \max_{1 \le j \le m} \left(\sum_{k=1}^n |\alpha_{jk}| \right) \|x\|_\infty,$$

and

$$\|T\| \le \max_{1 \le j \le m} \sum_{k=1}^n |\alpha_{jk}|.$$

The right hand side is the so called *maximum row sum*; we denote it by C. We find a unit vector $x \in \mathbb{C}^n$ which satisfies $\|Tx\|_\infty = C$. Suppose that the maximum row sum is attained at the j_0th row. Define

$$\xi_k = \begin{cases} |\alpha_{j_0,k}|\alpha_{j_0,k}^{-1} & \text{if } \alpha_{j_0,k} \ne 0, \\ 1 & \text{otherwise,} \end{cases} \qquad k = 1, \dots, n.$$

Then $\|x\|_\infty = 1$ and $\alpha_{j_0,k}\xi_k = |\alpha_{j_0,k}|$ for $k = 1, \dots, n$. Therefore

$$\left| \sum_{k=1}^n \alpha_{jk}\xi_k \right| \le \sum_{k=1}^n |\alpha_{jk}| \text{ if } j \ne j_0, \qquad \left| \sum_{k=1}^n \alpha_{j_0,k}\xi_k \right| = \sum_{k=1}^n |\alpha_{j_0,k}|,$$

which proves $\|Tx\|_\infty = C$, and $\|T\| = C$.

9.13 (i) Let $x = (\xi_1, \xi_2, \xi_3, \dots) \in \ell^1$, and let $(Tx)_i$ be the ith coordinate of Tx. We write $\|\cdot\|$ for the ℓ^1 norm. Then

$$\|Tx\| = \sum_{i=1}^\infty |(Tx)_i| = \sum_{i=1}^\infty \left| \sum_{j=1}^\infty \alpha_{ij}\xi_j \right| \le \sum_{i=1}^\infty \sum_{j=1}^\infty |\alpha_{ij}||\xi_j|$$

$$= \sum_{j=1}^\infty \left(\sum_{i=1}^\infty |\alpha_{ij}| \right) |\xi_j| = \sum_{j=1}^\infty c_j|\xi_j| \le c \sum_{j=1}^\infty |\xi_j| = c\|x\|.$$

This shows that $Tx \in \ell^1$ whenever $x \in \ell^1$. From the definition of T we conclude that T is linear on ℓ^1. The inequality $\|Tx\| \le c\|x\|$ shows that T is bounded with the operator norm $\|T\| \le c$.

(ii) We want to show that $\|T\| = \sup_{\|x\|=1} \|Tx\| = c$. Let $\varepsilon > 0$. Since $c = \sup_{j \in \mathbb{N}} c_j$, there exists an index j_0 such that $c - \varepsilon < c_{j_0} \le c$. Let $x = e_{j_0}$, where (e_k) is the standard Schauder basis of ℓ^1. Then $\|x\| = 1$ and

$$\|Tx\| = \sum_{i=1}^\infty |\alpha_{ij_0}| = c_{j_0}.$$

Thus for each $\varepsilon > 0$ there exists $x \in \ell^1$ of unit norm such that $c - \varepsilon < \|Tx\| \le c$. By the definition of suprema, $\|T\| = c$.

9.15 Linearity:

$$
\begin{aligned}
A(x + y) &= A(\xi_1 + \eta_1, \xi_2 + \eta_2, \xi_3 + \eta_3, \dots) \\
&= (0, \alpha_1(\xi_1 + \eta_1), \alpha_2(\xi_2 + \eta_2), \alpha_3(\xi_3 + \eta_3), \dots) \\
&= (0, \alpha_1\xi_1, \alpha_2\xi_2, \alpha_3\xi_3, \dots) + (0, \alpha_1\eta_1, \alpha_2\eta_2, \alpha_3\eta_3, \dots) = Ax + Ay, \\
A(\lambda x) &= A(\lambda\xi_1, \lambda\xi_2, \lambda\xi_3, \dots) = (0, \alpha_1\lambda\xi_1, \alpha_2\lambda\xi_2, \alpha_3\lambda\xi_3, \dots) \\
&= \lambda(0, \alpha_1\xi_1, \alpha_2\xi_2, \alpha_3\xi_3, \dots) = \lambda Ax
\end{aligned}
$$

Next, $\|Ax\| = \|(0, \alpha_1\xi_1, \alpha_2\xi_2, \alpha_3\xi_3, \dots)\| = \left(\sum_{n=1}^{\infty} |\alpha_n\xi_n|^2\right)^{1/2} \le c\|x\|$, where $c = \sup_i |\alpha_i|$; hence $Ax \in \ell^2$ for $x \in \ell^2$ and $\|A\| \le c$. We prove that $\|A\| = c$: Let $0 < \varepsilon < c$. Then there exists $k \in \mathbb{N}$ such that $|\alpha_k| > c - \varepsilon$. If e_k is the kth element of the standard Schauder basis for ℓ^2, then $\|e_k\| = 1$ and $\|A\| \ge \|Ae_k\| = |\alpha_k| > c - \varepsilon$.

Let $x = (\xi_n) \in \ell^2$ and $y = (\eta_n) \in \ell^2$. Then

$$
\begin{aligned}
\langle Ax, y \rangle &= \langle (0, \alpha_1\xi_1, \alpha_2\xi_2, \alpha_3\xi_3, \dots), (\eta_1, \eta_2, \eta_3, \eta_4, \dots) \rangle \\
&= \alpha_1\xi_1\overline{\eta}_2 + \alpha_2\xi_2\overline{\eta}_3 + \alpha_3\xi_3\overline{\eta}_4 + \dots \\
&= \langle (\xi_1, \xi_2, \xi_3, \dots), (\overline{\alpha}_1\eta_2, \overline{\alpha}_2\eta_3, \overline{\alpha}_3\eta_4, \dots) \rangle \\
&= \langle x, A^*y \rangle.
\end{aligned}
$$

Thus for all $y \in \ell^2$,

$$
A^*y = A^*(\eta_1, \eta_2, \eta_3, \dots) = (\overline{\alpha}_1\eta_2, \overline{\alpha}_2\eta_3, \overline{\alpha}_3\eta_4, \dots).
$$

9.17 If $f \ne 0$, then $N = f^{-1}(\{0\})$ is a closed subspace of H. We show that $N^{\perp}$ is a one-dimensional subspace of H. (Such a subspace N is called a *hyperplane*.) For this we choose a nonzero vector $b \in N^{\perp}$ and observe that, for each $x \in H$, $u := x - f(x)f(b)^{-1}b$ is in N. Then $\langle u, b \rangle = 0$, that is, $\langle u, b \rangle = \langle x, b \rangle - f(x)\|b\|^{-2}f(b) = 0$, and

$$
f(x) = \frac{f(b)}{\|b\|^2}\langle x, b \rangle = \langle x, a \rangle, \quad x \in H,
$$

where $a = \|b\|^{-2}f(b)\,b$.

The norm of f: $|f(x)| = |\langle x, a \rangle| \le \|a\|\|x\|$, and $\|f\| \le \|a\|$. The equality follows when we choose $x = a$. Suppose there is $a_1 \in H$ such that $f(x) = \langle x, a_1 \rangle$ for all $x \in H$. Then $\langle x, a - a_1 \rangle = 0$ for all $x \in H$; choosing $x = a - a_1$, we get $\|a - a_1\|^2 = 0$, that is, $a_1 = a$.

9.21 We observe the following equivalent statements:

$$
\begin{aligned}
x \in R(A^*)^{\perp} &\iff \langle x, A^*y \rangle \quad \text{for all } y \in H \\
&\iff \langle Ax, y \rangle \quad \text{for all } y \in H
\end{aligned}
$$

$$\Longleftrightarrow Ax = 0$$
$$\Longleftrightarrow x \in N(A)$$

So $R(A^*)^\perp = N(A)$, and

$$H = N(A) \oplus^\perp N(A)^\perp = N(A) \oplus^\perp R(A^*)^{\perp\perp} = N(A) \oplus^\perp \overline{R(A^*)}.$$

9.27 (i) First we observe that $(A^*)^{-1} = (A^{-1})^*$; this follows by taking adjoints of the equations $AB = I = BA$. If A is self-adjoint, then $(A^{-1})^* = (A^*)^{-1} = A^{-1}$, and A^{-1} is self-adjoint. Then for all y,

$$\langle A^{-1}y, y \rangle = \langle A^{-1}y, A(A^{-1}y) \rangle = \langle A(A^{-1}y), A^{-1}y \rangle = \langle Az, z \rangle \geq 0.$$

9.33 We use results of Problem 9.31.

(i) Let $P \leq Q$ and let $x \in R(P)$. Then $Px = x$, and we want to show that $Qx = x$: We have $\|x\|^2 = \|Px\|^2 = \langle Px, x \rangle \leq \langle Qx, x \rangle = \|Qx\|^2 \leq \|x\|^2$. This proves $\|Qx\| = \|x\|$, which implies $\|x\|^2 = \|Qx + (x - Qx)\|^2 = \|Qx\|^2 + \|Qx - x\|^2$; so $\|Qx - x\|^2 = 0$ and $Qx = x$. Hence $R(P) \subset R(Q)$. If $R(P) \subset R(Q)$, then for each $x \in H$, $Px = QPx$, that is, $P = QP$. If $P = QP$, then $P = PQ$ by taking adjoints. Finally, if $PQ = P$, then for all x

$$\langle Px, x \rangle = \|Px\|^2 = \|PQx\|^2 \leq \|Qx\|^2 = \langle Qx, x \rangle.$$

(ii) If $PQ = QP$, then PQ is self-adjoint and $(PQ)^2 = P^2Q^2 = PQ$. Conversely, if PQ is an orthogonal projection, then PQ is self-adjoint, that is $QP = (PQ)^* = PQ$. Note that $R(PQ) = R(P) \cap R(Q)$.

(iii) If $Q - P$ is a projection, then

$$\langle Qx, x \rangle - \langle Px, x \rangle = \langle (Q - P)x, x \rangle = \|(Q - P)x\|^2 \geq 0$$

for all x. By (i), $PQ = P$. Conversely, let $PQ = P$. Then also $QP = P$ (adjoints), and

$$(Q - P)^2 = Q - QP - PQ + P = Q - P.$$

We have $R(Q - P) = R(Q) \cap R(P)^\perp$.

(iv) If $P + Q$ is a projection, then $P + Q = (P + Q)^2 = P + PQ + QP + Q$, that is, $PQ = -QP$. Hence $PQ = -QP = -QPQ = -Q(-QP) = QP$, and $PQ = 0$. Conversely, let $PQ = 0$; then also $QP = 0$, and $(P + Q)^2 = P + PQ + QP + Q = P + Q$. We have $R(P + Q) = R(P) + R(Q)$.

9.39 Let $e_1, e_2, e_3, \ldots$ be the standard Schauder basis in ℓ^p. If $f \in (\ell^p)'$, set $\alpha_n = f(e_n)$ and $a = (\alpha_n)$. To show that $a \in \ell^q$, we define $x = (\xi_n)$ to satisfy $\alpha_n \xi_n = |\alpha_n|^q = |\xi_n|^p$ by setting

$$\xi_n = \begin{cases} \alpha_n^{-1}|\alpha_n|^q & \text{if } \alpha_n \neq 0, \\ 0 & \text{if } \alpha_n = 0 \end{cases}$$

(check). Then

$$\sum_{k=1}^{n} |\alpha_k|^q = \sum_{k=1}^{n} \alpha_k \xi_k = f\left(\sum_{k=1}^{n} \xi_k e_k\right) \le \|f\| \left(\sum_{k=1}^{n} |\xi_k|^p\right)^{1/p} = \|f\| \left(\sum_{k=1}^{n} |\alpha_k|^q\right)^{1/p},$$

and

$$\left(\sum_{k=1}^{n} |\alpha_k|^q\right)^{1/q} \le \|f\| \quad \text{for all } n.$$

Hence $a \in \ell^q$ with $\|a\|_q \le \|f\|$.

Conversely, Hölder's inequality ensures that each $a = (\alpha_n) \in \ell^q$ defines a bounded linear functional f on ℓ^p by $f(x) = \sum_{n=1}^{\infty} \alpha_n \xi_n$ for all $x = (\xi_n)$ in ℓ^p, and that $\|f\| \le \|a\|_q$.

9.43 (a) First assume that a matrix $[\alpha_{ij}]$ with the given properties exists. Write $c_i = \sup \sum_{j=1}^{\infty} |\alpha_{ij}|$ $(i = 1, 2, \dots)$; then $c = \sup_i c_i$. Let $x = (\xi_j) \in \mathsf{c}_0$. Then

$$\sum_{j=1}^{\infty} |\alpha_{ij}| |\xi_j| \le \|x\|_{\infty} \sum_{j=1}^{\infty} |\alpha_{ij}| = c_i \|x\|_{\infty}.$$

Then $\eta_i = \sum_{j=1}^{\infty} \alpha_{ij}\xi_j$ converges, and $\sup_i |\eta_i| \le c_i \|x\|_{\infty} \le c\|x\|_{\infty}$ for all $x \in \mathsf{c}_0$. Hence $y = Ax \in \ell^{\infty}$ with $\|Ax\|_{\infty} \le c\|x\|_{\infty}$, and $\|A\| \le c$.

To prove the reverse inequality, take an arbitrary $\varepsilon > 0$. Then there exists i_0 such that $c \ge c_{i_0} > c - \frac{1}{2}\varepsilon$ (definition of supremum). Further, there exists $N \in \mathbb{N}$ such that $\sum_{j=N+1}^{\infty} |\alpha_{i_0,j}| < \frac{1}{2}\varepsilon$. We make a clever choice of $x \in \mathsf{c}_0$ which will satisfy $\|Tx\|_{\infty}/\|x\|_{\infty} > c - \varepsilon$. Define

$$\xi_j = \begin{cases} |\alpha_{i_0,j}|/\alpha_{i_0,j} & \text{if } \alpha_{i_0,j} \ne 0, \\ 1 & \text{if } \alpha_{i_0,j} = 0, \end{cases} \qquad j = 1, \dots, N,$$

and set $x = (\xi_1, \dots, \xi_N, 0, 0, \dots)$. Then $x \in \mathsf{c}_0$, $\|x\|_{\infty} = 1$ and $\alpha_{i_0,j}\xi_j = |\alpha_{i_0,j}|$ for $j = 1, \dots, N$. We calculate:

$$\|Ax\|_{\infty} = \|(|\alpha_{i_0,1}|, |\alpha_{i_0,2}|, \dots, |\alpha_{i_0,N}|, 0, 0, \dots)\|_{\infty}$$
$$= \sum_{j=1}^{N} |\alpha_{i_0,j}| = \sum_{j=1}^{\infty} |\alpha_{i_0,j}| - \sum_{j=N+1}^{\infty} |\alpha_{i_0,j}|$$
$$> c_{i_0} - \tfrac{1}{2}\varepsilon > (c - \tfrac{1}{2}\varepsilon) - \tfrac{1}{2}\varepsilon = c - \varepsilon.$$

Thus $c \ge \|A\| \ge \|Ax\|_{\infty} > c - \varepsilon$; this proves $\|A\| = \sup\{\|Au\|_{\infty} : \|u\|_{\infty} = 1\} = c$.

(b) Conversely, suppose that A is a bounded linear operator on c_0 to ℓ^{∞}. It is known that $e_1 = (1, 0, 0, \dots)$, $e_2 = (0, 1, 0, \dots)$, $\dots$ is a Schauder basis in c_0. If

$x = (\xi_1, \xi_2, \dots) \in \mathbf{c}_0$, we show that $Ax = \sum_{i=1}^{\infty} \xi_i A e_i$ in ℓ^∞. Indeed,

$$Ax = A\left(\sum_{i=1}^{\infty} \xi_i e_i\right) = A\left(\lim_{n\to\infty} \sum_{i=1}^{n} \xi_i e_i\right) = \lim_{n\to\infty} A\left(\sum_{i=1}^{n} \xi_i e_i\right)$$

$$= \lim_{n\to\infty} \sum_{i=1}^{n} \xi_i A e_i = \sum_{i=1}^{\infty} \xi_i A e_i \qquad \text{(F.2)}$$

by the continuity and linearity of A. Define the infinite matrix M by

$$M := [Ae_1 | Ae_2 | Ae_3 | \dots] = \begin{bmatrix} \alpha_{11} & \alpha_{12} & \alpha_{13} & \dots \\ \alpha_{21} & \alpha_{22} & \alpha_{23} & \dots \\ \alpha_{31} & \alpha_{32} & \alpha_{33} & \dots \\ \dots & \dots & \dots & \dots \end{bmatrix},$$

where each Ae_i is written as a column. Then $Ax = Mx$ (infinite matrix product) with $Mx \in \ell^\infty$ for all $x \in \mathbf{c}_0$ in view of (F.2).

We show that for each i, $c_i := \sum_{j=1}^{\infty} |\alpha_{ij}| < \infty$. We do this using the uniform boundedness principle. For each fixed n we define a functional $f_{in}(x) = \sum_{i=1}^{n} \alpha_{ij} \xi_j$ for all $x = (\xi_1, \xi_2, \dots) \in \mathbf{c}_0$ (cut off after the nth term). It is routine to verify that f_{in} is linear and bounded with the norm $\|f_{in}\| = \sum_{j=1}^{n} |\alpha_{ij}|$. Since $\lim_{n\to\infty} f_{in}(x)$ exists for each $x \in \mathbf{c}_0$, $(f_{in})_{n=1}^{\infty}$ converges to a bounded linear functional f_i on $\mathbf{c}_0$ (the Banach-Steinhaus theorem) with $\|f_i\| = \lim_{n\to\infty} \|f_{in}\| = \sum_{j=1}^{\infty} |\alpha_{ij}|$.

Next we prove that $c := \sup_i c_i < \infty$. Using the notation of the preceding paragraph, we have $f_i(x) = \sum_{j=1}^{\infty} \alpha_{ij} \xi_j$ for each $x \in \mathbf{c}_0$, with each f_i a bounded linear functional on $\mathbf{c}_0$. By hypothesis, $Ax = (f_1(x), f_2(x), \dots) \in \ell^\infty$, that is, $\sup_i |f_i(x)| < \infty$. Thus the sequence (f_i) is pointwise bounded, and therefore uniformly bounded by the uniform boundedness principle:

$$c = \sup_i c_i = \sup_i \sum_{j=1}^{\infty} |\alpha_{ij}| = \sup_i \|f_i\| < \infty.$$

9.51 By Problem 9.46, $A'' \in L(X'', Y'')$ is invertible. Let $\pi \colon X \to X''$ and $\sigma \colon Y \to Y''$ be the James embeddings. Show that the following diagram commutes, that is, show that $\sigma A = A'' \pi$:

$$\begin{CD} X @>\pi>> X'' \\ @VAVV @VVA''V \\ Y @>>\sigma> Y'' \end{CD}$$

Use this identity to show that $g(A) = \inf\{\|Ax\| : \|x\| = 1\} > 0$. Apply a corollary to the Hahn–Banach theorem to prove that A is surjective.

Chapter 10

10.1 The range of $\lambda I - A$ is closed: Let $(\lambda I - A)x_n \to y$. If $\|x_n\| \not\to \infty$, then a suitable subsequence of (x_n) converges to some x, so that $y = (\lambda I - A)x$. If $\|x_n\| \to \infty$, we extract a suitable subsequence to reach a contradiction with the injectivity of $\lambda I - A$. Further, $H = N(\lambda I - A) \oplus^{\perp} \overline{R(\lambda I - A)} = R(\lambda I - A)$ proves the surjectivity.

10.3 Consider the closed unit ball K in $N(\lambda I - A)$. Using the compactness of A, we conclude that K is compact. Hence $N(\lambda I - A)$ is finite dimensional by Theorem 7.16.

Chapter 11

11.1 This is a result of classical analysis solved by methods of functional analysis. Let $\mathbf{c}_0$ be the space of all complex sequences $x = (\xi_1, \xi_2, \dots)$ convergent to 0 equipped with the supremum norm. For each $n \in \mathbb{N}$ define a (linear) functional on $\mathbf{c}_0$ by

$$f_n(x) = \sum_{k=1}^{n} \alpha_k \xi_k, \quad x \in \mathbf{c}_0.$$

The following estimate shows that each f_n is bounded:

$$|f_n(x)| \leq \sum_{k=1}^{n} |\alpha_k| |\xi_k| \leq \left(\sum_{k=1}^{n} |\alpha_k| \right) \|x\|, \quad x \in \mathbf{c}_0.$$

The family $\{f_n : n \in \mathbb{N}\}$ of bounded linear functionals $f_n \colon \mathbf{c}_0 \to \mathbb{C}$ is pointwise bounded as $\lim_{n\to\infty} f_n(x)$ exists for each $x \in \mathbf{c}_0$ by hypothesis. Since $\mathbf{c}_0$ is a Banach space, the uniform boundedness principle applies to $\{f_n\}$ to ensure the existence of a constant $M > 0$ such that $\|f_n\| \leq M$ for all n.

We need to show that $\|f_n\| = \sum_{k=1}^{n} |\alpha_k|$: Set $x_n = (\xi_1, \dots, \xi_n, 0, 0, \dots) \in \mathbf{c}_0$, where $\xi_k = |\alpha_k|/\alpha_k$ if $\alpha_k \neq 0$, and $\alpha_k = 1$ otherwise. Then $\|x_n\| = 1$ and $|f_n(x_n)| = \sum_{k=1}^{n} \alpha_k \xi_k = \sum_{k=1}^{n} |\alpha_k|$. Thus $\sum_{k=1}^{n} |\alpha_k| = \|f_n\| \leq M$ for all n, that is, $\sum_{k=1}^{\infty} |\alpha_k|$ converges.

11.3 For each $x \in X$ define $\hat{x} \colon X' \to \mathbb{C}$ by $\hat{x}(x') = x'(x)$ for all $x' \in X'$ (James embedding). Apply the principle of uniform boundedness to the family $\{\hat{x} : x \in A\} \subset X''$.

11.7 Since $x_n \rightharpoonup x$, the sequence (x_n) is bounded in norm (uniform boundedness principle), that is, $\|x_n\| \leq M$ for some $M > 0$ and all n. Further, $\langle x_n - x, y \rangle \to 0$. Then

$$\begin{aligned} |\langle x_n, y_n \rangle - \langle x, y \rangle| &= |\langle x_n - x, y \rangle + \langle x_n, y_n - y \rangle| \\ &\leq |\langle x_n - x, y \rangle| + \|x_n\| \|y_n - y\| \leq |\langle x_n - x, y \rangle| + M\|y_n - y\| \to 0. \end{aligned}$$

11.9 Let $S\colon X \to R(T)$ be the codomain restriction of T, that is, $Sx = Tx$ for all $x \in X$. Then S is linear, bounded and *surjective*. We note that $R(T)$ is a closed subspace of a Banach space Y, and therefore also a Banach space. Thus we can apply the open mapping theorem to $S\colon X \to R(T)$: There exists $\alpha > 0$ such that

$$\alpha B_{R(T)} \subset T(B_X). \tag{F.3}$$

Let $y \in R(T)$, $y \neq 0$. The vector $\frac{1}{2}\alpha\|y\|^{-1}y$ belongs to $\alpha B_{R(T)}$. According to (F.3), there exists $u \in B_X$ such that

$$\frac{\alpha y}{2\|y\|} = Tu, \quad \text{and} \quad y = \frac{2\|y\|}{\alpha} Tu = T\Big(\frac{2\|y\|}{\alpha}u\Big).$$

Set $x = 2\alpha^{-1}\|y\|u$. Then $\|x\| < 2\alpha^{-1}\|y\|$ (as $\|u\| < 1$). Then $y = Tx$ and $\|x\| \leq c\|y\|$, where $c = 2\alpha^{-1}$ is a constant.

ALTERNATIVE: Define the operator $\widetilde{T}\colon X/N(T) \to R(T)$ on the quotient space $X/N(T)$ by $\widetilde{T}(x + N(T)) = Tx$. Apply the Banach inverse mapping theorem.

11.13 (i) We prove that the differential operator $D\colon X \to Y$ is closed if both spaces are equipped with the uniform norm. For this we need to show that for any sequence (x_n) in X and $x \in Y$,

$$\big(\|x_n - x\| \to 0 \text{ and } \|Dx_n - y\| \to 0\big) \implies \big(x \in X \text{ and } y = Dx\big).$$

Since $x_n \to x$ and $x_n' \to y$ uniformly on $[0,1]$, x and y are continuous. So

$$\left|\int_0^t x_n'(s)\,ds - \int_0^t y(s)\,ds\right| \leq \int_0^t |x_n'(s) - y(s)|\,ds \leq \|Dx_n - y\| \to 0,$$

that is, $\int_0^t x_n'(s)\,ds \to \int_0^t y(s)\,ds$, and

$$x(t) - x(0) = \lim_{n\to\infty}(x_n(t) - x_n(0)) = \lim_{n\to\infty}\int_0^t x_n'(s)\,ds = \int_0^t y(s)\,ds.$$

Hence x is differentiable with $x' = y$; since y is continuous, x belongs to X. This proves that D is closed. In view of Problem 9.7, D is not bounded.

The conclusion of the closed graph theorem fails since the space X is *not* complete (a sequence of continuously differentiable functions may converge uniformly on $[0,1]$ to a function x which is not continuously differentiable).

(ii) Let $(X, \|\cdot\|)$ be a separable Banach space and $\{x_\alpha : \alpha \in D\}$ a Hamel basis for X, normalized so that $\|x_\alpha\| = 1$. Apply Baire's theorem to show that D is uncountable. If $x = \sum_{\alpha \in D}\xi_\alpha x_\alpha$ (with only finitely many ξ_α nonzero), set $\|x\|_1 = \sum_{\alpha \in D}|\xi_\alpha|$. Then $\|\cdot\|_1$ is a norm in X with $\|x\| \leq \|x\|_1$. Let Y be the space X with the norm $\|\cdot\|_1$. Then Y is not separable since $\|x_\alpha - x_\beta\|_1 = 2$ whenever $\alpha \neq \beta$, and D is uncountable. Consider the mapping $T := \mathrm{id}\colon X \to Y$. Then T is closed since $T^{-1}\colon Y \to X$ is closed ($\|T^{-1}\| \leq 1$). But T cannot be continuous: If it were, T would be a homeomorphism, and Y would be separable since X is separable. The conclusion of the closed graph theorem fails since Y is not complete.

11.15 P is the projection on M along N if and only if $Px \in M$ and $x - Px \in N$ for all $x \in X$. Then $P \colon X \to X$ is linear, but not necessarily bounded. Suppose that M, N are closed. We apply the closed graph theorem. Let $x_n \to x$ and $Px_n \to y$. Then $y - Px = \lim_{n\to\infty} P(x_n - x) \in M$ as M is closed. On the other hand, $x_n - Px_n \to x - y \in N$ as N is closed. Hence $y - Px = (y-x) + (x-Px) \in N$, and $y - Px \in M \cap N = \{0\}$. Therefore $y = Px$, and P is bounded by the closed graph theorem.

Conversely, if $X = M \oplus N$ and the projection P onto M along N is bounded, then $N = N(P)$ and $M = N(I - P)$. Hence both subspaces are closed.

Chapter 12

12.3 Let $A_n \in \Sigma$ for all $n \in \mathbb{N}$ and let $A = \bigcup_{n=1}^{\infty} A_n$. Suppose first that all A_n are countable; then A is countable (a countable union of countable sets), and so $A \in \Sigma$. Next suppose that at least one of the given sets, say A_1, is not countable; then A_1^c is countable. Since $A_1 \subset A$, we have $A^c \subset A_1^c$, that is, A^c is countable, and $A \in \Sigma$. Since $\emptyset$ is countable (zero elements), $\emptyset \in \Sigma$.

12.5 We give a solution for $\mathbb{R}^k$. Write $\mathcal{B}$ for the σ-algebra of Borel sets on $\mathbb{R}^k$, and $\mathcal{A}$ for the least σ-algebra containing all semiclosed k-cells. Each semiclosed k-cell A is a Borel set: $A = G \cap F$, where

$$G = \{(x_1, \ldots, x_k) : x_i > a_i, \ i = 1, \ldots, k\} \quad \text{is open,}$$
$$F = \{(x_1, \ldots, x_k) : x_i \leq b_i, \ i = 1, \ldots, k\} \quad \text{is closed.}$$

Since $\mathcal{A}$ is the least σ-algebra containing semiclosed k-cells, we have $\mathcal{A} \subset \mathcal{B}$.

For the reverse inclusion we first show that *every open set $G \subset \mathbb{R}^k$ is a countable union of semiclosed k-cells*: For each open set $G \subset \mathbb{R}^k$ and each point $c \in G$ we can find an open k-cell B with rational vertices such that $c \in B \subset G$ (give details). Hence there is a countable family $\mathcal{F}$ of open k-cells which forms a base for the Euclidean topology of $\mathbb{R}^k$. By a property of a base, every open set in $\mathbb{R}^k$ is the union of some subfamily of $\mathcal{F}$, that is, a countable union of open k-cells. But every open k-cell is a countable union of semiclosed k-cells (for instance,

$$(a_1, b_1) \times \ldots \times (a_k, b_k)$$
$$= \bigcup_{n=1}^{\infty} (a_1, b_1 - (b_1 - a_1)/n] \times \cdots \times (a_k, b_k - (b_k - a_k)/n].)$$

As every open set is a countable union of semiclosed k-cells, $\mathcal{A}$ contains all open sets. Since $\mathcal{B}$ is the least σ-algebra which contains all open sets, $\mathcal{B} \subset \mathcal{A}$. Therefore $\mathcal{A} = \mathcal{B}$.

12.7 (i) $B \setminus A = (B^c \,\dot\cup\, A)^c$. (ii) If $A_n \in \mathcal{D}$, set $B_1 = A_1$ and $B_n = A_n \setminus (B_1 \,\dot\cup\, \ldots \,\dot\cup\, B_{n-1})$ if $n \geq 2$. Then the sets B_n are disjoint, belong to $\mathcal{D}$ for all n, and $\bigcup_{k=1}^{n} B_k = \bigcup_{k=1}^{n} A_k$ for all n.

12.11 First we show that if $a, c \in [0, \infty]$, then $a \le a + c$. If $a = \infty$, then $a = \infty = a + c$, and $a \le a + c$ holds as $\infty \le \infty$. If $a < \infty$ and $c < \infty$, then $a \le a + c$ by the laws of inequalities for real numbers. If $a < \infty$ and $c = \infty$, then $a + c = \infty$ and $a \le \infty = a + c$.

Let $A \subset B$ for $A, B \in \Sigma$. Then $B = A \,\dot\cup\, (B \setminus A)$, and $\mu(B) = \mu(A) + \mu(B \setminus A)$ by the additivity of μ. By the first part of the proof, $\mu(B) = \mu(A) + \mu(B \setminus A) \ge \mu(A)$.

12.15 We need to prove that if (B_n) is a sequence of disjoint Σ-measurable sets in S, then $\mu\left(\bigcup_{n=1}^{\infty} B_n\right) = \sum_{n=1}^{\infty} \mu(B_n)$. For this we start from the given sequence (B_n) and construct a new sequence (A_n): $A_1 = B_1$, $A_n = B_1 \cup B_2 \cup \cdots \cup B_n$, $n \ge 2$. By the definition of a σ-algebra, each A_n is in Σ. Since μ is finitely additive and the B_n are pairwise disjoint, we have

$$\mu(A_n) = \sum_{k=1}^{n} \mu(B_k), \quad n \in \mathbb{N}. \tag{F.4}$$

Let $B = \bigcup_{n=1}^{\infty} B_n$. The sequence (A_n) is expanding and $A_n \nearrow B$. By hypothesis, $\mu(A_n) \to \mu(B)$. Applying (F.4), we get

$$\sum_{k=1}^{\infty} \mu(B_k) = \lim_{n \to \infty} \sum_{k=1}^{n} \mu(B_k) = \lim_{n \to \infty} \mu(A_n) = \mu(B).$$

This proves that μ is countably additive, and therefore a measure on Σ.

12.21 Let $A_1 = \{(x, y) \in \mathbb{R}^2 : x > 1\}$, $A_2 = \{(x, y) \in \mathbb{R}^2 : y < 1/x\}$ and $A_3 = \{(x, y) \in \mathbb{R}^2 : y \ge 0\}$. Then $A = A_1 \cap A_2 \cap A_3$, where A_1 and A_2 are open and A_3 is closed. This proves that A is a Borel set. For each $n \ge 2$ let U_n be the rectangle $(n - 1, n] \times [0, 1/n)$. Then $U = \bigcup_{n=2}^{\infty} U_n$ is a Borel set (the countable union of Borel sets), and $U \subset A$. Since the U_n are disjoint, the Lebesgue measure of U is equal to

$$m(U) = \sum_{n=2} m(U_n) = \sum_{n=2} \text{area}\,(U_n) = \sum_{n=2} \frac{1}{n} = \infty.$$

Hence $\infty \ge m(A) \ge m(U) = \infty$, that is, $m(A) = \infty$.

12.25 Let u be a fixed vector in $\mathbb{R}^k$. By Problem 12.24,

$$E \in \mathcal{B}(\mathbb{R}^k) \implies E + u \in \mathcal{B}(\mathbb{R}^k).$$

Following the suggestion, we define a set function μ on $\mathcal{B}(\mathbb{R}^k)$ by

$$\mu(E) = m(E + u), \quad E \in \mathcal{B}(\mathbb{R}^k).$$

To show that μ is a measure, we first observe that $\mu(\emptyset) = m(\emptyset + u) = m(\emptyset) = 0$.

Suppose that E_n are disjoint Borel sets on $\mathbb{R}^k$. Then the sets $E_n + u$ are also Borel, and disjoint: Suppose that $x \in (E_n + u) \cap (E_m + u)$. Then $x - u \in E_m \cap E_n$, which is false if $m \neq n$. Hence $(E_n + u) \cap (E_m + u) = \emptyset$ if $m \neq n$. Hence

$$\mu\left(\bigcup_{n=1}^{\infty} E_n\right) = m\left(\left(\bigcup_{n=1}^{\infty} E_n\right) + u\right) = m\left(\bigcup_{n=1}^{\infty}(E_n + u)\right) = \sum_{n=1}^{\infty} m(E_n + u) = \sum_{n=1}^{\infty} \mu(E_n).$$

Finally we show that μ coincides with the Euclidean volume on semiclosed k-cells. If

$$A = (a_1, b_1] \times (a_2, b_2] \times \cdots \times (a_k, b_k],$$

then

$$\begin{aligned}
\mu(A) = m(A + u) &= m((a_1 + u, b_1 + u] \times (a_2 + u, b_2 + u] \times \cdots \times (a_k + u, b_k + u]) \\
&= \text{vol}((a_1 + u, b_1 + u] \times (a_2 + u, b_2 + u] \times \cdots \times (a_k + u, b_k + u]) \\
&= (b_1 - a_1) \cdot (b_2 - a_2) \ldots (b_k - a_k) \\
&= \text{vol}((a_1, b_1] \times (a_2, b_2] \times \cdots \times (a_k, b_k]) \\
&= \text{vol}(A).
\end{aligned}$$

Since the Lebesgue measure m is the unique measure on $\mathcal{B}(\mathbb{R}^k)$ which coincides with the Euclidean volume on semiclosed k-cells, we have $\mu = m$.

12.27 (i) The mapping $\tau \colon \{0, 1\}^{\mathbb{N}} \to C$ defined by $\tau((s_n)) = \sum_{n=1}^{\infty} s_n \, 2/3^n$ is a bijection, while the set $\{0, 1\}^{\mathbb{N}}$ has cardinality $2^{\aleph_0} = \mathfrak{c}$.

(ii) C is closed and bounded.

(iii) $C \subset C_k$ for all k, while the Lebesgue measure of C_k is $(2/3)^k$.

12.29 We assume that $\bar{\mu}$ is well defined (proof as per the suggestion).

$\overline{\Sigma}$ is a σ-algebra: Any set $E \in \Sigma$ is in $\overline{\Sigma}$ since $E \subset E \subset E$ and $\mu(E \setminus E) = \mu(\emptyset) = 0$. In particular, $\emptyset \in \overline{\Sigma}$.

Let $E \in \overline{\Sigma}$. Then there exist $A, B \in \Sigma$ such that $A \subset E \subset B$ and $\mu(B \setminus A) = 0$. Since Σ is a σ-algebra, $A^c, B^c \in \Sigma$. Further, $B^c \subset E^c \subset A^c$, and $A^c \setminus B^c = A^c \cap (B^c)^c = B \cap A^c = B \setminus A$, where $\mu(A^c \setminus B^c) = \mu(B \setminus A) = 0$. Hence $E^c \in \overline{\Sigma}$.

Let $E_n \in \overline{\Sigma}$ for $n \in \mathbb{N}$ and let $E = \bigcup_n E_n$. For each n there exist $A_n, B_n \in \Sigma$ such that $A_n \subset E_n \subset B_n$ and $\mu(B_n \setminus A_n) = 0$. Then $A = \bigcup_n A_n \in \Sigma$, $B = \bigcup_n B_n \in \Sigma$ and $A \subset E \subset B$. First we show that

$$B \setminus A \subset \bigcup_{n=1}^{\infty}(B_n \setminus A_n). \tag{F.5}$$

Let $x \in B \setminus A$. Then $x \in B$ and $x \in A^c$. By the definition of union, there exists n_0 such that $x \in B_{n_0}$. By de Morgan's law, $x \in A^c = \left(\bigcup_k A_k\right)^c = \bigcap_k A_k^c \subset A_{n_0}^c$.

Hence $x \in B_{n_0} \cap A_{n_0}^c = B_{n_0} \setminus A_{n_0} \subset \bigcup_n (B_n \setminus A_n)$. This proves (F.5). Hence

$$\mu(B \setminus A) \le \mu\left(\bigcup_n (B_n \setminus A_n)\right) \le \sum_n \mu(B_n \setminus A_n) = 0$$

by the monotonicity and the countable subadditivity of μ. Then $\mu(B \setminus A) = 0$ and $E \in \overline{\Sigma}$.

$\Sigma \subset \overline{\Sigma}$ and $\overline{\mu}(A) = \mu(A)$ if $A \in \Sigma$: If $E \in \Sigma$, then $E \subset E \subset E$ with $\mu(E \setminus E) = \mu(\emptyset) = 0$. So $E \in \overline{\Sigma}$ and $\overline{\mu}(E) = \mu(E)$ by the definition of $\overline{\mu}$. In particular, $\overline{\mu}(\emptyset) = \mu(\emptyset) = 0$.

$\overline{\mu}$ is countably additive on $\overline{\Sigma}$: Let $E_n \in \overline{\Sigma}$, $n \in \mathbb{N}$, be disjoint sets. For each $n \in \mathbb{N}$ there exist sets $A_n, B_n \in \Sigma$ such that $A_n \subset E_n \subset B_n$ and $\mu(B_n \setminus A_n) = 0$. As above, we write $E = \bigcup_n E_n$, $A = \bigcup_n A_n \in \Sigma$, $B = \bigcup_n B_n \in \Sigma$. We have shown previously that $\mu(B \setminus A) = 0$. Since E_n are pairwise disjoint, so are the A_n as $A_n \cap A_m \subset E_n \cap E_m = \emptyset$ if $m \ne n$. By the definition of $\overline{\mu}$ and the countable additivity of μ,

$$\overline{\mu}(E) = \mu(A) = \mu\left(\bigcup_n A_n\right) = \sum_n \mu(A_n) = \sum_n \overline{\mu}(E_n).$$

$(S, \overline{\Sigma}, \overline{\mu})$ is measure-complete: We consider a set $E \in \overline{\Sigma}$ of measure zero. Then there exist $A, B \in \Sigma$ such that $A \subset E \subset B$, where $\mu(B \setminus A) = 0$ and $\overline{\mu}(E) = \mu(A) = \mu(B) = 0$. Let F be an arbitrary subset of A. Then $\emptyset \subset F \subset E \subset B$, where $\mu(B \setminus \emptyset) = \mu(B) = 0$. This proves that $F \in \overline{\Sigma}$, and the space is measure-complete.

Chapter 13

13.3 There exist sets A_n ($n = 0, 1, 2, \dots$) of measure zero such that $f_n(t) \to f(t)$ for all $t \in A_0^c$, and that $f_n(t) = g_n(t)$ for all $t \in A_n^c$. The set $D = \bigcup_{n=0}^{\infty} A_n$ is of measure zero as $\mu(D) \le \sum_{n=0}^{\infty} \mu(A_n) = 0$ (subadditivity of measure). If $t \in D^c$, then $t \in A_n^c$ for all $n = 0, 1, 2, \dots$, and $g_n(t) = f_n(t) \to f(t)$ as $n \to \infty$. This means that $g_n \to f$ μ-a.e.

13.5 A function $f \colon \mathbb{N} \to \mathbb{C}$ is integrable and simple if it takes only finitely many nonzero values, each on a set of finite measure. This happens if and only if there exists $n \in \mathbb{N}$ such that $f(k) = 0$ for all $k > n$. In this case $f = \sum_{k=1}^{n} f(k)\chi_{\{k\}}$, and the integral of f is given by

$$\int_{\mathbb{N}} f \, d\nu = \sum_{k=1}^{n} f(k).$$

Assume that $f \colon \mathbb{N} \to \mathbb{C}$ satisfies $\sum_{n=1}^{\infty} |f(n)| < \infty$. We show that f is integrable in $(\mathbb{N}, \mathcal{P}(\mathbb{N}), \nu)$. For each n define f_n by

$$f_n(k) = \begin{cases} f(k), & 1 \le k \le n, \\ 0, & k > n. \end{cases} \tag{F.6}$$

Then each f_n is an integrable simple function. We show that (f_n) is an approximating sequence for f:

(a) $f_n(k) \to f(k)$ for all $k \in \mathbb{N}$;

(b) if $n < m$, then

$$\|f_m - f_n\|_{L^1} = \int_{\mathbb{N}} |f_m - f_n| \, d\nu = \sum_{k=1}^{n} |f_m(k) - f_n(k)| + \sum_{k=n+1}^{m} |f_m(k)|$$

$$= \sum_{k=n+1}^{m} |f(k)| \le \sum_{k=n+1}^{\infty} |f(k)| \to 0 \text{ as } n = \min\{m, n\} \to \infty.$$

Thus (f_n) is Cauchy relative to $\|\cdot\|_{L^1}$. From the definition of the integral,

$$\int_{\mathbb{N}} f \, d\nu = \lim_{n \to \infty} \int_{\mathbb{N}} f_n \, d\nu = \lim_{n \to \infty} \sum_{k=1}^{n} f(k) = \sum_{k=1}^{\infty} f(k).$$

Assume that $f \in L^1(\nu) = L^1(\mathbb{N}, \mathcal{P}(\mathbb{N}), \nu)$. Then also $|f| \in L^1(\nu)$. For each n define f_n by (F.6). Since $|f_n| \le |f|$ on $\mathbb{N}$, the monotonicity of the integral yields

$$\sum_{k=1}^{n} |f(k)| = \int_{\mathbb{N}} |f_n| \, d\nu \le \int_{\mathbb{N}} |f| \, d\nu = \text{const.}$$

Hence $\sum_{k=1}^{\infty} |f(k)| < \infty$.

13.9 Let $f \in L^1(\mu)$, and let (f_n) be an approximating sequence for f. We show that $(\text{Re } f_n), (\text{Im } f_n)$ are approximating sequences for $\text{Re } f$ and $\text{Im } f$, respectively. From $f_n \to f$ μ-a.e. it follows that $\text{Re } f_n \to \text{Re } f$ and $\text{Im } f_n \to \text{Im } f$ μ-a.e. Since $z \mapsto \text{Re } z$ is linear, we have

$$\|\text{Re } f_n - \text{Re } f_m\|_{L^1} = \|\text{Re}(f_n - f_m)\|_{L^1} = \int_S |\text{Re}(f_n - f_m)| \, d\mu \le \int_S |f_n - f_m| \, d\mu,$$

that is, $\|\text{Re } f_n - \text{Re } f_m\|_{L^1} \le \|f_n - f_m\|_{L^1}$, and $(\text{Re } f_n)$ is Cauchy in the $L^1(\mu)$-norm. The proof for $(\text{Im } f_n)$ is similar.
The converse is clear.

13.11 Define $f = u + iv$. Since $u, v \in L^1(\mu)$, we have $f \in L^1(\mu)$. The Lebesgue integral is an absolute integral, that is, if $f \in L^1(\mu)$, then also $|f| \in L^1(\mu)$. Hence $|f| = \sqrt{u^2 + v^2} \in L^1(\mu)$.

In the measure space $((0, 1), \Lambda(0, 1), m)$ consider the functions $u(t) = t^{-1/2}$ and $v(t) = t^{-3/4}$. Then neither $u^2(t) = t^{-1}$ nor $v^2(t) = t^{-3/2}$ is integrable, but $\sqrt{t^{-1} + t^{-3/2}}$ is.

13.15 Write $B_n = A_1 \cup \cdots \cup A_n$. Since the A_k are disjoint, we have $\chi_{B_n} = \chi_{A_1} + \cdots + \chi_{A_n}$, and

$$\int_{B_n} f \, d\mu = \int_S f \chi_{B_n} \, d\mu = \int_S f(\chi_{A_1} + \cdots + \chi_{A_n}) \, d\mu = \int_{A_1} f \, d\mu + \cdots + \int_{A_n} f \, d\mu.$$

We observe that $B_n \nearrow A$ and $f\chi_{B_n} \to f\chi_A$, while $|f\chi_{B_n}| \le |f\chi_A| \le |f| \in L^1(\mu)$. By Lebesgue's dominated convergence theorem, $\int_{B_n} f \, d\mu = \int_S f\chi_{B_n} \, d\mu \to \int_S f\chi_A \, d\mu = \int_A f \, d\mu$, that is,

$$\int_A f \, d\mu = \lim_{n \to \infty} \int_{B_n} f \, d\mu = \lim_{n \to \infty} \sum_{k=1}^n \int_{A_k} f \, d\mu = \sum_{k=1}^\infty \int_{A_k} f \, d\mu.$$

13.19 It is routine to verify that Σ_A and Σ_B are σ-algebras and subfamilies of Σ. Let f satisfy the conditions of the problem, and let G be an open subset of the complex plane. Then

$$t \in f^{-1}(G) \iff t \in S = A \cup B \text{ and } f(t) \in G$$
$$\iff [t \in A \text{ and } f_A(t) \in G] \text{ or } [t \in B \text{ and } f_B(t) \in G]$$
$$\iff t \in f_A^{-1}(G) \text{ or } t \in f_B^{-1}(G),$$

that is, $f^{-1}(G) = f_A^{-1}(G) \cup f_B^{-1}(G) \in \Sigma_A \cup \Sigma_B \subset \Sigma$. Hence f is measurable.

13.21 Assume that g is monotonic on $[a, b]$. By Problem 2.27, g is quasicontinuous. Then g is Lebesgue integrable, being the uniform limit on $[a, b]$ of step functions.

Alternatively, we use the fact that a quasicontinuous function is continuous outside a countable set, while $[a, b]$ has finite Lebesgue measure.

13.25 By hypothesis there exists a Lebesgue measurable set of measure 0 such that f is continuous and bounded on $B := E \setminus A$. Then E is the disjoint union of measurable sets $E = A \dot\cup B$. Let f_A (f_B) be the restriction of f to A (B). We show that f is Lebesgue measurable on E. If G is an open subset of $\mathbb{C}$, then $U := f_B^{-1}(G)$ is open in B and $V := f_A^{-1}(G) \subset A$. We observe that U, V are Lebesgue measurable (V as a subset of a set of measure 0), and that $f^{-1}(G) = U \cup V$ is measurable. If $c > 0$ is a constant such that $|f| \le c$ on B, set $g = c\chi_B$. Then $|f| \le g$ almost everywhere in E, and g is an integrable dominant for f on E. This proves that f is Lebesgue integrable on E.

13.27 Assume first that f is measurable. For each $n \in \mathbb{N}$ the set

$$A_n = \{t \in S : |f(t)| \le n\} = |f|^{-1}([0, n])$$

is measurable. Set $f_n = |f|\chi_{A_n}$. Then all f_n are measurable, $0 \le f_n \le |f|$, and $f_n \nearrow |f|$ pointwise on S. By Lebesgue's dominated convergence theorem,

$$\int_S f_n \, d\mu \nearrow \int_S |f| \, d\mu, \quad \text{or} \quad \int_S (|f| - f_n) \, d\mu \searrow 0.$$

Let $\varepsilon > 0$ be given. Then there exists $n \in \mathbb{N}$ such that $0 \le \int_S (|f| - f_n) \, d\mu < \frac{1}{2}\varepsilon$. Set $\delta = \varepsilon/(2n)$.

Let $A \in \Sigma$ be such that $\mu(A) < \delta$. Then

$$\int_A |f| \, d\mu = \int_A (f_n + (|f| - f_n)) \, d\mu = \int_A f_n \, d\mu + \int_A (|f| - f_n) \, d\mu$$

$$\le \int_{A \cap A_n} |f| \, d\mu + \int_S (|f| - f_n) \, d\mu < n\mu(A) + \frac{1}{2}\varepsilon$$

$$\le n\delta + \frac{1}{2}\varepsilon = \frac{1}{2}\varepsilon + \frac{1}{2}\varepsilon = \varepsilon.$$

If f is only μ-measurable, there exists a measurable function g such that $f = g$ μ-a.e. The preceding argument is applicable to g. The result then follows as $\int_A f \, d\mu = \int_A g \, d\mu$ for any measurable set A.

13.29 Initially we assume that $E = S$. Define $f_n = |f|\chi_{E_n}$. Then each f_n is integrable on S, $f_n \nearrow |f|$ pointwise on S, and $\int_S f_n \, d\mu = \int_{E_n} |f| \, d\mu \le c$ for some constant $c > 0$. By the monotone convergence theorem, $|f|$ is integrable. We also have $g_n = f\chi_{E_n} \to f$ pointwise on S, where each g_n is μ-integrable, and therefore μ-measurable. Hence f is μ-measurable. Since $|f|$ is μ-integrable, so is f. Finally, $|g_n| = f_n \le |f|$ for all n, and Lebesgue's dominated convergence theorem implies that $\int_{E_n} f \, d\mu = \int_S g_n \, dmu \to \int_S f \, d\mu$.

To prove the result for a general measurable E, replace f in the preceding argument by $f\chi_E$.

13.31 Write $f(t, x) = 1/(1 + tx)$, with $t \in (-1, \infty)$ a parameter, and x the integration variable. Then, for any fixed $t \in (-1, \infty)$, $x \mapsto 1/(1 + tx)$ is Lebesgue integrable on $(0, 1)$ (bounded and continuous on $(0, 1)$). Further, for each $t \in (-1, \infty)$, the partial derivative of $f(t, x)$ with respect to t exists:

$$\frac{\partial}{\partial t} f(t, x) = -\frac{x}{(1 + tx)^2}.$$

Choose points t_1, t_2 such that $-1 < t_1 < t_2 < \infty$, and consider $t \in (t_1, t_2)$. Then for any $x \in (0, 1)$ and any $t \in (t_1, t_2)$,

$$\left| \frac{-x}{(1 + tx)^2} \right| = \frac{x}{(1 + tx)^2} \le \frac{x}{(1 + t_1 x)^2} =: g(t).$$

as $t > t_1 > -1$ implies $xt > xt_1 > -x$ and $1 + xt > 1 + xt_1 > 1 - x > 0$. The function g is a Lebesgue integrable dominant for $(\partial/\partial t)f$ on $(0,1)$, and the hypotheses of Theorem 13.47 are fulfilled for $t \in (t_1, t_2)$. Hence

$$F'(t) = \frac{d}{dt} \int_0^1 \frac{dx}{1+tx} = \int_0^1 \frac{\partial}{\partial t} \frac{1}{1+tx}\, dx = \int_0^1 \frac{-x\, dx}{(1+tx)^2}.$$

Using the Fundamental theorem of calculus for the Lebesgue integral,

$$F(t) = \int_0^1 \frac{dx}{1+tx} = \frac{\log(1+t)}{t}, \quad t > -1,$$

and

$$F'(t) = \frac{1}{t(1+t)} - \frac{\log(1+t)}{t^2}, \quad t > -1.$$

Thus

$$\int_0^1 \frac{x\, dx}{(1+x)^2} = -F'(1) = -\frac{1}{2} + \log 2.$$

We now repeat the problem for the function $h(t,x) = -x/(1+tx)^2$ and

$$\frac{\partial}{\partial t} h(t,x) = \frac{2x^2}{(1+tx)^3}, \quad t > -1.$$

For $-1 < t_1 < t < t_2 < \infty$ we have an integrable dominant $k(x) = 2x^2/(1+t_1 x)^3$ for $(\partial/\partial t)h(t,x)$ on $(0,1)$. Then for all $t > -1$,

$$F''(t) = \frac{d^2}{dt^2} \int_0^1 \frac{dx}{1+tx} = \frac{d}{dt} \int_0^1 \frac{-x\, dx}{(1+tx)^2} = \int_0^1 \frac{\partial}{\partial t} \frac{-x\, dx}{(1+tx)^2} = \int_0^1 \frac{2x^2\, dx}{(1+tx)^3}.$$

Further,

$$F''(t) = \frac{d}{dt}\left(\frac{1}{t(1+t)} - \frac{\log(1+t)}{t^2} \right) = \frac{2\log(1+t)}{t^3} - \frac{2}{t^2(1+t)} - \frac{1}{t(1+t)^2}$$

and

$$\int_0^1 \frac{x^2\, dx}{(1+x)^3} = \frac{1}{2} F''(1) = \log 2 - \frac{5}{8}.$$

13.33 For each fixed t, the function $s \mapsto \exp(-ts)\sin s$ is the imaginary part of $f(s,t) = \exp((-t+i)s)$. For a fixed $t > 0$, $s \mapsto f(s,t)$ is continuous for all $s \in (0, \infty)$, and

$$|\exp((-t+i)s)| = \exp(-ts),$$

where $s \mapsto \exp(-ts)$ is an integrable dominant for $f(s,t)$ on $(0,\infty)$. Thus $f(s,t)$ is integrable for $s \in (0,\infty)$. By the compatibility of the Lebesgue and Newton integrals,

$$\int_0^\infty \exp((-t+i)s)\,ds = \left[\frac{\exp((-t+i)s)}{t-i}\right]_0^\infty = \frac{1}{t-i} = \frac{t+i}{t^2+1}.$$

Taking the imaginary part, we get

$$\int_0^\infty \exp(-ts)\sin s\,ds = \frac{1}{t^2+1}.$$

We have to check the conditions of Theorem 13.47 for the differentiation under the integral sign:

(i) $f(t,s) \in L^1(0,\infty)$ for $t \in (0,\infty)$ by the preceding argument.

(ii) $\partial f/\partial t = -s\exp(-st)\sin s$ exists for all $t \in (0,\infty)$.

(iii) We restrict t to a finite interval $t \in (t_1, t_2) \subset (0,\infty)$, and seek an integrable dominant for $(\partial/\partial t)f(s,t) = -s\exp(-st)\sin s$ independent of t:

$$|-s\exp(-st)\sin s| \le s\exp(-st) \le s\exp(-t_1 s) = O(s^{-2}) \text{ as } s \to \infty.$$

(This is confirmed by l'Hôpital's rule: $\lim_{s\to\infty} s^3 \exp(-t_1 s) = 0$.) Since $g(s) = s^{-2}$ is integrable on $(0,\infty)$, the differentiation under the integral sign is applicable and

$$\int_0^\infty s\exp(-ts)\sin s\,ds = -\int_0^\infty \frac{\partial}{\partial t}\exp(-ts)\sin s\,ds = -\frac{d}{dt}\int_0^\infty \exp(-ts)\sin s\,ds$$

$$= -\frac{d}{dt}\frac{1}{t^2+1} = \frac{2t}{(t^2+1)^2}.$$

Since every $t > 0$ can be fitted into some interval (t_1, t_2) as above, the result holds for all $t > 0$.

Chapter 14

14.1 In order to prove that f is not absolutely continuous on $[0,1]$ it is enough to show that for each $\delta > 0$ there exists a finite family of nonoverlapping intervals $[a_k, b_k]$ of total length less that δ such that the sum of the terms $|f(b_k) - f(a_k)|$ is greater than or equal to 1.

To prove this set $t_k = 1/((k+\frac{1}{2})\pi)$, $k = 1, 2, \dots$ Then $f(t_k) = t_k \sin(1/t_k) = t_k \cos(k\pi) = (-1)^{k+1}/((k+\frac{1}{2})\pi)$. We observe that

$$|f(t_n) - f(t_{n+1})| + |f(t_{n+2}) - f(t_{n+3})| + \cdots$$

$$= \frac{1}{\pi}\left(\frac{1}{n+\frac{1}{2}} + \frac{1}{n+1+\frac{1}{2}} + \frac{1}{n+2+\frac{1}{2}} + \frac{1}{n+3+\frac{1}{2}} + \cdots\right),$$

which is infinite. Given $\delta > 0$, choose n such that $t_n < \delta$. Then any finite family of intervals $[t_{n+1}, t_n]$, $[t_{n+3}, t_{n+2}]$, $\dots$, $[t_{n+p+1}, t_{n+p}]$ has a total length less than δ. Choosing p sufficiently large, we obtain $|f(t_n) - f(t_{n+1})| + \cdots + |f(t_{n+p}) - f(t_{n+p+1})| \geq 1$.

To see that f is absolutely continuous on any interval $[\varepsilon, 1]$ we observe that $f'(t)$ exists and is Lebesgue integrable on $[\varepsilon, 1]$.

14.3 Suppose that $|f'(t)| \leq M$ for all $t \in (a, b)$ and some $M > 0$. Let $\varepsilon > 0$ be given, and let $\delta = \varepsilon/M$. Let $[a_k, b_k]$ $(k = 1, \dots, n)$ be a family of nonoverlapping subintervals of (a, b) of total length less than δ. By the mean value theorem, for each k there exists c_k between a_k and b_k such that $f(b_k) - f(a_k) = f'(c_k)(b_k - a_k)$. Then

$$\sum_{k=1}^{n} |f(b_k) - f(a_k)| \leq M \sum_{k=1}^{n} (b_k - a_k) < M\delta = \varepsilon.$$

14.5 We know that the integral is absolutely continuous (Problem 13.27): If f is integrable on (a, b), then for each $\varepsilon > 0$ there exists $\delta > 0$ such that $\int_A |f(t)| \, dt < \varepsilon$ whenever A is a Lebesgue measurable subset of (a, b) of Lebesgue measure less than δ. Let $[u_k, v_k]$ $(k = 1, \dots, n)$ be a family of nonoverlapping subintervals of (a, b) such that $A = \bigcup_{k=1}^{n} [u_k, v_k]$ has a total length less than δ. Then $m(A) < \delta$, and

$$\sum_{k=1}^{n} |F(v_k) - F(u_k)| = \sum_{k=1}^{n} \left| \int_c^{v_k} f(t) \, dt - \int_c^{u_k} f(t) \, dt \right|$$

$$= \sum_{k=1}^{n} \left| \int_{u_k}^{v_k} f(t) \, dt \right| \leq \sum_{k=1}^{n} \int_{u_k}^{v_k} |f(t)| \, dt$$

$$= \int_A |f(t)| \, dt < \varepsilon.$$

Select sequences (a_n), (b_n) such that $a_n \searrow a$, $b_n \nearrow b$. By Lebesgue's Theorem 14.6 on differentiating integrals, $F'(t) = f(t)$ for all $t \in [a_n, b_n] \setminus B_n$, where B_n is a subset of $[a_n, b_n]$ of Lebesgue measure zero. Then $B = \bigcup_{n=1}^{\infty} B_n$ is a subset of (a, b) of measure zero, and $F'(t) = f(t)$ for all $t \in (a, b) \setminus B$. Finally, since $[a_n, b_n] \nearrow (a, b)$,

$$\int_a^b f(t) \, dt = \lim_{n \to \infty} \int_{a_n}^{b_n} f(t) \, dt = \lim_{n \to \infty} (F(b_n) - F(a_n)) = F(b-) - F(a+)$$

by Theorem 14.7.

14.7 The functions $F(x) = \frac{1}{2}x^2$ and $G(x) = \frac{1}{2}x^2 \operatorname{sgn} x$ are generalized primitives to x and $|x|$, respectively, and both $F(1) - F(-1)$ and $G(1) - G(-1)$ exist.

14.11 For each $n \in \mathbb{N}$ the function f_n is continuous and therefore Lebesgue measurable on $(1, \infty)$. For each fixed $x > 1$,

$$\lim_{n \to \infty} f_n(x) = \lim_{n \to \infty} \frac{n + i}{nx^2 + i} = \lim_{n \to \infty} \frac{1 + i/n}{x^2 + i/n} = \frac{1}{x^2} \text{ as } n \to \infty.$$

To apply Lebesgue's dominated convergence theorem we need to find an integrable dominant for the sequence (f_n) on $(1, \infty)$. Find an upper estimate assuming $x > 1$ and $n \geq 2$. From $|nx^2 + i| \geq nx^2 - 1 \geq \frac{1}{2}nx^2$ we get

$$|f_n(x)| = \frac{|n + i|}{|nx^2 + i|} \leq \frac{n+1}{nx^2 - 1} \leq \frac{2n}{\frac{1}{2}nx^2} = \frac{4}{x^2},$$

where $4x^{-2}$ is an integrable dominant for (f_n) on $(1, \infty)$ (absolutely Newton integrable). Then

$$\int_1^\infty \frac{dx}{x^2} = (N) \int_1^\infty \frac{dx}{x^2} = \lim_{t \to \infty} (-t^{-1}) - \lim_{t \to 1+} (-t^{-1}) = 1.$$

By Legesgue's dominated convergence thoerem, $\int_1^\infty f_n(t)\, dt \to \int_1^\infty f(t)\, dt = 1$.

14.17 At $x = 1$ the function has a limit calculated by l'Hôpital's rule:

$$\lim_{x \to 1} \frac{\log x}{x^2 - 1} = \lim_{x \to 1} \frac{x^{-1}}{2x} = \lim_{x \to 1} \frac{1}{2x^2} = \tfrac{1}{2}.$$

When we set $f(1) = \frac{1}{2}$, the function becomes continuous at $x = 1$. The function is then continuous and therefore Lebesgue measurable on $(0, \infty)$. As $t \to 0+$ and $t \to \infty$, the function diverges to ∞. Using l'Hôpital's rule again we show that

$$f(x) = O(1/x^{1/2}) \text{ as } x \to 0+, \quad f(x) = O(1/x^{3/4}) \text{ as } x \to \infty$$

by first showing that $f(x) = o(1/x^{1/2})$ as $x \to 0+$ and $f(x) = o(1/x^{3/4})$ as $x \to \infty$. Then $1/x^{1/2}$ is an integrable dominant for f on some interval $(0, \delta)$, and $1/x^{3/4}$ is an integrable dominant for f on some interval (K, ∞). On the compact interval $[\delta, K]$ the function f is continuous and therefore Lebesgue integrable. Overall, f is Lebesgue integrable on $(0, \infty)$.

(The problem does not ask for the evaluation of the integral but as a point of interest, the value is $\pi^2/4$.)

14.21 Select a point $c \in (a, b)$ and for any $x \in (a, b)$ define $F(x) = \int_c^x f(t)\, dt$, $G(x) = \int_c^x g(t)\, dt$ (Lebesgue integrals). Then $(FG)' = Fg + fG$, and the Newton integral of $Fg + fG$ on (a, b) exists. Since the functions Fg and fG are continuous on (a, b), they both have the Lebesgue (and Newton) integral over any compact subinterval of (a, b). Chose real sequences (a_n) and (b_n) such that $a_n \searrow a$ and $b_n \nearrow b$. Since Fg is Newton integrable by hypothesis, the limit of $(N)\int_{a_n}^{b_n} Fg$ as $n \to \infty$ exists. Then also the limit of $(N)\int_{a_n}^{b_n} fG$ as $n \to \infty$ exists since

$$(N) \int_{a_n}^{b_n} fG = (FG)(b_n) - (FG)(a_n) - (N) \int_{a_n}^{b_n} Fg.$$

By Problem 14.20, fG is Newton integrable on (a, b), and the result follows on taking the limit as $n \to \infty$.

14.23

(i) integrable (absolutely Newton integrable), value -1,

(ii) integrable if $p < -1$ (absolutely Newton integrable), value $-1/(p+1)$

(iii) integrable if $q > -1$ (absolutely Newton integrable), value $1/(q+1)$

(iv) integrable (absolutely Newton integrable), value $-1/4$

(v) integrable; the only problem point is $t = 0$, where $\cos(t^{-1})t^{-1/2} = O(t^{-1/2})$ as $t \to 0+$

(vi) integrable: substitution transform the integral to $-\int_0^1 \log x \cdot \sin x \, dx$, where $|\log x|$ is an integrable dominant for the integrand on $(0,1)$

(vii) integrable: at the problem point $t = 1$ we have $t^2(1 - t^4)^{-1/2} = O((1 - t)^{-1/2})$ as $t \to 1-$

(viii) not integrable: at the point $t = 0$ we have $1/(t^3 + t^2)^{1/2} \sim 1/t$ as $t \to 0+$, where $1/t$ is not integrable in any interval $(0, \delta)$

(ix) integrable: the integrand is of order $O(t^{-3/2})$ as $t \to \infty$ and $O(t^{-1/2})$ as $t \to 0+$

(x) integrable: using Taylor's expansion of the integrand in a neighbourhood of $\pi/2$ or otherwise show that the integrand is of order $O((x - \pi/2)^{-1/3})$ as $x \to \pi/2-$; value $-3(1 - \sin 1)^{1/3} + 3$

(xi) integrable (absolutely Newton integrable—a primitive $F(x) = 2\sqrt{\log x}$ exists on $(1, \mathrm{e}))$, value 2

(xii) integrable: $t^{-3/2} \sin t = O(t^{-3/2})$ as $t \to \infty$

14.25 (i) Write $f(x) = 1/(x^\alpha \log^\beta x)$. Show that, as $x \to \infty$, the following happens: If $\alpha > 1$, then $f(x) = O(x^{-\gamma})$ for $1 < \gamma < \alpha$; if $\alpha < 1$, then $x^{-\gamma} = O(f(x))$ for $0 < \gamma < 1$. (ii) Substitute $x = \mathrm{e}^t$.

Chapter 15

15.1 The outer measure λ is monotonic, countably subadditive, and satisfies $\lambda(A) \le \rho(A)$ for all $A \in \mathcal{K}$. First we observe that $\lambda(\{x\}) = 0$ for all $x \in \mathbb{R}$:

$$0 \le \lambda(\{x\}) \le \lambda((x - \tfrac{1}{n}, x + \tfrac{1}{n})) \le \rho((x - \tfrac{1}{n}, x + \tfrac{1}{n})) = \tfrac{4}{n^2} \text{ for all } n.$$

Then $\lambda([a, b]) = \lambda((a, b))$ since $\lambda((a, b)) \le \lambda([a, b]) \le \lambda(\{a\}) + \lambda((a, b)) + \lambda(\{b\}) = \lambda((a, b))$. For any partition $a = t_0 < t_1 < \cdots < t_n = b$ of the interval $[a, b]$ we have $[a, b] = \bigcup_{k=1}^n [t_{k-1}, t_k]$ and

$$\lambda([a, b]) \le \sum_{k=1}^n \lambda([t_{k-1}, t_k]) = \sum_{k=1}^n \lambda((t_{k-1}, t_k)) \le \sum_{k=1}^n \rho((t_{k-1}, t_k)) = \sum_{k=1}^n (t_k - t_{k-1})^2.$$

If the partition is equidistant, we have

$$0 \le \lambda([a,b]) \le \sum_{k=1}^{n} \left(\frac{b-a}{n}\right)^2 = \frac{(b-a)^2}{n} \quad \text{for any } n.$$

Hence $\lambda([a,b]) = 0$. From this we conclude that $\lambda = 0$.

15.3 λ is the counting measure on S: $\lambda(\emptyset) = 0$, $\lambda(A) = \mathrm{card}\,(A)$ if $A \subset S$ is finite, and $\lambda(A) = \infty$ if $A \subset S$ is inifinite.

15.5 (i) The intersection of two nails is always an interval of the same type or $\emptyset$. For the complement of two nails we check that we get either $\emptyset$, an interval or the disjoint union of two intervals.

(ii) The intersection of two nails in $\mathbb{R}^k$ is a (possibly empty) nail:

$$I \cap J = (I_1 \times \cdots \times I_k) \cap (J_1 \times \cdots \times J_k) = (I_1 \cap J_1) \times \cdots \times (I_k \cap J_k).$$

For the set difference we use mathematical induction. The case $k = 1$ is part (i). Suppose that for some $k \ge 1$, the difference of any two nails in $\mathbb{R}^k$ is a finite disjoint union of nails. For two nails I, J in $\mathbb{R}^{k+1}$ we write $I = I_1 \times I_2$ and $J = J_1 \times J_2$, where I_1, J_1 are one-dimensional nails, and I_2, J_2 are nails in $\mathbb{R}^k$. Then

$$I \setminus J = (I_1 \times I_2) \cap (J_1 \times J_2)^c = [(I_1 \setminus J_1) \times I_2] \cup [(I_1 \cap J_1) \times (I_2 \setminus J_2)]$$

is a disjoint union. By the induction hypothesis, $I_1 \setminus J_1 = K_1 \cup K_2$ and $I_2 \setminus J_2 = \bigcup_{m=1}^{p} L_m$ are disjoint unions of suitable nails. Hence

$$I \setminus J = (K_1 \times J_2) \cup (K_2 \times J_2) \cup ((I_1 \cap J_1) \times L_1) \cup \cdots \cup ((I_1 \cap J_1) \times L_p).$$

15.9 (i) $\Longrightarrow$ (ii): Let $\widetilde{\mathcal{N}}_k$ be the family of all sets which can be expressed as finite disjoint unions of nails. We extend ρ to $\widetilde{\mathcal{N}}_k$ by writing $\widetilde{\rho}(A) = \sum_{i=1}^{p} \rho(A_i)$ if $A = \bigcup_{i=1}^{p} A_i$ (disjoint union of nails). Then $\widetilde{\rho}$ is countably additive on $\widetilde{\mathcal{N}}_k$. Let $A_n \searrow A$ for $A_n, A \in \mathcal{N}_k$. Write $B_n = A_1 \setminus A_n$, $C_n = B_n \setminus (B_1 \cup \cdots \cup B_{n-1})$, and $C = A_1 \setminus A$. Then $\bigcup_{n=1}^{\infty} C_n = C$ is a disjoint union of sets in $\widetilde{\mathcal{N}}_k$, and $\widetilde{\rho}(C) = \sum_{p=1}^{\infty} \widetilde{\rho}(C_p) = \lim_n \sum_{p=1}^{n} \widetilde{\rho}(C_p) = \lim_n \widetilde{\rho}(B_n) = \rho(A_1) - \lim_n \rho(A_n)$; on the other hand, $\widetilde{\rho}(C) = \rho(A_1) - \rho(A)$, and the result follows.

(ii) $\Longrightarrow$ (iii): True since (iii) is a special case of (ii) with $A = \emptyset \in \mathcal{N}_k$.

(iii) $\Longrightarrow$ (iv): Let $A = (a_1, b_1] \times \cdots \times (a_k, b_k] \in \mathcal{N}_k$ and let $\varepsilon > 0$. For each sufficiently large n set $B_n = (a_1, b_1 + \frac{1}{n}] \times \cdots \times (a_k, b_k + \frac{1}{n}]$, and

$$P_{ni} = (a_1, b_1] \times \cdots \times (b_i, b_i + \tfrac{1}{n}] \times \cdots \times (a_k, b_k], \quad i = 1, \ldots, k.$$

Then $B_n = A \cup \left(\bigcup_{i=1}^{k} P_{ni}\right)$. Observe that $P_{ni} \searrow \emptyset$ as $n \to \infty$ for $i = 1, \ldots, k$. By hypothesis there exists N such that $\sum_{i=1}^{k} \rho(P_{ni}) < \varepsilon$ if $n \ge N$. Set $B = B_N$. Then $A \subset B^\circ$ and

$$\rho(B) \le \rho(A) + \sum_{i=1}^{k} \rho(P_{Ni}) < \rho(A) + \varepsilon.$$

A similar procedure applied to k-cells $C_n = (a_1 + \frac{1}{n}, b_1] \times \cdots \times (a_k + \frac{1}{n}, b_k]$ will show that there exist $C \in \mathcal{N}_k$ such that $\overline{C} \subset A$ and $\rho(A) < \rho(C) + \varepsilon$. This proves that ρ is a Stieltjes gauge.
(iv) $\Longrightarrow$ (i): Theorem 15.27.

15.11 The space $\mathbb{R}^k$ can be expressed as a disjoint union of measurable bounded sets; hence $A = \bigcup_{n=1}^{\infty} A_n$, where A_n are disjoint, measurable and bounded. Given $\varepsilon > 0$, there exists N such that $\sum_{n=N+1} m(A_n) < \varepsilon$. Then $B = \bigcup_{n=1}^{N} A_n$ is a bounded measurable subset of A, and $m(A \setminus B) = m(\bigcup_{n=N+1}^{\infty} A_n) \le \sum_{n=N+1} m(A_n) < \varepsilon$.

15.13 (i) Let $\varepsilon > 0$ be given. Assume first that $m(A) < \infty$. By the definition of the Lebesgue outer measure, there exists a countable cover of A by open k-cells B_n such that $m(A) \le \sum_n \text{vol}(B_n) < m(A) + \varepsilon$. Then $G = \bigcup_{n=1}^{\infty} B_n$ is an open superset of A satisfying $m(A) \le m(G) < m(A) + \varepsilon$. Hence $0 \le m(G \setminus A) = m(G) - m(A) < \varepsilon$.
Let $m(A) = \infty$. Since $(\mathbb{R}^k, \Lambda, m)$ is σ-finite measure space, $A = \bigcup_{n=1}^{\infty} A_n$, where A_n are disjoint measurable sets of finite measure. For each n there exists an open superset G_n of A_n such that $m(G_n \setminus A_n) < (\frac{1}{2})^n \varepsilon$. Then $G = \bigcup_{n=1}^{\infty} G_n$ is an open superset of A satisfying $G \setminus A \subset \bigcup_{n=1}^{\infty}(G_n \setminus A_n)$ (see (F.5)). Hence

$$m(G \setminus A) \le \sum_{n=1}^{\infty} m(G_n \setminus A_n) < \sum_{n=1}^{\infty} (\tfrac{1}{2})^n \varepsilon = \varepsilon.$$

(ii) If $m(A) = \infty$ and G is an open superset of A, then $m(G) = \infty$, and the result follows. If $m(A) < \infty$, by the first part for each $\varepsilon > 0$ there exists an open superset G of A such that $m(G \setminus A) < \varepsilon$. Then $m(A) \le m(G) = m(A) + m(G \setminus A) < m(A) + \varepsilon$.

15.17 Since A is bounded, $\lambda(A) < \infty$. Let $\varepsilon > 0$ be given. By the definition of the outer measure there is a sequence of open k-cells B_n of diameter less than 1 such that $\lambda(A) \le \sum_{n=1}^{\infty} \text{vol}(B_n) < \lambda(A) + \frac{1}{2}\varepsilon$. Let $G = \bigcup_{n=1}^{\infty} B_n$. Then G is open and bounded, so that $m(G) < \infty$. Let F be a closed subset of A such that $\lambda(A \setminus F) < \frac{1}{2}\varepsilon$. After a short calculation we get $m(G \setminus F) < \varepsilon$, and $A \in \Lambda^k$ by Problem 15.16.

15.19 Let K be a compact subset of $\mathbb{R}^k$. Given $\varepsilon > 0$, there exists an open set $G \supset A$ such that $\lambda(G \setminus A) < \varepsilon$. Then $A^c \cap K \supset G^c \cap K$ and

$$(A^c \cap K) \setminus (G^c \cap K) = (A^c \cap K) \cap (G \cup K^c) = G \cap A^c \cap K \subset G \cap A^c = G \setminus A.$$

Thus $B := A^c \cap K$ is a bounded set such that for each $\varepsilon > 0$ there exists a closed set $F := G^c \cap K$ such that $B \supset F$ and $\lambda(B \setminus F) \le \lambda(G \setminus A) < \varepsilon$. Then $B \in \Lambda^k$ by Problem 15.17. Further, $A^c \in \Lambda^k$ by Problem 15.18, and $A \in \Lambda^k$.

15.25 Let $\bigcup_{k=1}^{\infty} E_k = E$ be a disjoint union of measurable sets. Given $\varepsilon > 0$, for each k there is a measurable partition $\{A_{kj}\}_j$ of E_k such that $|\mu|(E_k) - \varepsilon 2^{-k} < \sum_{j=1}^{\infty} |\mu(A_{kj})|$. Since $E = \bigcup_{k,j=1}^{\infty} A_{kj}$, we have

$$\sum_k |\mu|(E_k) - \varepsilon \leq \sum_{k,j} |\mu(A_{kj})| \leq |\mu|(E).$$

So $\sum_{k=1}^{\infty} |\mu|(E_k) \leq \mu|(E)$. For the reverse inequality let $\{B_n\}$ be a measurable partition of E. For each fixed n, $\{B_n \cap E_k\}_k$ is a measurable partition of B_n; so

$$\sum_n |\mu(B_n)| = \sum_n \left| \sum_k \mu(B_n \cap E_k) \right| \leq \sum_n \sum_k |\mu(B_n \cap E_k)|$$

$$= \sum_k \sum_n |\mu(B_n \cap E_k)| \leq \sum_k |\mu|(E_k).$$

Taking the supremum over all measurable partitions $\{B_n\}$ of E, we obtain $|\mu|(E) \leq \sum_k |\mu|(E_k)$.

15.27 Let $E \in \Sigma$ and let $\varepsilon > 0$ be given. There exists a measurable partition $\{A_k\}$ of E such that

$$|\mu|(E) - \varepsilon < \sum_k |\mu(A_k)| = \sum_k \left| \int_{A_k} f \, d\mu \right| \leq \sum_k \int_{A_k} |f| \, d\mu = \int_E |f| \, d\mu.$$

Since ε was arbitrary, $|\mu|(E) \leq \int_E |f| \, d\mu$.

Chapter 16

16.1 (i) For any $m \in \mathbb{N}$, $\mathcal{B}^m$ is the least σ-algebra containing $\mathcal{N}_m$. If $A \in \mathcal{N}_k$ and $B \in \mathcal{N}_s$, then $A \times B \in \mathcal{N}_{k+s}$. Hence $\mathcal{B}^{k+s} \subset \mathcal{B}^k \otimes \mathcal{B}^s$. For the reverse inclusion we show that $\mathcal{B}^{k+s}$ contains all measurable rectangles $A \times B$, where $A \in \mathcal{B}^k$ and $B \in \mathcal{B}^s$. The family $\mathcal{A}^k = \{A \subset \mathbb{R}^k : A \times \mathbb{R}^s \in \mathcal{B}^{k+s}\}$ is a σ-algebra containing open sets, so $\mathcal{A}^k \supset \mathcal{B}^k$. Similarly $\mathcal{A}^s = \{B \subset \mathbb{R}^s : \mathbb{R}^k \times B \in \mathcal{B}^{k+s}\} \supset \mathcal{B}^s$. Thus for any $A \in \mathcal{B}^k$ and $B \in \mathcal{B}^s$, $A \times B = (A \times \mathbb{R}^s) \cap (\mathbb{R}^k \times B) \in \mathcal{B}^{k+s}$.
(ii) Recall that $\Lambda^m = \overline{\mathcal{B}^m}$. From the previous result we get

$$\overline{\Lambda^k \otimes \Lambda^s} = \overline{\mathcal{B}^k \otimes \mathcal{B}^s} = \overline{\mathcal{B}^{k+1}} = \Lambda^{k+s},$$

where the first equality requires some elaboration. It can be shown that the product space $\Lambda^k \otimes \Lambda^s$ is not measure-complete.

16.5 (i) Construct sequences (g_n), (h_n) of step functions with $g_n \leq g_{n+1} \leq f \leq h_{n+1} \leq h_n$ and $\int_I (h_n - g_n) < 1/n$ for all n. Each step function is Lebesgue integrable, and its elementary integral is the Lebesgue integral. Hence we can apply the monotonic sequence theorem to conclude that $g_n \to g$ and $h_n \to h$ a.e. with $g \leq f \leq h$ a.e. From $\int_I (h - g) = 0$ we get $g = f = h$ a.e. So f is Lebesgue integrable. From the definition of supremum it follows that there exists an increasing sequence (u_n) of step functions such that $\int_I u_n \nearrow (\mathcal{R}) \int_I f(t)\,dt$. Hence $\int_I f(t)\,dt = (\mathcal{R}) \int_I f(t)\,dt$ by the monotonic sequence theorem.

(ii) Let f be Riemann integrable on I. By part (i) there exist sequences (g_n) and (h_n) of step functions such that $g_n \nearrow f$ on $I \setminus A$, and $h_n \searrow f$ on $I \setminus B$ where A, B are sets of Lebesgue measure zero. Hence f is lower semicontinuous on $I \setminus A$ and upper semicontinuous on $I \setminus B$, that is, f is continuous on $I \setminus (A \cup B)$. (A real valued function f is *lower semicontinuous* at a point t_0 if for every $\varepsilon > 0$ there exists an open neighbourhood U of t_0 such that $f(t_0) - \varepsilon < f(t) \le f(t_0)$ whenever $t \in U$. Analogously for *upper semicontinuous*.)

One way to prove the converse is to show that every bounded lower (upper) semicontinuous function on I is the limit of an increasing (decreasing) sequence of step functions.

(iii) Any bounded monotonic function on $[a, b]$ is continuous on $[a, b] \setminus C$, where C is countable, and hence of Lebesgue measure zero.

16.9 Equivalently we may consider the double integral on $\Delta = (0, 1) \times (a, b)$. The function $f(x, y) = x^y = \exp(y \log x)$ is continuous on Δ: $(x, y) \mapsto y$ and $(x, y) \mapsto \log x$ are continuous on Δ; the product $\varphi(x, y) = y \log x$ is continuous on Δ, and so is the composite function $f = \exp \circ \varphi$. Hence f is Lebesgue measurable on Δ. Tonelli's theorem:

$$\int_a^b \left(\int_0^1 |f(x, y)|\,dx \right) dy = \int_a^b \left(\int_0^1 x^y\,dx \right) dy = \int_a^b \left[\frac{x^{y+1}}{y+1} \right]_{x=0}^{x=1} dy$$

$$= \int_a^b \frac{dy}{y+1} = \log \frac{b+1}{a+1}$$

(the consistency of the Newton and Lebesgue integral). Fubini's theorem:

$$\int_0^1 \left(\int_a^b \exp(y \log x)\,dy \right) dx = \int_0^1 \left[\frac{\exp(y \log x)}{\log x} \right]_{y=a}^{y=b} dx$$

$$= \int_0^1 \frac{x^b - x^a}{\log x}\,dx = \log \frac{b+1}{a+1}.$$

(In the process we have proved that $(x^b - x^a)/\log x$ is Lebesgue integrable on $(0, 1)$.)

16.13 The function $f(x, y) = \exp(-x^2 - y^2)$ is positive and symmetric in the x and y axes; so $\int_{\mathbb{R}^2} f\,dx\,dy = 4 \int_\Omega f\,dx\,dy$, where $\Omega = (0, \infty) \times (0, \infty)$.

For the polar substitution set $u(r, \theta) = (r \cos \theta, r \sin \theta)$. The strip $R = (0, \infty) \times (0, \frac{1}{2}\pi)$ in the (r, θ) plane is mapped under u bijectively onto Ω in the (x, y) plane; u is a diffeomorphism with the absolute value of the Jacobian equal to r. Hence

$$\int_{\mathbb{R}_+^2} e^{-x^2 - y^2}\,dx\,dy = \int_R e^{-r^2} r\,dr\,d\theta$$

provided one of the integrals exists. The integral on the right exists by Tonelli's theorem:

$$\int_0^{\pi/2} \left(\int_0^\infty e^{-r^2} r\, dr \right) d\theta = \int_0^{\pi/2} \left(\int_0^\infty e^{-s} \tfrac{1}{2} ds \right) d\theta = \int_0^{\pi/2} \tfrac{1}{2}\, d\theta = \tfrac{1}{4}\pi$$

Fubini's theorem:

$$\int_{\mathbb{R}_+^2} e^{-x^2-y^2}\, dx\, dy = \int_0^\infty e^{-x^2}\, dx \int_0^\infty e^{-y^2}\, dy = \left(\int_0^\infty e^{-s^2}\, ds \right)^2.$$

Hence

$$\int_0^\infty e^{-x^2}\, dx = \frac{\sqrt{\pi}}{2}.$$

16.15 Write $f(x,y) = (1+y)^{-1}(1+x^2y)^{-1}$. Let $\Omega = (0,\infty) \times (0,\infty)$. Tonelli's theorem: (i) $(\Omega, \Lambda(\Omega), m)$ is σ-finite. (ii) f is continuous in (x,y) on Ω, and hence f is Lebesgue measurable on Ω. (iii) Repeated integral:

$$\int_0^\infty \left(\int_0^\infty |f(x,y)|\, dx \right) dy = \int_0^\infty \left(\int_0^\infty f(x,y)\, dx \right) dy.$$

The inner integral (the Newton integral with the substitution $u = \sqrt{y}\, x$):

$$\int_0^\infty \frac{1}{(1+y)(1+x^2y)}\, dx = \frac{1}{1+y}(N)\int_0^\infty \frac{dx}{1+x^2y} = \frac{1}{1+y}(N)\int_0^\infty \frac{1}{1+u^2} \frac{du}{\sqrt{y}}$$

$$= \frac{1}{\sqrt{y}(1+y)} \lim_{t\to\infty} (\arctan(t) - \arctan(0)) = \frac{1}{\sqrt{y}(1+y)} \frac{\pi}{2}.$$

The function $1/(\sqrt{y}(1+y))$ is nonnegative; a primitive $G(y) = 2\arctan(\sqrt{y})$. Then

$$\int_{(0,\infty)} \frac{1}{\sqrt{y}(1+y)}\, dy = (N)\int_0^\infty \frac{1}{\sqrt{y}(1+y)}\, dy = \lim_{t\to\infty} (G(t) - G(0)) = \pi.$$

The repeated integral exists, and f is Lebesgue integrable on Ω by Tonelli's theorem.

Fubini's theorem (note that the integrand is nonnegative):

$$\iint_\Omega \frac{1}{(1+y)(1+x^2y)}\, dx dy = \int_0^\infty \left(\frac{1}{(1+y)(1+x^2y)}\, dx \right) dy$$

$$= \int_0^\infty \frac{\pi}{2} \int_0^\infty \frac{1}{\sqrt{y}(1+y)}\, dy = \frac{\pi^2}{2}.$$

Changing the order of integration according to Fubini's theorem, we have

$$\iint_\Omega (\cdots)\, dx\, dy = \int_0^\infty \left(\int_0^\infty \frac{1}{(1+y)(1+x^2y)}\, dy \right) dx = \int_0^\infty \frac{2\log x}{x^2-1}\, dx.$$

Hence

$$\int_0^\infty \frac{\log x}{x^2 - 1} \, dx = \frac{\pi^2}{4}.$$

16.17 (i) Integrable, value $-\pi/4$. (ii) Integrable, value 2. (iii) Integrable for $0 < p < 2$; value $\pi^2/6$ for $p = 1$.

Chapter 17

17.1 The function f is μ-measurable being the μ-a.e. limit of μ-measurable functions f_n (Theorem 13.41). From $f_n \to f$ μ-a.e. and $|f_n| \leq g$ μ-a.e. we deduce that $|f| \leq g$ μ-a.e. Then f_n and f are in $L^p(\mu)$ as $|f_n|^p \leq g^p$ and $|f|^p \leq g^p$. Also $f_n - f \in L^1(\mu)$, and $|f_n - f|^p \leq 2^p(|f_n| + |f|) \leq 2^{p+1}g^p$. Thus $|f_n - f|^p$ is a dominated sequence of $L^1(\mu)$ functions convergent to 0 μ-a.e. By the ordinary Lebesgue's dominated convergence theorem, $\int_S |f_n - f|^p \, d\mu \to 0$, which implies $\|f_n - f\|_{L^p} \to 0$.

17.3 Let $p = s/r$; then $p \geq 1$. If $f \in L^s$, then $|f|^r \in L^p$ as $pr = s$. Since S is a finite measure space, the constant function 1 is in L^q, where $q = p/(p-1)$ is the conjugate index of p. Applying Hölder's theorem to $|f|^r \in L^p$ and $1 \in L^q$, we observe that $|f|^r = |f|^r \cdot 1 \in L^1$ and

$$\int_S |f|^r \, d\mu \leq \| |f|^r \|_{L^p} \|1\|_{L^q}.$$

This proves $f \in L^r$, and

$$\|f\|_{L^r} \leq (\|1\|_{L^q})^{1/r} (\| |f|^r \|_{L^p})^{1/r} = \mu(S)^{\frac{1}{r} - \frac{1}{s}} \|f\|_{L^s}.$$

17.5 From the definition of infimum it follows that there is a minimizing sequence (A_n) of sets of measure zero such that

$$\|f\|_{L^\infty} = \lim_{n \to \infty} \sup_{A_n^c} |f|.$$

Set $A = \bigcup_{k=1}^\infty A_k$. Then A has measure zero, and $\|f\|_{L^\infty} \leq \sup_{A^c} |f| \leq \sup_{A_n^c} |f|$. The result follows on taking the limit as $n \to \infty$.

17.9 Suggestion: The continuous functions are dense in $(L^p[0,1], \| \cdot \|_{L^p})$, and the polynomials with rational coefficients are dense in $(C[0,1], \|\cdot\|_\infty)$.

17.11 Assume first that f is measurable and $\|f\|_{L^\infty} > 0$. For any $\alpha \in (0, \|f\|_{L^\infty})$ set $A_\alpha = |f|^{-1}[\alpha, \infty)$. Then A_α is measurable, $\mu(A_\alpha) > 0$, and

$$\alpha^p \mu(A_\alpha) \leq \int_{A_\alpha} |f|^p \, d\mu \leq \int_S |f|^p \, d\mu \leq \|f\|_{L^\infty}^p \mu(S),$$

$$\alpha \mu(A_\alpha)^{1/p} \leq \|f\|_{L^p} \leq \|f\|_{L^\infty} \mu(S)^{1/p},$$

$$\alpha \leq \lim_{p \to \infty} \|f\|_{L^p} \leq \|f\|_{L^\infty}.$$

The result follows since α satisfying $0 < \alpha < \|f\|_{L^\infty}$ was arbitrary. The general case is left to the reader.

Bibliography

[1] Amir, D. (1986). *Characterizations of Inner Product Spaces, Operator Theory: Advances and Applications*, Vol. 20 (Birkhäuser, Basel).

[2] Baire, R. (1899). Sur les fonctions de variables réeles, *Annali di Matematica* **3**.

[3] Banach, S. (1932). *Théorie des Opérations Linéaires* (Warsaw), reprinted by Chelsea Publ. Co., New York 1955.

[4] Barnsley, M. (1993). *Fractals Everywhere, 2nd ed.* (Academic Press, New York).

[5] Barsnley, M. (2006). *Superfractals* (Cambridge University Press, Cambridge).

[6] Bartle, R. G. (1995). *The Elements of Integration and Lebesgue Measure* (J. Wiley Classics, New York).

[7] Bartle, R. G. (2001). *A Modern Theory of Integration* (American Mathematical Society, Providence).

[8] Bartle, R. G. and Sherbet, D. R. (2000). *Introduction to Real Analysis, 3rd ed.* (Wiley, New York).

[9] Botsko, M. W. (1989). The use of full covers in real analysis, *American Mathematical Monthly* **96**, pp. 328–333.

[10] Botsko, M. W. (2003). An elementary proof of Lebesgue's differentiation theorem, *American Mathematical Monthly* **110**, pp. 834–838.

[11] Brown, A. L. and Page, A. (1970). *Elements of Functional Analysis* (Van Nostrand Reinhold, New York).

[12] Cain, G. L. (1994). *Introduction to General Topology* (Addison-Wesley, New York).

[13] Čech, E. (1969). *Point Sets* (Academic Press, New York).

[14] Černý, I. and Rokyta, M. (1998). *Differential and Integral Calculus of One Real Variable* (Karolinum, Prague).

[15] Cohn, D. L. (1980). *Measure Theory* (Birkhäuser, Stuttgart).

[16] Dieudonné, J. (1969). *Foundations of Modern Analysis* (Academic Press, New York).

[17] Fréchet, M. (1906). Sur quelqes points du calcul fonctionel, *Rendiconti del Circolo Matematico di Palermo* **22**, pp. 1–74.

[18] Gemignani, M. C. (1972). *Elementary Topology* (Addison-Wesley, New York).

[19] Gohberg, I., Goldberg, S. and Kaashoek, M. A. (2003). *Basic Classes of Linear Operators* (Birkhäuser, Basel).

[20] Gordon, R. A. (1994). *The Integrals of Lebesgue, Denjoy, Perron and Henstock*, GSM 4 (American Mathematical Society, Providence).

[21] Hagood, J. W. (2003/2004). The Lebesgue differentiation theorem via nonoverlapping interval covers, *Real Analysis Exchange* **29**, pp. 953–956.

[22] Hausdorff, F. (1914 (reprinted by Chelsea 1949)). *Grundzüge der Mengenlehre* (Leipzig).

[23] Jarnik, V. (1976a). *Differential Calculus II (in Czech)* (Academia, Prague).

[24] Jarnik, V. (1976b). *Integral Calculus II (in Czech)* (Academia, Prague).

[25] Jordan, P. and von Neumann, J. (1936). On inner products in linear metric spaces, *Annals of Mathematics (2)* **36**, pp. 719–723.

[26] Kasriel, R. H. (1971). *Undergraduate Topology* (W. B. Saunders, Philadelphia).

[27] Koliha, J. J. (2003). Lebesgue through Newton integral, *Gazette of the Australian Mathematical Society* **30**, pp. 261–264.

[28] Koliha, J. J. (2006). Fundamental theorem of calculus for Lebesgue integration, *American Mathematical Monthly* **115**, pp. 551–554.

[29] Kreyszig, E. (1978). *Introductory Functional Analysis with Applications* (J. Wiley, New York).

[30] Kurtz, D. S. and Swartz, C. W. (2004). *Theories of Integration* (World Scientific, Singapore).

[31] Lang, S. (1993). *Real and Functional Analysis, 3nd edition* (Springer, Berlin).

[32] Lebesgue, H. (1902). Intégrale, longueur, aire, *Ann. Mat. Pura Appl.* **7**, pp. 231–259.

[33] Lebesgue, H. (1904). *Leçons sur l'Integration et la Recherche des Fonctions Primitives* (Gauthier-Villars, Paris).

[34] Lebesgue, H. (1966). *Measure and the Integral* (Holden-Day, San Francisco).

[35] Lin, S. Y. T. and Lin, Y. F. (1974). *Set Theory: An Intuitive Approach* (Houghton Mifflin, Boston).

[36] Mikusiński, J. and Mikusiński, P. (1993). *An Introduction to Analysis. From Number to Integral* (Wiley, New York).

[37] Pitts, C. G. C. (1972). *Introduction to Metric Spaces* (Oliver and Boyd, Edinburgh).

[38] Pryce, J. D. (1973). *Basic Methods of Functional Analysis* (Hutchinson University Library, London).

[39] Rassias, T. M. (1984). New characterizations of inner product spaces, *Bull. Sci. Math. (2)* **108**, pp. 95–99.

[40] Riesz, F. and Sz.-Nagy, B. (1955). *Functional Analysis* (Frederik Ungar, New York).

[41] Sutherland, W. A. (1975). *Introduction to Metric and Topological Spaces* (Clarendon Press, Oxford).

[42] Swartz, C. (1994). *Measure, Integration and Function Spaces* (World Scien-

tific).

[43] Taylor, A. E. and Lay, D. C. (1980). *Introduction to Functional Analysis, 2nd ed.* (Wiley, New York).

[44] Volintiru, C. (2000/2001). A proof of the fundamental theorem of calculus using Hausdorff measures, *Real Analysis Exchange* **26**, pp. 381–390.

Index